Stratification of tropical forests as seen in leaf structure: Part 2

Tasks for vegetation science 21

HELMUT LIETH
University of Osnabrück, F.R.G.

HAROLD A. MOONEY
Stanford University, Stanford Calif., U.S.A.

Stratification of tropical forests as seen in leaf structure

Part 2

by

B. ROLLET

CH. HÖGERMANN

and

I. ROTH

Kluwer Academic Publishers

Dordrecht / Boston / London

Library of Congress Cataloging in Publication Data

Rollet, B. (Bernard)
Stratification of tropical forests as seen in leaf structure, part 2 / by B. Rollett, Ch. Högermann, and I. Roth.
p. cm. -- (Tasks for vegetation science ; 21)
Includes bibliographical references.

ISBN-13: 978-94-010-7335-6

1. Leaves--Anatomy. 2. Leaves--Tropics--Anatomy. 3. Rain forest plants--Anatomy. 4. Rain forest plants--Ecology. I. Högermann, Ch. II. Roth, Ingrid. III. Title. IV. Series.
QK649.R64 1989
581.4'97--dc20 89-15548

ISBN-13: 978-94-010-7335-6 e-ISBN-13: 978-94-009-1872-6
DOI: 10.1007/ 978-94-009-1872-6

Published by Kluwer Academic Publishers,
P.O. Box 17, 3300 AA Dordrecht, The Netherlands.

Kluwer Academic Publishers incorporates the publishing programmes of D. Reidel, Martinus Nijhoff, Dr W. Junk and MTP Press.

Sold and distributed in the U.S.A. and Canada
by Kluwer Academic Publishers,
101 Philip Drive, Norwell, MA 02061, U.S.A.

In all other countries, sold and distributed
by Kluwer Academic Publishers Group,
P.O. Box 322, 3300 AH Dordrecht, The Netherlands.

CONTENTS

FOREWORD BY THE SERIES EDITOR AND THE VOLUME EDITOR

Since the start of the Task for Vegetation Science series (T:VS), 21 volumes have appeared. The series has found its proper place in the scientific community. Its initial goal was first to provide an outlet for monographic treatments of innovative approaches in vegetation sciences; secondly, the documentation of special data sets important for future practical or theoretical treatments and, thirdly, the translation into English of important treatments of vegetation science related books published in other languages.

The series has volumes published in all three categories. The content of the volumes is often interdisciplinary in nature and emphasizes topics where vegetation science is used to benefit other fields. Several of these volumes were especially successful on the market.

One of the basic requirements for each author/editor of a volume is to provide a clean manuscript in English. In the past, this was a serious obstacle for many projects which would otherwise have been accepted in our series. In many cases, the mother tongue of the editor was not English, and his style and grammar had to be corrected by many people who were polyglots or not from the scientific field. This might have caused some unusual expressions or unfamiliar terminology. However, this is a major problem of our time. If the members of the scientific community request to be monolingual, they must accept the fact that style and grammar of that language vary from country to country, that mistakes occur, and that sentence structures suffer. As a series editor, one can smooth some of the problems but not all, and it does not seem appropriate to override the personal style of a respected colleague one has invited to serve as the key person for an individual volume. Furthermore, it seems financially irresponsible to burden volumes with the cost of a complete linguistic correction when the scientific content is understandable.

The three volumes by Roth (1984, 1987) and Rollet *et al.* (this volume) were included in our series because they provided unique information from the tropical rain forest, a vegetation type from which the least information of this kind is available. The data compiled in these volumes will be used in modelling transfer rates of various kinds in the future. It is unlikely that anywhere in the tropics will it again be possible to collect such a wealth of data. The volumes have been lauded for that fact but, on the other hand, they have been criticized for various reasons. We have therefore asked Professor Roth to answer the criticism by reviewers where necessary, especially those comments where the discussion could improve or clarify objectives. In the following paragraphs, she answers the reviewers on those points which we agreed upon:

"Proctor and Garwood criticize that the author divided the woody plants into three height categories (above 30 m, 10—29 m, and below 10 m). But, as usual, the people who made the inventory also introduced the classification into height categories. I only received the material from Dr. Rollet, who collected it. The abovementioned classification is the one generally used in tropical forestry. The cloud forest of Rancho Grande was classified in a similar way by other forestry experts. Proctor also criticizes that individuals of less than 10 cm dbh are excluded from the inventory.

In the introduction of T:VS.17, I emphasized that most samples were taken from trees with an average diameter of between 10 cm and 1 m or more at breast height. But then I emphasized on page two that the principal objective of the dendrological studies was the collection of wood samples for technological purposes. The inventory was made to study the properties of the wood and the economic return of the forest when cut. No

B. Rollet et al.: Stratification of Tropical Forests as seen in Leaf Structure, Part 2, pp. vii—x.

ecological interest was involved in these studies. The collection of fruits was only an adventitious by-product. But, with reference to the treelets and shrubs with a dbh of less than 10 cm, the reader finds a special note on page two of the book stating: 'However, some trees do not even reach 20 cm dhb in the adult stage, such as some Rubiaceae, Capparidaceae, and Melastomaceae. These trees were also included, however, in an attempt to *cover the maximum number of genera and species*.' This means that trees and shrubs with a dbh of less than 10 cm were also included. Not included were herbs, forbs, tree seedlings, vines and epiphytes. The objective of the primary collection was trees, not the whole flora.

A further point mentioned by Proctor is gaps and phases. Many authors emphasize that the natural regeneration of tropical forests occurs in gaps. This seems to be the paradigm of the present time. Rollet (Bois et forêt tropiques no. 201:3—34: no. 202:19—34, 1983. "La régeneration naturelle dans les trouvées. Un processus général de la dynamique des forêts tropicales humides") thinks that there is a grossly unbalanced exaggeration of the importance of gaps.

Concerning phases in tree life: the literature of the last two decades (in English) seems to present these phases as a new finding and makes proposals for an updated terminology. Foresters were aware of juvenile and senescent individuals in their forests long before 1970. But here again, the inventory was made for other purposes, and not to study regeneration. For more details, we refer to Rollet, B. (1969) "Etudes quantitatives d'une forêt dense humide sempervirente de plaine de la Guyane Vénézuélienne." Trav. Lab. Forest. Toulouse, *T.* 1, Vol. 8, Art. 1:1—36.

In his critical review of Volume 17, Proctor also claims a statement which he thinks is a contradiction to another assertion. On page 15, I say that 'The fact that the height of release has little influence on dispersal distance provides one of the reasons why *explosive* dispersal mechanisms are not confined to trees or other plants which attain a great height' and (on page 139) that '*Autochory* seems to be advantageous when the seeds fall from a great height, in this way they may be removed further from the parent tree'. There is a difference between autochory in general and explosive mechanisms. As is easily checked in any book dealing with fruit dispersal, true explosive mechanisms are actually rare and extremely specialized. There are quite a few herbs which show explosive mechanisms (see Roth 1977, 'Fruits of Angiosperms', Encyclopedia of Plant Anatomy). But fruits or seeds are usually 'shot forth' a few meters. In the interior of the forest, this is an adequate dispersal distance. The world record in 'shooting' is held by Hura crepitans at 45 m. But this is a species of dry and hot habitats where the vegetation is open and where, at times, a strong wind is blowing. Most autochorous plants are less effective. Autochory is a general conception of dispersal types in which no other agents are involved. All plants with dehiscing fruits are considered to be autochorous (see also page 14 of volume 17). However, many of the autochorous mechanisms are not effective. The fruits just open and drop the seeds. In a herb, the seeds remain close to the mother plant. The same is true for a high tree, when no air movement takes place. However, the environmental conditions in the upper crown regions are different from those in the forest interior. At times, we have to take into account the strong winds, in which case fruits or seeds may be taken far away. Furthermore, fruits or seeds falling down from high crowns may become deviated. For these reasons, even a not very effective autochorous method may become effective in high trees, especially when dispersal takes place during the dry season.

Finally, concerning the complaint of Proctor and Garwood that the tables on page 39 are in the wrong place, I completely agree. I asked the publishers to put them in the right place, but they refused for financial reasons. The computer printing has its disadvantages — there is little possibility for any corrections to be made after the appearance of the first proofs, except when the author pays for them.

Answering the questions and criticisms of Nancy Garwood: fruiting times were observed by Dr. Rollet and a team of experts over several years. On page 1 and in the bibliography, the publications of Rollet concerning this topic are cited. However, I do not understand of what 'critical' importance the preservation of fruits should be for descriptive purposes. Fruits and seeds usually have strong cell walls and the photographs show that the tissue types are perfectly preserved. The author has 48 years' experience, in the preservation and processing of plant material for anatomical studies. Another misunderstanding of my publication is the assumption that it would be useful for *field* identification. Who could carry the microscope to the forest, make slides and preparations there to identify fruits? particularly when the person in question is perhaps *unable to distinguish different tissue types, even when the legend underneath explains them*?

Furthermore, I never consider a morphological-anatomical work on fruits and seeds of Venezuelan Guiana a 'substitute' for a study on fruits on the Dutch Guianan flora. Of what use should this comparison be? Garwood misses convincing explanations why some species of Sapotaceae have a higher number of individuals and others have not. Many factors are involved here, of which many are not known. Garwood admits that most wind-dispersed species are tall. Was this not to be expected? Furthermore, the classification of height categories has some disadvantages: when we regard a tree of 30 m height as tall, a tree of 29 m height is not short. The overall statistics are thus of limited value. Results have to be considered with more care — certain trends in certain families or genera are of importance. The fact that the number of seeds is reduced towards the upper strata is of phylogenetic value. What the statistics show is that anemochory drastically diminishes towards the lower strata, as had been expected, and that zoochory prevails in the forest has never been denied. It is, however, surprising that autochorous and auto-zoochorous methods increase in number towards the canopy. Instead of confronting the height categories with the number of species (and individuals) and their dispersal mechanisms (Table 1 of Garwood), which is misleading, it would be much more informative to confront height categories and *percentages* of species (and individuals) with their dispersal mechanisms. In this case it is possible really to follow the increase or diminution of certain dispersal types in the vertical direction.

For those who are working in tropical forests, it is already *commonplace* that stratification exists in the forest. There is a microclimatic gradient, particularly concerning humidity and illumination. While, in the undergrowth, illumination is a limiting factor, a high degree of humidity prevails; in the crown region, by contrast, illumination is often very strong and drought may be considerable when strong winds are blowing. Leaves of herbs, low treelets and shrubs are thus adapted to humidity and are often 'artists' in capturing low light intensities, while leaves in the crown region are more xeric, leathery and have a well developed palisade parenchyma. It is also well known that animal life is more vigorous in the lower forest layers and additionally it shows a certain 'stratification' concerning the different species preferring a certain stratum (ground animals, etc.). Plants have adapted to these gradients in the forest, even in their dispersal types. Anemochory, for example, is only advantageous in high trees. That autochory also seems to be more successful in high trees than in low herbs is likewise a result of the air movement and possibly also of the deviation of seeds or fruits by obstacles. There is no 'stratification hypothesis', but gradation in the forest concerning leaf morphology and anatomy as well as dispersal due to a microclimatic gradient, is just a fact (as I can demonstrate in my next publication 'Leaf structure of a Venezuelan cloud forest as related to microclimate'.)

In the review of the book on leaf structure, Professor Ernst commented on my use of 'ridiculous decimal points', as he phrased it. He forgets that plant anatomists read four digit numbers from the microdrive of their microscopes and these are calibrated in mμ. These numbers converted into the CGS System yield the figures presented in the book. These numbers are familiar to plant anatomists and they are very valid, indeed, usually down to at least the third figure on the dial.

"The main value of this study for ecosystems analyses lies in the completeness of the material, which allows comparisons in many ways. Up to the present time, an area of such a large extension has never been studied morphologically and anatomically (barks, leaves, fruits and seeds). The advantage was the inventory in which not only the number of species, but also the number of individuals was known. Usually, conclusions are drawn after general information is available. The results obtained from a thorough study of all species and individuals showed contradictions with the usual dogma. It is unlikely that ecologists will be able to conduct studies of this type in the future."

So much for the responses for the volume editor.

The publisher and the series editors admit that they too were included in the criticism of the reviewers to some extent and that some of the critique directed at Professor Roth should have been directed at the series editor or the publisher. We are ready to apologize for any mistakes we made. Our main objective with the T:VS series is service to the scientific community at the lowest possible cost. The success of the series on the market, including the volumes by Professor Roth, seems to justify our decisions. We hope, therefore, that the current volume will be as successful as the previous ones. Since this volume completes the documentation of the entire project, it may even stimulate more users to buy the complementary T:VS Volumes 6 and 17. We also hope that our readers understand from this foreword that we as editors, and publisher as well, respect the comments of the reviewers of our books. Whatever we can adopt from their suggestions will be done. We hope to keep the series going in the future on the same principles and with the same care as at present, and thank all readers and users of our volumes for their help.

H. Lieth, *Series editor*

I. Roth, *Volume editor*

W. R. Peters, *Kluwer Academic Publisher*

INTRODUCTION

This volume is the last contribution of a series of studies concerned with the plant material of one and the same area of Venezuelan Guiana. The studies originated through a collaboration with the forest engineer Dr. B. Rollet, the FAO expert in forest inventory who collected the material of tree barks, leaves, fruits and seeds in Venezuelan Guiana around the "Rio Grande", "El Paraiso", and "El Dorado" camps. In the first place, tree barks of about 280 species of dicotyledons belonging to 48 families were studied (family by family) by Roth in separate publications which mainly appeared in Acta Botānica Venezuelica and in Acta Biológica Venezuelica (see the bibliography in Roth 1981). A comprehensive survey of bark structures, principally from the area of Venezuelan Guiana, was given in "Structural patterns of tropical barks" by Roth (1981). Secondly, leaf structure of 42 species belonging to 28 families of dicotyledons, chiefly trees but also herbs and lianas, of the same region, was studied with the aim of estimating leaf consistency and texture ("Anatomía y textura foliar de plantas de la Guayana Venezolana", Roth 1977). Some material of the leaf collection of Dr. Rollet from the same region was left to the disposal of Dr. M. Pyykkö for morphological and anatomical studies (Pyykkö 1979). An exhaustive survey of the leaf anatomy of 232 species belonging to 48 different families from the collection of Dr. Rollet in Venezuelan Guiana, was given by Roth (1984) ("Stratification of tropical forests as seen in leaf structure"). Lastly, morphology, anatomy, and dispersal of fruits and seeds of 58 families with 375 species from the same collection in Venezuelan Guiana was investigated by Roth (1987) ("Stratification of a tropical forest as seen in dispersal types"). Wood structure was studied separately by the forestry department of the Universidad Central de Venezuela in Mérida.

With the present book, a further gap concerning leaf morphology and leaf venation, as well as some structural peculiarities of physiological importance, is closed so that an exhaustive survey of bark and leaf morphology and anatomy as well as of fruit and seed structure of the plants of a certain well-known area is herewith given. Not only were hundreds of species studied, but structural characteristics were related to "forest stratification", i.e. to the different microclimatic conditions in the forest, as the height of the trees and shrubs studied was known. It is of common knowledge that in the lower forest layers, light is a limiting factor, while humidity is sufficiently available, whereas in the crown region, insolation is strong, but drought may play an important part. In the undergrowth and lower layers, we thus find the hygromorphic shade leaf, while in the uppermost layers and in emergent trees, the sun leaf with some xeromorphic features prevails. Furthermore, zoochory predominates in the lower strata, whereas autochory and wind dispersal are more common to the upper storey and in emergent trees. In order to distinguish the different storeys, the trees and shrubs were separated into three categories:

A — very tall trees of 30 m height or more,
a — trees between 10 m and 29 m in height, and
aa — small trees and shrubs of less than 10 m in height.

This subdivision into height categories was made by the inventory experts. Additionally, mature leaves of juvenile plants of different heights (e.g. 1 m, 5 m, 7 m, 12 m etc.) were available so that a comparison between mature leaves of one and the same species at different stages of tree development and at different forest strata could be made. We owe this knowledge to the initiation of an inventory carried out by the Ministry of Agricul-

B. Rollet et al.: Stratification of Tropical Forests as seen in Leaf Structure, Part 2, pp. xi–xv.

ture in Venezuela (MAC) together with the FAO and UNDP through which the commercially important trees in the studied region were identified. A strip 25 m wide and about 67 km long was enumerated roughly east-west through the forest for trees and vines 10 cm dbh and above, thus covering a 155 ha area with about 68 000 trees (see Map 1). The main value of this inventory lies in its completeness, as not only were all species growing on a certain area identified, but also the number of individuals of each species was recorded. Barks, leaves, and fruits of trees, treelets, and shrubs were collected among emergents, in the canopy, and in the undergrowth, recording the total height of each collected individual. Most species were identified by Dr. Julian Steyermark, Instituto Botanico, Caracas. Others were sent to specialists in the U.S.A., England, France, The Netherlands, Belgium, Sweden, Colombia, and Brazil. Some samples, however, are still in the process of identification or could not be identified with any certainty, such as species of the families Burseraceae, Sapotaceae, Myrtaceae, and Rosaceae.

The tropical humid forest from which leaf samples were collected corresponds to a dense evergreen rain forest with a short dry season between January and April which Beard would classify as a "Seasonal rain forest". Geographically, the forest is located on the ancient shield of Guiana at the massif of Imataca (south of the Orinoco river and east of the junction with the Caroní river) which is composed of rocks of the Precambrian Era, essentially of granites and gneisses (see Map 2). This extreme northern part of the Guianan shield has an undulated ground relief which fluctuates between 150 and 500 m in altitude. Precipitation amounts to about 2000 mm or more. The mean annual temperature oscillates between 18 and 24°. Approximately 10% or 60 species of the total number of species collected shed their leaves, but as defoliation is somewhat irregular, the forest maintains its evergreen aspect throughout the year. One can object to any tentative stratification of stands into height classes, i.e. into physical layers above the forest floor. Having in mind the performance of the different tree species in terms of maximum height or diameter, one will generally accept that in undisturbed lowland tropical forest, trees above 30 m total height (roughly corresponding to 30—40 cm in diameter) belong to the main canopy or emergents, whereas trees below 10 m total height (roughly less than 10 cm in diameter) belong to the undergrowth. These two extreme subpopulations, below 10 m and above 30 m total height, will give an ecological basis for morphological comparisons of leaves. The in-between stratum (10 to 30 m) will reflect intermediate effects, especially concerning leaf size and leaf consistence. However, it must be admitted that the height category below 10 m was not completely enumerated for individuals. Herbs, tree seedlings, vines, and epiphytes were not systematically covered. The objective was trees, not the whole flora. The original goal of the inventory was not an ecological study, but the collection of wood samples for technological purposes. These samples were processed in the Instituto Forestal in Mérida, Venezuela, and in the Instituto Tecnológico Forestal of the University of Mérida. For further information, the reader is referred to B. Rollet "Inventario forestal en la Guayana Venezolana", MAC, Informe no. 3, parte 2, Caracas 1967.

The main value of the anatomical and morphological studies in question lies in the large material of plant species (several hundred), in its diversity (about 50 different families), as well as in the fact that not only the number of species is known, but also the number of individuals present in a well-known geographical region. There are no other areas of tropical forest in the world in which bark and leaf structure as well as fruit and seed anatomy and dispersal has been studied thoroughly of such a large plant material of one and the same region, in which even the number of individuals is known. Unfortunately, there are only few plant anatomists left and it is, therefore, not to be expected that a comparably large investigation of this type will be repeated in another tropical region in the near future. Furthermore, we think that anatomical studies of this type should still be of great value to ecologists and physiologists, the more as practically extremely little is known of the inner struc-

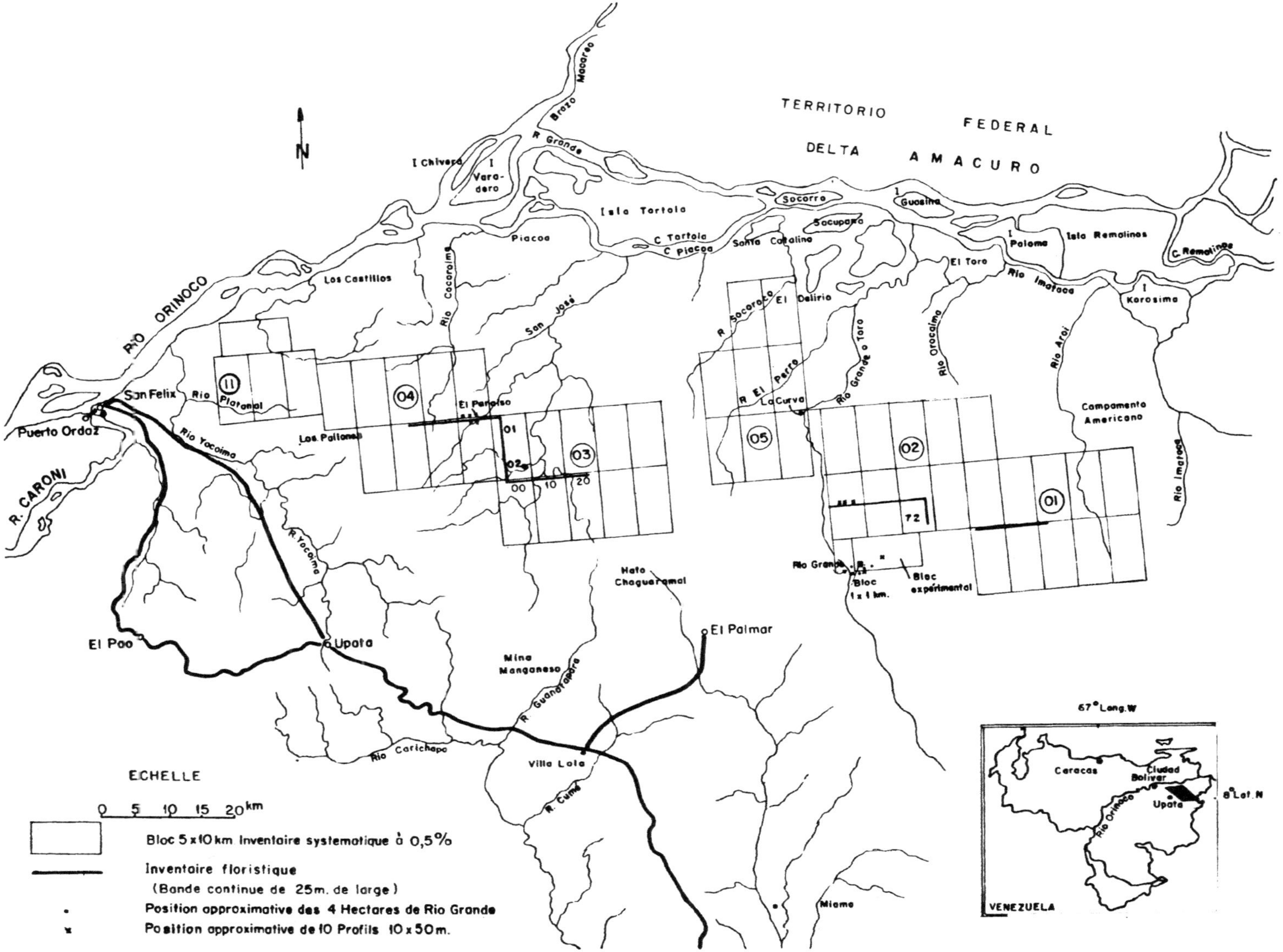

Map 1. Source of the map: B. Rollet. Bois For. Trop. no. 124 (1969)

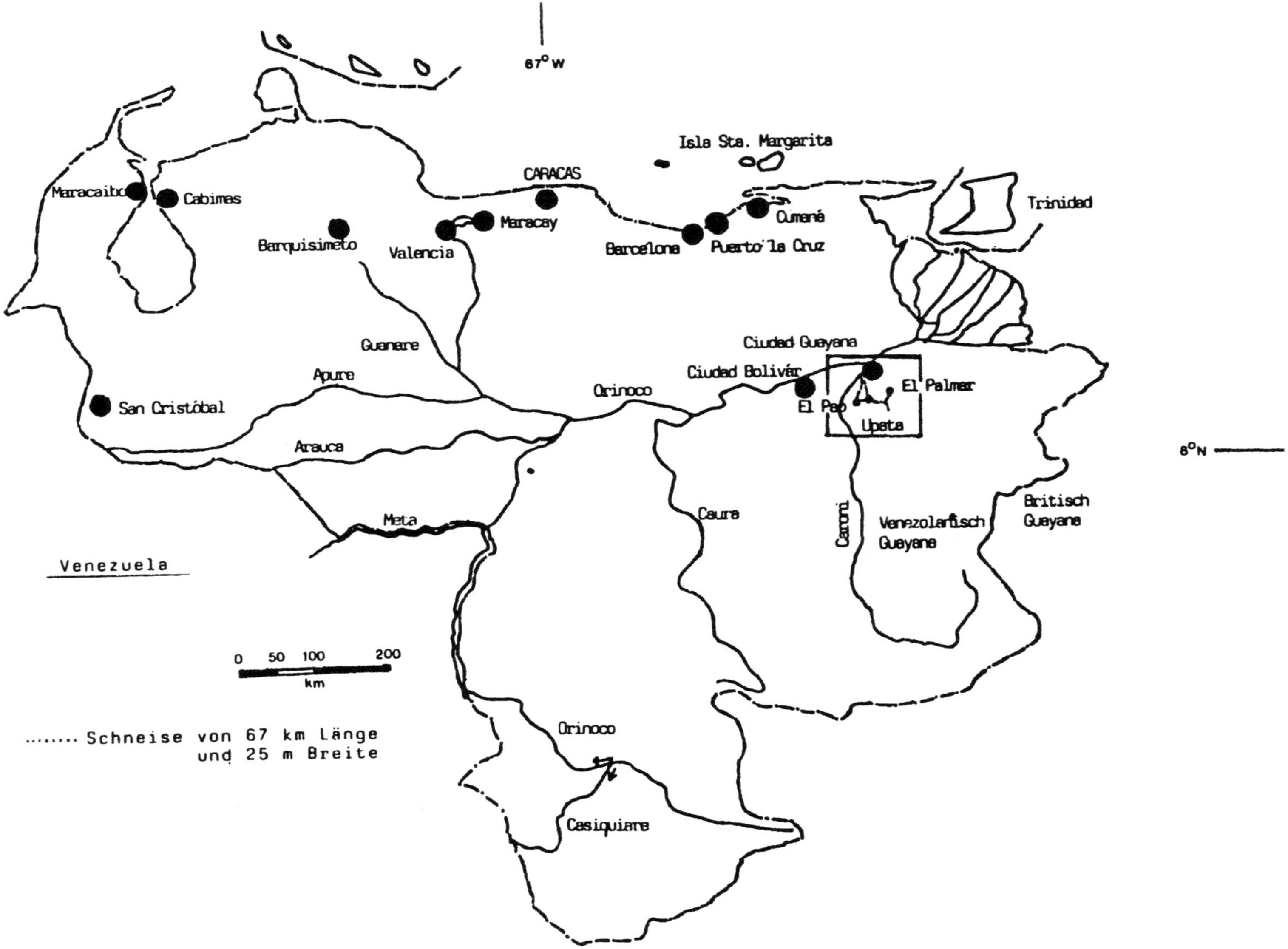

Map 2. Source of the map: Ministerio de Agricultura y Cria, Caracas 1961

ture of humid tropical forest plants. In our modern times, the tropical forests will be more rapidly destroyed than the structure and usefulness of their plants will be studied. We thus do not understand that some ecologists still object against our studies or devaluate them in a strange way. The only purpose we have is to show ecologists and physiologists that there occur interesting relations between the inner and outer structure of plant organs, their function, and their environment.

The aim of the present contribution was to complete the studies of leaf structure of the plants growing in a certain well-known are in Venezuelan Guiana. As leaf anatomy has already been investigated by Roth (1984), leaf morphology as well as leaf venation are presented in this book besides some other interesting structures concerning surface sculpturation, hydathodes and certain structures for the gaseous exchange of leaves.

Ingrid Roth November 1989

B. ROLLET

LEAF MORPHOLOGY

CONTENTS

1. INTRODUCTION

A description and an interpretation of leaf morphology are attempted for the tropical rain forest of Imataca, Venezuela with some comparisons being made with rain forests outside Venezuela and with other formations.

Roth gave an anatomical description and an ecological interpretation for the tropical rain forest, essentially according to the situation of the leaves above the ground (herbaceous stratum, undergrowth, codominant and emergent trees), i.e. taking account of the decreasing influence of humidity and the increasing influence of light from the bottom to the top of the trees. We shall retain

B. Rollet et al.: Stratification of Tropical Forests as seen in Leaf Structure, Part 2, pp. 1—75.

the term stratum because of its evocative value and the convenience of the expression. Actually, tree crowns merge gradually in the whole profile without evidencing clear-cut layers.

Roth demonstrated very clearly the anatomical differences that are induced by the intense light above the crowns. Sun leaves are provided with several well developed layers of vertically elongated palisade cells at the upper surface, whereas palisade layers are reduced in number or elongation in shade leaves.

A second factor in addition to light is humidity (water availability in the soil and atmospheric humidity which controls transpiration). Hygromorphic leaves are readily distinguished from xeromorphic leaves.

Although xeromorphic conditions and intense light often occur together, and although hygromorphic conditions often coexist with low light intensity, these ecological factors are not necessarily coupled. Leaves can retain some hygromorphic characters in crowns exposed to full light. I. Roth showed that some species exhibit at an early stage characteristics that are in conflict with the environmental conditions and therefore are genetically bound.

The occurrence of families, genera or species generally bound to montane conditions is sometimes surprising in the tropical lowlands and likewise the occurrence in the rain forest of deciduous or xeromorphic species usually bound to dry forest types. Their anatomical study discloses beyond any doubt their original status and the occurrence of these species outside their normal habitat cast some light on the floristic vicissitudes during the geological times, especially the Quaternary and Holocene.

2. MATERIAL COLLECTED

The site and circumstances for the collecting of the botanical material have been described by Roth. It seems convenient to add two comments.

(1) The objective of botanical collecting in Imataca was originally the determination of the species occurring in a commercial forest inventory dealing with all trees above 40 cm DBH. This inventory was carried out during three years by a dendrological team with the help of two or three climbers. The work was painful, time-consuming, dangerous and expensive. It is still an occasional practice in tropical botany and forestry and is therefore of special interest.

Collecting on big trees was progressively extended to the undergrowth species and was mainly concentrated on two regions of Imataca, Rio Grande and El Paraiso, some 65 km apart immediately south of the lower Orinoco River, east of the junction with Caroni River. About 600 trees belonging to 300 species were located and marked along two access-trails, and monitored until obtaining a complete collection (leaves, flowers and fruits on the same tree). Most of the collected species were in the adult phase but we endeavoured to collect also intermediary and juvenile stages.

Roth could deal with 205 adult species and 103 species at various stages of development, 232 species in all. I studied 197 species in the adult phase and 112 at intermediary and juvenile stages, 240 species in all. For every species, 30 leaves were collected from the same tree. Compound leaves were collected complete and were not sampled for leaflets. In many cases the collection was duplicated or triplicated on a second or a third tree with 30 leaves or only 12 leaves.

(2) A strip 25 m wide and about 67 km long was surveyed into 12.5 × 50 m plots for trees 10 cm DBH and over. The objective was essentially the study of the architecture and floristics of the forest. This 155 ha inventory gives the distribution of frequencies of trees by 10 cm diameter classes (about 400 species between 10 cm and 220 cm DBH), and is used to get a statistical image of the leaf morphology of the forest. Thus the somewhat qualitative and insufficient representativeness of our collection on 240 species is corrected to a certain extent by considering the relative numerical importance of the various species and improving the evaluation of their proportions.

3. METHODOLOGY

Studied characters

The following coded characters were considered for every species.

Leaf types
1. Simple
2. Compound even-pinnate
3. 〃 odd-pinnate
4. 〃 2 — even-pinnate
5. 〃 2 — odd-pinnate
6. 〃 palmate

Leaf arrangements
1. Alternate distichous
2. Alternate
3. Opposite distichous
4. 〃 decussate
5. Whorled

Leaf margin
1. Entire
2. Not entire

Leaf forms
1. Lanceolate
2. Ovate
3. Elliptic lanceolate
4. Oblong
5. Spatulate-obovate
6. Obovate

Consistence
1. Very coriaceous
2. Metallic sound
3. Leather
4. Leather, somewhat breakable
5. Papyraceous
6. Weaker than papyraceous
7. Soft
8. Very soft
9. Very fibrous
10. Herbaceous
11. Succulent, fleshy

Lawrence's terminology (1951) is used for the form classification and applies for leaves and leaflets. Other codes (linear, rhomboid, etc.) have been used for leaflets of compound leaves. The orbicular form is present in only one species of *Coccoloba* (v. n. Uvero).

Measurements

In most cases 30 leaves were collected for every species on the full length of several twigs of one tree in order to cover the whole range of leaf size. As far as possible only leaves that were undamaged by insects were collected. If this was not possible, the leaf areas were corrected for their browsed portion.

Leaves were dried and measured with 12% humidity. Later on and for other formations, lengths widths and weight were measured on green leaves also to evaluate retractibility and succulence.

For a given species, total length and total width to the nearest mm were measured on each of the 30 leaves of a species sample and also petiole and acumen length, width of the base of the acumen, leaf area, and leaf weight to the nearest centigram (including petiole). Petioles were weighed separately only when they were obviously important and heavy.

Leaf areas were measured using grids of points, a process which I find to be less time consuming than a planimeter or drawing on paper, cutting and weighing. Four grids were used: one point every 4 cm^2 (2 × 2 cm) for the biggest leaves; one point per cm^2 for most of the leaves larger than 20 cm^2; 4 points per cm^2 for leaves between 5 and 20 cm^2; one point every 4 mm^2 (2 × 2 mm) for leaves less than 5 cm^2.

Photoelectric machines are available but they frequently fail, especially when leaves are not perfectly flat. On the other hand leaflets of bipinnate leaves often partially cover one another, which results in important underestimations if a photoelectric process is used. Air-dry leaves were weighed individually. Water content was checked for 12 species (30 leaves each) and was found to be close to 12%.

When dealing with leaf biometrics various difficulties must be overcome and some cautions are needed.

- During leaf sampling, young leaves must be avoided because they are much thinner and lighter than adult leaves, especially when leaves are flushing after a deciduous period of the tree in dry or semi-deciduous forests. Leaves with a colour obviously different from the colour of adult leaves, and with a weaker consistency should not be collected.
- Leaves must be laid flat before drying, folded leaves will be more tedious to meausre. It is necessary collect leaves with complete petioles and to put an ink mark at the limit blade-petiole for a correct measurement of retracbility on length. Some leaves do not remain flat after drying: the margin may become undulated or the whole leaf may take on a saddle-shape in which case it must be cut into two parts along the principal nerve.
- Some species are badly damaged in their

leaves by insects, especially *Meliaceae*, to the point that an adequate correction for weight is extremely time consuming. Some species are attacked by galls and weighing the leaves may be useless. Their presence may be a good characteristic for some species.

— Epiphylls are rarely abundant in lowland rain forests, at least when annual rainfall does not exceed 3 m, but they are profuse in montane rain forests and may have an incidence on weight measurements.

Direct area measurements in case of bipinnate leaves is tedious because of the number and minuteness of the leaflets. An indirect measurement by weighing (with or without the rachis) is the obvious answer, perhaps after one or two days to ease the deciduousness of the leaflets.

Leaf weighing is usually carried out by including the petiole, especially in biomass measurements and for the evaluation of leaf area indices. However the computation of linear regressions $S = a + bX$ with S = area and X = weight may give high values for the coefficient a. When the weight of the petioles greatly exceeds the weight of the blades, it may be convenient to measure them separately, e.g. for *Sterculia, Cecropia, Didymopanax*. In the case of compound leaves, weights refer to leaflets.

The major problem is to avoid biases by correct sampling. Replicates with a second or third tree usually show quite significant differences (see Appendices 1 and 3), and variability is very high within a tree or between trees of the same species. In other words, it is quite risky to consider that the mean size of 30 leaves originating from one tree could be an unbiased estimate of the size of the species: the size of the tree greatly affects the mean area of the leaves and since the populations of the various species are quite different, only large inventories of trees and a wide sampling of the leaves can approach some representativeness. A 155 ha inventory will be used to improve the reliability of the results. It is practically impossible to avoid systematic errors when collected material is limited, and it is easy to imagine the inadequacy of the results based on fragmentary data.

The most lengthy measurements concern compound leaves, especially bipinnate leaves; a complete direct area measurement is unrealistic: on the other hand a sampling on leaflets should not be handled without care: the leaflet size is often correlated with its position in the leaves; the lower leaflets are often smaller but the larger ones may be at the upper and or in the medium portion of the leaf. Their number is not independent of their size. In bipinnate leaves the number of leaflets per pinna is not independent of the number of the pinnae; thence there arises too complicated a computation of the variability of the leaf size as a product equal to the mean size of leaflet × mean number of leaflet per pinna × mean number of pinnae. The variance of a product is not a simple expression.

Finally it is not certain whether one should deal with leaflets as units, a system used by Raunkiaer in his classification of leaf sizes, or with the whole compound leaf. According to the procedure chosen quite different results are obtained.

Incidence of drip-tips

Drip-tips are apices that are well individualized, long and narrow. Their area is negligible with respect to the blade area.

Lengths of drip-tips and the ratio of total length to total width of the blade are interesting ecological criteria, but the overall appraisal of leaf form in the Lawrence system does not need to consider the length of the drip-tip. However, the form factor of the leaf K may be computed in two ways; $K = S/(L \times l)$ where L = total length including drip-tip and l = total width or $K' = S/(L' \times l)$ where L' = length without drip-tip length. Both will be considered; K may be conspicuously smaller than K'.

Accuracy of leaf area measurements

The accuracy of a planimeter is between 1 and 2%. Bouchon (1973) studied the accuracy of area measurements using grids of points. It depends upon the length of the line which delimits the area and upon the fineness of the grid. However the amount of time necessary to count the points is proportional to this number and a compromise is necessary between the accuracy and the amount of time required.

In practice the type of grid is selected according to the size to be measured in such a way that at least 30 points are counted per leaf. If this is not so, the grid is cast a second time at random over the leaf for another count and the average of the two counts is retained. A point falling exactly on the margin of the leaf is given a score of 0.5. Moreover one or two leaves per species are remeasured and departures from the first measurement (either plus or minus) are tabulated. The distribution of these paired measurements can be used for an estimate of the error. Thus, based on 105 differences on leaves belonging to the 20–182 cm^2 size class of Raunkiaer (mesophyll leaves), the distribution of absolute errors is slightly skewed with probably a small excess of value 1, the mean deviation being + 7/105 = 0.067 cm^2.

Values of differences	−3	−2	−1	0	+1	+2	+3	+4	Total
Frequency	3	7	25	28	32	8	1	1	105

The relative error on a 100 cm^2 area is therefore negligible. Relative errors on lengths and widths are rather high but since 30 leaves are used the mean error on K (form factor) is about 0.5% for leaves of the mesophyll class, i.e. the third decimal of K is not significant. The same applies to the ratio L/l.

Indirect measurements of leaf area through weight have been used for a long time for field crops (Watson, 1937) or for forests using regression analysis and allometric relations between leaf area and more readily accessible related parameters. Since we are not dealing with leaf area index (LAI) and biomass measurements in this study, these techniques will not be considered.

4. TYPES OF LEAVES

Among the 183 species studied, 126 are provided with simple leaves (68.85%), 45 are pinnate (24.59%), 8 are bipinnate (4.37%) and 4 are palmate (2.19%). The proportion of simple leaves is therefore quite close to two thirds (see illustrations).

These proportions show only a gross spectrum of leaf types. Other figures will be given later for several tropical rain forests of South America.

The gross spectrum can be corrected by taking account of the number of trees occurring in each species, and also the different sizes of the trees. As a first approximation it can be accepted that the total number of leaves of a tree is related to the size of the crown, i.e. to the square or cube of its diameter. The crown diameter is (roughly) proportional to DBH.

Hence a possible weighing owing to the 155 ha inventory which gives the number of trees per species and 10 cm DBH classes. Weighing can be applied using the proportion of number of trees in a class to the grand total. Any other kind of weighing can be preferred either using basal area (or squared DBH) by species or their timber volumes, or $(DBH)^3$.

Using basal areas by species, the corrected spectrum of leaf types yields 68.51% simple leaves, 25.83% pinnate and palmate, 5.65% bipinnate i.e. virtually the same proportion of simple leaves compared with the gross spectrum and a slightly increased proportion of bipinnate leaves. Weighing by $(DBH)^3$ would give 66.70%, 27.23% and 6.07%, respectively.

Compound leaves: Importance and significance

Givnish (1978) questioned the importance and significance of compound leaves in tropical forests and two particularly significative statements are quoted (p. 372):

> Compound leaves appear to be adaptive in at least two sorts of environmental contexts: in warm seasonally arid situations that favour the deciduous habit, and in light gap and early successional vegetation where rapid upward growth and competition for light favour the cheap throwaway branch . . .

and further

> If most emergents begin growth as gap colonizers, this may

help explain the high incidence of compound leaves among such species.

Concerning the importance of compound leaves in tropical forests some elements of an answer can be supplied on three points:

- The proportion of compound leaves in several forests of Tropical America.
- Are compound leaves more frequent in natural gaps than in the neighbouring forest?
- Are species with compound leaves more deciduous than species with simple leaves?

Proportions of compound leaves in the strata and in different formations: The Venezuelan Guiana Imataca range

The interpretation is based on the inventory of 155 ha based on transects running more or less East—West and North—South, with 67 km total lengh and 67777 enumerated trees over 10 cm DBH.

The proportion of compound leaves will be expressed in two ways; considering the *number of species* and the *number of trees* over 10 cm DBH. In both cases, six diameter classes will be considered, corresponding to a rough stratification in height of the forest: 10 to 19 cm DBH; 20 to 29; 30 to 39; 40 to 49; 50 to 59 and above 60 cm DBH (Table 1).

The proportion of compound leaves among all trees ⩾ 10 cm DBH is 31.10% (Table 1). The proportion varies to some extent in the different "strata": 31% among trees between 10 and 29 cm; 29% among trees 30 to 49 cm; 37% among trees ⩾ 60 cm DBH.

This last proportion (37%) among trees bigger than 60 cm DBH is very significantly different from the proportion (31%) found among trees smaller than 60 cm DBH (χ^2 19.5); the same is true between trees above and below 50 cm DBH; but the difference of proportions is not significant between trees 20 to 29 cm DBH and 30 to 39 cm DBH.

Difference of proportions between trees 10 to 29 cm DBH (32.8%) and trees 30 to 49 cm DBH (28.8%) is very significant.

These results are rather surprising, because one would expect that the proportion of compound leaves would decrease progressively as one moves from the emergent trees to the undergrowth.

If now the *proportion of species* with compound leaves is considered (Table 1), the increase is more progressive from small diameter (less than 30%) to big diameter: it is 42.7% for species ⩾60 cm DBH (Table 1).

Cain *et al.* (1956) found 37% of species with compound leaves among trees 30 m and higher and 27% of species in the undergrowth between 8 m and 30 m in height in a rainforest near Belém (Pará State, Brazil). This result can be compared with three others originating also from Brazilian Amazonia, Jari, Jabuti and Curua Una. They have been expressed in the same way as Imataca.

The rainforest of Jari is located about 400 km

Table 1. Proportion of compound leaves in 10 cm diameter classes (Imataca) 155 ha.

	Number of species							Number of trees						
Type of leaves	10–	20–	30–	40–	50–	⩾60	Total	10–	20–	30–	40–	50–	⩾60	Total
Pinnate	75	73	57	49	40	38		9028	3376	1669	865	305	282	
Bipinnate	16	14	14	13	10	12		2323	1001	342	117	50	134	
Bifoliolate	03	04	03	02	02	03		79	48	56	28	19	36	
Trifoliolate	02	01	01	03	02	02		16	04	04	08	06	14	
Palmate	05	07	06	06	04	04		543	243	77	32	12	24	
Palm trees	07	05	02	01				215	187	18	01			
TOTAL	108	104	83	74	58	59	127	12914	4859	2166	1051	392	450	21112
GRAND TOTAL	361	304	262	209	142	138	397	38592	15597	7523	3625	1224	1216	67777
% COMPOUND	29.9	34.2	31.7	35.4	40.8	42.7	32.0	31.6	31.1	28.8	29.0	32.0	37.0	31.1

West of Belém immediately North of the Amazon in the vicinity of Almeirim. On a small sample of 3.4 ha. the proportion of species with compound leaves is regularly increasing from trees 20 cm DBH to trees 60 cm DBH and above (Table 2). When expressed in *number of trees*, the proportion of compound leaves drops in the diameter class 40—59 cm, a result difficult to explain and possibly due to the small size of the sample.

The rainforest of Jabuti is located about 180 km South-East of Belém. A 6 ha sample distributed over 3 000 ha shows a fairly regular increase of the proportion of compound leaves from the undergrowth to the emergent trees, with more irregularities when this proportion is expressed in number of trees (Table 3). Curiously the diameter class above 55 cm DBH shows a larger proportion of species with compound leaves than the corresponding number of trees: 50 trees with simple leaves totalize 36 species whereas 41 trees with compound leaves are distributed among 16 species only; this indicates a high spatial concentration of trees with compound leaves.

The rainforest of Curua Una is located in the vicinity of Santarém about 700 km West of Belém. A 5 ha sample randomized in a 100 ha block (1 × 1 km) shows a lower proportion of species with compound leaves than in the four examples above. The tendency suggested by Cain *et al.* whereby the proportion of species with compound leaves should increase from the bottom strata of the forest to the top of the emergents does not apply here (Table 4).

The tendency suggested by Cain exists in 3 cases out of 5 when expressed in species, only in 2 cases out of 4 when expressed in number of trees (and even so with irregulaties). Therefore one must be careful before generalizing a hypothesis: it seems that there are many local variations which reflect the variability of the floristic compositions in the rainforests.

According to our examples in South America, the rainforests show a proportion of compound leaves between 20 and 45% among the emergent trees and possibly less in the undergrowth 25 to 37%, but still high. We can draw the following conclusion: *the tropical lowland rainforests do not show a constant proportion in compound leaves but a high proportion (unknown in temperate forests) is certainly one of their characteristics.*

It is even highly probable that the proportions mentioned for tropical America are exceeded in Central Africa. An inventory on 100 000 ha in Kango (Gabon) gives for trees ⩾ 10 cm DBH 106 species with compound leaves (of which 66 are legumes) out of a total of 308 species, i.e. 34.4%. Weighing by gross volumes of timber instead of basal area as before gives a proportion

Table 2. Proportion of compound leaves in 20 cm diameter classes (Jari) 3.4 ha.

	Number of species				Number of trees			
Diameter cm	20—39	40—59	⩾60	⩾20	20—39	40—59	⩾60	⩾20
Compound leaves	43	16	15	49	347	30	27	404
TOTAL	131	47	29	146	916	123	70	1071
% Compound leaves	32.8	34.0	51.7	33.6	37.9	24.4	38.6	37.7

Table 3. Proportion of compound leaves in 10 cm diameter classes (Jabuti) 6 ha.

	Number of species					Number of trees				
Diameter cm	15—24	25—	35—	45—	⩾55	15—24	25—	35—	45—	⩾55
Compound leaves	47	32	28	20	16	246	107	52	34	41
TOTAL	158	107	76	54	52	902	420	167	85	91
% Compound leaves	29.7	29.9	36.8	37.0	30.8	27.3	25.5	31.1	40.0	45.5

Table 4. Proportion of compound leaves. Curua Una 5 ha.

	Number of species			Number of trees		
Diameter cm	5–9	10–19	⩾60	5–9	10–19	⩾60
Compound leaves	37	28	08	727	389	19
TOTAL	146	113	37	2772	1396	87
% Compound leaves	25.3	24.8	21.6	26.2	27.9	21.8

of 51.5% in compound leaves, a much higher figure than any previously mentioned.

For the sake of comparison, we would like to give the proportion of compound leaves in two quite different forest types, a semideciduous tropical forest and a montane rainforest.

The semideciduous rainforest of Caimital is located in the vicinity of Barinas in the western Llanos of Venezuela: in a 100 ha block, (1 × 1 km) the proportion of species with compound leaves is 41.6% in the undergrowth, and 51% among the emergents, and respectively 32.0% and 43.8% if the proportion is expressed in number of trees (Table 5). Such proportions are fairly higher than what has been mentioned for lowland evergreen rainforests in Venezuela or Brazilian Amazonia.

The montane tropical rainforest La Carbonera is located some 40 km from Merida in the Venezuela Andes at an altitude between 2250 and 2450 m. The proportion of compound leaves found on 16 ha sampled in a 360 ha forest is 16 to 17%; this figure is much lower than in any previously mentioned lowland tropical forests, with only a slight difference between undergrowth and emergents (Table 6). The proportion varies irregularly in the different strata but is conspicuously low among emergent trees (4.6%) because of the high percentage of a conifer *Podocarpus rospigliosii* among trees 60 cm DBH and above.

Table 5. Proportion of compound leaves. Caimital Venezuela 100 ha.

	Number of species		Number of trees	
Diameter cm	20–29	⩾60	20–29	⩾60
Compound leaves	32	25	1459	711
TOTAL	77	49	4560	1622
% Compound leaves	41.6	51.0	32.0	43.8

In summary among the tropical forest formations studied, the lowland rainforest, the semideciduous rainforest and the montane rainforest are fairly well distinguishable concerning their proportions in compound leaves: the montane rainforest is the poorest; the lowland evergreen rainforest follows with large irregularities from one country to another while the semideciduous rainforest is the richest formation in compound leaves.

Abundance of compound leaves in natural gaps of tropical forests

Is there a greater proportion of compound leaves in gaps than in the neighbouring forest?

The question can be put in different terms: do light-demanding species show a higher proportion of compound leaves than tolerant species? (See below: compound leaves and behaviour of species.)

The lowland evergreen rainforest of Curua Una. Brazilian Amazonia.

Natural gaps were sampled in a 100 ha-block (1 × 1 km) with 3307 quadrant (1 × 1 m) for all individuals 1 m high and above; 726 individuals were found with compound leaves among a total of 4736, i.e. 15.3%; these 726 individuals were composed of 52 species out of a grand total of 182 species, i.e. 28.6% of species with compound leaves.

A similar inventory is available for the neighbouring forest but only for individuals 5 cm DBH and above. The populations of diameter classes 5–9 cm and 10–19 cm have been mentioned in Table 4: their proportion of compound leaves is higher than 15.3%. For example, the portion of the forest between 5 and 9.9 cm diameter with

Table 6. Proportion of compound leaves. La Carbonera Venezuela 16 ha.

	Number of species						Number of trees					
Diameter cm	20–29	30–	40–	50–	≥60	≥20	20–29	30–	40–	50–	≥60	≥20
Compound leaves	12	09	09	05	06	12	369	225	103	32	14	743
TOTAL	68	56	51	44	38	76	2490	1139	479	283	302	4693
% Compound leaves	17.6	16.0	17.6	11.4	15.8	15.8	14.8	19.7	21.5	11.3	4.6	15.8

compound leaves is 26.2% (727 trees out of 2772). gaps are significantly *less* abundant in individuals with compound leaves $\chi^2 = 11.0$d.f. = 1.

Thus it is difficult to agree with Givnish that the successional phase created by natural gaps is a highly favourable niche for species with compound leaves. These species are supposed to be generally quick-growing light demanders, and owing to these advantages happen to be in excellent conditions to become the emergents of the forest. But, in fact, the compound leaved species do not follow this process of selection. They do not preferably invade gaps but their survival is probably better there, and the fact that they are in general long-living species determine a relative enrichment of the emergent stratum in species with compound leaves.

Since it has been shown that the resuts of Cain *et al.* cannot be generalized without caution, the enrichment of the canopies in species with compound leaves is more a tendency than a rule in the tropical rainforests.

The montane rainforest La Carbonera. Venezuela We have already mentioned the relative paucity of species with compound leaves in the montane forest. A comparison is carried out between a small sample of eleven gaps and the undergrowth of the neighbouring forest concerning the proportion of compound leaves among individuals higher than 1 m and less than 10 cm DBH in La Carbonera (Table 7).

The proportion of compound leaves is not significantly higher in gaps than in the neighbouring undergrowth, a result similar to the situation at Curua Una.

Compound leaves and behaviour of the species Is there a higher proportion of compound leaves among light-demanding species than among tolerant species?

Results will be expressed in number of species and in number of trees as before.

Criteria for light demanders and tolerant species A classification of species according to their behaviour was proposed on the basis of the mentioned Imataca inventory and using the different types of distribution of trees in the diameter-classes by species (Rollet, 1974).

Tolerant species
- Undergrowth (1)
- Edificators (2)
- p. p. species represented only by trees in the smaller diameter classes (3)

Light-demanders
- Erratic distribution (4)
- Bell-shaped (5)

species of intermediate behaviour
- S shaped distribution (6)
- Flattened distribution (7)

A first comparison will use the criteria in the most restricted meaning (stricto sensu), i.e. category (1) versus categories (4) + (5). The second comparison will use the criteria in a wider sense (lato sensu) i.e. will compare categories (1) + (3) versus categories (4) + (5) + (7).

The four tables grouped as Table 8 show that

Table 7. Comparison of the number of trees with compound leaves. La Carbonera.

	Gaps	Undergrowth	TOTAL
Compound leaves	11	18	29
Simple leaves	304	497	801
TOTAL	315	515	830
% Compound leaves	3.5	3.5	3.5

$\chi^2 = 0.039$ d.f. = 1

Table 8. Proportion of compound leaves among tolerant and light demanding species, Venezuelan Guiana.

	Number of species ⩾ 10 cm DBH			Number of trees ⩾ 10 cm DBH		
	Simple leaves	Compound	TOTAL	Simple	Compound	TOTAL
Tolerant species stricto sensu	44	16	60	11833	4297	16130
Light demanders id.	56	47	103	1224	1142	2366
TOTAL	100	63	163	13057	5439	18497
Tolerant species lato sensu	80	27	107	12049	4344	16392
Light demanders id.	66	51	117	2490	1495	3985
TOTAL	146	78	224	14539	5839	20378

the number of trees (or species) of light-demanders has always a larger proportion of compound leaves. For example, stricto sensu there are nearly as many light-demanders with compound leaves (1142) as with simple leaves (1224), whereas there are only 4297 trees with compound leaves versus 11833 trees with simple leaves among tolerant species. The differences are significant or very significant.

Values of χ^2

CRITERIA	Number of species	Number of trees
Stricto Sensu	4.97	463.8
Lato Sensu	7.51	190.9

In conclusion, the character of simple leaves versus compound leaves is not independant of the character of tolerant or light-demanding species. *Light-demanders have a significantly higher proportion of compound leaves.* This proportion is 50% among light-demanders, 25% among tolerant species (percentages expressed either in number of species or number of trees).

Compound leaves and deciduous species Are species with compound leaves more deciduous than species with simple leaves?

Venezuelan Guiana: Trees ⩾ 10 cm DBH from the 155 ha inventory are separated into four categories: evergreen with simple leaves, evergreen with compound leaves; deciduous with simple leaves, deciduous with compound leaves (Table 9).

The paired characteristic simple leaf—compound leaf is not independent of the paired characteristic evergreen—deciduous. The proportion of deciduous leaves among compound leaves (8.33%) is significantly higher than the same proportion among simple leaves (7.66%) although the difference appears very small. This is due to the high number of observations, which makes the chi-square test very sensitive.

Alternatively there is 33% of compound leaves among deciduous leaves and only 31% among evergreen leaves.

Compound leaves are (slightly but) *significantly more deciduous than simple leaves.* Actually Imataca rainforest with about 8% deciduous trees should be called seasonal rainforest, according to Beard's terminology.

Caimital. Venezuela: The two pairs of characteristics are considered in two diameter classes: 20 to 29 cm and 60 cm and above, both in number of species and number of trees. In number of species the percentage of compound leaves is independent of the deciduous—evergreen characteristic (Table 10) $\chi^2 = 3.50$ (corrected for small frequencies).

In number of trees, compound leaves are distinctly more deciduous (55.9%) than the simple leaves (33.1%). See Table 10, $\chi^2 = 212.8$.

The former result concerning species appears abnormal only because of the low frequencies: the difference in deciduousness is not statistically

Table 9. Proportion of compound leaves and deciduousness, Venezuelan Guiana.

	Number of trees ≥ 10 cm DBH			
	Evergreen	Deciduous	TOTAL	% Deciduousness
Simple leaves	43090	3575	46665	7.66
Compound leaves	19353	1759	21112	8.33
TOTAL	62443	5334	67777	7.87
% Compound leaves	30.99	32.97	31.15	$\chi^2 = 8.92$ d.f = 1

Table 10. Proportion of compound leaves and deciduousness, Caimital.

	Number of species 20—29 cm DBH			Number of trees 20—29 cm DBH			%
	Evergreen	Deciduous	TOTAL	Evergreen	Deciduous	TOTAL	Deciduousness
Simple leaves	29	16	45	2075	1026	3101	33.1
Compound leaves	12	20	32	644	815	1459	59.9
TOTAL	41	36	77	2719	1841	4560	40.4

significant but quite perceptible: 20 deciduous species out of 32 species with compound leaves; only 16 deciduous species out of 45 species with simple leaves.

The same results are obtained from Table 11 for trees above 60 cm DBH: statistically non-significant for species with $\chi^2 = 3.44$, the difference is highly significant for trees $\chi^2 = 342.7$.

Eight species are deciduous out of 24 species with simple leaves (one third); 17 species are deciduous out of 25 species with compound leaves (about two thirds). The difference in percentages is even higher with trees: 31.9% versus 78.3%.

In short, compound leaves are definitely more deciduous than simple leaves in a semi-deciduous forest, and rather deciduous in evergreen rainforests.

The statement applies to a lesser degree to the lowland rainforests of Imataca only because the number of observations is high so that the chi-square test becomes very sensitive.

Arrangements of leaves on branches. Imataca

Out of 183 species, more than half have alternate leaves and less than 20% are opposite. The whorled arrangement is exceptional, e.g. *Tabebuia*

Table 11. Proportion of compound leaves and deciduousness, Caimital.

	Number of species ≥ 60 cm DBH			Number of trees ≥ 60 cm DBH			%
	Evergreen	Deciduous	TOTAL	Evergreen	Deciduous	TOTAL	Deciduousness
Simple leaves	16	8	24	620	291	911	31.9
Compound leaves	8	17	25	154	557	711	78.3
TOTAL	24	25	49	774	848	1622	52.3

stenocalyx. The distribution of the 183 species according to leaf types (simple, pinnate, bipinnate) and the arrangement (alternate, alternate distichous, opposite, opposite decussate) is found in Table 12.

5. *LEAF FORMS*. DESCRIPTION. BIOMETRICS

Introduction

Morphogenesis will not be considered and we suggest consulting the thesis by Cusset and the contributions of Jeune for an introduction to the subject and for bibliographical orientations.

Twenty-three centuries ago Theophrastus questioned the significance of various leaf forms of ivy.

Concerning leaf forms Ashby (1949) posed four questions:

- How leaf form is determined by local differences of cell division and growth?
- What is the chemical and physical basis of these differences?
- What is the influence of the environment on leaf form?
- What is the influence of age and maturity of the plant on leaf form?

Ashby did not attempt to answer the first two questions. To say that some genetic reason is involved in leaf form is of little help, even within one species where varieties may show very different leaf forms.

For a century many works have contributed to the study of the influence of the environment on leaf form. Ashby mentions only the principal phenotypic factors: water supply (water and aerial leaves of water plants and rheophytes), mineral nutrition (Goebel's experiments in 1909), influence of temperature, light, and day length.

Ashby studied in greater detail the influence of physiological age. The observation of shoots of a flowering species shows that the successive leaves are not similar along the stem. These differences do not occur haphazardly; there is a regular gradation from one node to another. This aspect of leaf morphology is related to the findings on architectural models and growth rhythms.

The biometrical examination of a large sample of leaves collected at random on a species shows various leaf forms, sizes and consistency belonging to various populations of leaves that obviously developed in different conditions. This aspect will not be dealt with here.

Leaf form will be studied first, independently of the size of the leaf. Its characterization will follow the classic terminology (Lawrence, 1951) according to patterns which have been defined previously: lanceolate, elliptic lanceolate, oblong, ovate, obovate, etc.

When a large number of species is processed, it becomes evident that many intermediate forms exist which are difficult to grade into one form rather than another unless new forms are adopted.

The observation of many leaves belonging to the same species shows their high variability around the mean (Appendix 2, Column 8).

We will concentrate on two principal measurements (total length L and total width l) to charac-

Table 12. Frequencies of leaf types and arrangements, Imataca.

	Alternate	Distichous	Opposite	Decussate	TOTAL
Simple leaves	41	54	19	12	126
Pinnate leaves	2	45	1	1	49
Bipinnate leaves	–	7	–	1	8
TOTAL	43	106	20	14	183

terize leaf form quantitatively, using especially the ration L/l, the form factor $K = S/L \times l$ (S being the leaf area) and a slightly modified form factor using the length without the drip-tip length. A third morphological character will be added: the drip-tip length.

Ratio between length and width of the leaf

Does the ratio L/l vary during the life time of a species from the seedling to the adult phase?

Is the distribution of L/l data of some use in the classification of leaf forms?

1. *Influence of the size of trees on L/l*

Simple and compound leaves are pooled; L/l is calculated for each leaf or leaflet (30 leaves per species). See the means of the sample and the range in Appendix 3. These means are sorted according to the total height of the tree, into 5 height classes. The mean ratios of each height class are calculated and compared. They vary rather irregularly. Table 13 is a summary of the means and variability.

In a first step, species are compared whenever several individuals of different size are available: each line of Table 14 is a species. **Bold type** is used to show the few L/l values of saplings and young trees that are smaller than the corresponding tree value; in general the L/l value of a seedling is higher than for a sapling, and the latter higher than for a tree.

Some species have been studied on only one individual of any size, some others on two (or three) trees without any juvenile stage or trees smaller than 20 m. See Appendix 1.

Mean values of L/l by height classes are improved by pooling all the data available in Table 15. The following comments are relevant to Table 15:

- The irregularities of L/l average values shown in Table 13 disappear in Table 15.

Table 13. Values of L/l ratios in function of total height (only species with tress and juvenile or young stages).

Total height m	<1	1–4.9	5–9.9	10–19.9	⩾20
Means of L/l	2.75	2.91	2.79	2.93	2.57
Standard deviation	0.771	0.477	0.380	0.568	0.610
Coefficient of variation %	28.0	16.4	13.6	19.4	23.7
Number of individuals	9	50	20	16	60

Table 14. Comparison of means of the ratio *L/l* by species and height classes.

Total height m						Total height m					
<1	1–4.9	5–9.9	10–19.9	(1) ⩾20	(2) ⩾20	<1	1–4.9	5–9.9	10–19.9	(1) ⩾20	(2) ⩾20
1.61++				1.86			3.52++		3.00		
	2.77++			2.05			2.88++			2.48	
	4.39++			2.42					3.25++	2.16	
6.77++	2.79–			2.69			1.99–			2.02	
	3.16++			2.71	2.69–	4.35++	2.88				
	2.79++						2.83++			2.44	
	3.31++			1.87			2.54++			2.15	

Table 14. (continued)

Total height m						Total height m					
<1	1–4.9	5–9.9	10–19.9	(1) ⩾20	(2) ⩾20	<1	1–4.9	5–9.9	10–19.9	(1) ⩾20	(2) ⩾20
			2.41++	2.11				2.92++		2.12	
	3.93++			2.94				2.54−	2.46		
	3.65++			2.73			3.06++			2.66	
		2.58++	2.12				**2.67−**			2.71	
	2.46+			2.61			2.23++				
	3.28++			2.67	2.39++		2.36++	2.10			
	3.37−						2.89−	**2.40++**		2.81	
	3.00++			2.32	2.39−		2.72++			2.40	
	2.70++			2.45	2.52−		3.90++			2.35	
	3.35−			3.31			2.92++			2.14	
	2.79+			2.42	2.51−		3.50++			2.80	2.87−
	2.75++	2.70++		2.32	2.69++		**2.62++**	2.99			
		2.97++		2.60		**2.72++**				2.97	
	2.66−			2.75	2.66−			2.80−	2.69		
3.04++				2.26	2.58++	3.11++				2.80	3.04++
		2.92−		2.78					2.99++	2.14	
		3.29++		4.20					**2.84−**	2.89	3.43++
		3.44++		2.98	3.24+		**3.04++**			3.65	
	2.32							**2.57++**	2.79		
	2.73++					**2.69−**				2.87	
	3.37++		4.33				3.12−			3.05	
	2.93++			2.51	2.25++			2.47++	2.99++	1.84	
	3.26++			2.07							
	2.53++			2.12	2.55++						
	2.73++			2.27							
		2.46	3.29++	2.55							
2.91++				2.71	2.07++						
		2.78++	3.56++	2.28	2.14−						
	2.53++		2.84−	2.74	2.37++						
	2.28−			2.13							
	3.22++			2.70	3.07++						
2.23+				2.00							
2.13++				3.07	3.22−						
	2.77++			3.04							
	2.34++	2.88									
	3.12++			2.11							
		3.68+		3.47							
	2.45−										
	2.31++	2.96	3.30++	2.43							
		2.41++	2.00								

−: Difference non significant,
+: Difference significant, $p = 0.05$
++: Difference very significant, $p = 0.01$.

— *L/l* ratios decrease progressively from 3.115 (individuals less than 1 m high) to 2.470 (all trees above 20 m high).
— Differences between *L/l* means are not significant between height classes 10 to 19 m and above 20 m; 5—9 and 10—19 m; 1—4, 9 m and 5—9 m. But they are very significant between height classes 5—9 and above 20 m,

Table 15. Values of L/l ratios in function of total height (all data together).

Total height m	<1	1—4.9	5—9.9	10—19.9	⩾20 (1) to (4)	⩾20 (2)	⩾20 (3)	⩾20 (4)
Mean of L/l	3.115	2.820	2.802	2.670	2.470	2.513	2.346	2.427
Standard deviation	1.381	0.514	0.377	0.637	0.515	0.448	0.420	0.502
Coefficient of variation %	44.3	18.2	13.4	23.8	20.8	17.8	17.9	20.7
Number of individuals	12	75	28	29	223	52	34	77

and between 1—4.9 and above 20 m. Therefore, the elongation of leaves in the undergrowth is progressive between the various height classes and becomes very significant between stratum 1—10 m and stratum ⩾20 m. Observations below 1 m high are too few in number to show a significant difference between height classes <1 m and 1 to 4.9 m.

- Differences between *L/l* averages for trees ⩾20 m in columns (2) and (3) are not significant; they are just significant between (1) and (3) with $t = 2.07$.
- *L/l* values by species are already averages; this obscures the variability of *L/l* within and between species. Coefficients of variation have not been computed but could be deduced from tables giving the ratio range as a function of the number of observations. The range of *L/l* is indicated by species in Appendix 3.

Exceptionally high or low values of *L/l* are noteworthy: 7.83 for *Inga lateriflora* (v. n. Guamo cinta $H = 2$ m); 6.77 for *Rollinia multiflora* (v. n. Anoncillo $H = 0.3$ m); 1.09 for *Coccoloba sp* (v. n. Uvero $H = 10$ m); 1.20 for *Piper reticulatum* (v. n. Anisillo $H = 3$ m).

2. *Use of the ratio L/l to characterize leaf forms*

We now present two kinds of statistics for the ratios *L/l*.

- The distribution of individual values of *L/l* (by individual leaf) in Table 16. Species are grouped into six leaf form categories: (1 lanceolate, 2 ovate, etc.) for simple leaves only. Compound leaves were pooled into two categories (pinnate and palmate); eight species with bipinnate leaves were excluded because of the difficulty of measuring *L/l* on minute leaflets and the dubious meaning of such measurements.
- The distribution of mean values of *L/l* by species in the same six categories (Table 17) with the total distribution for simple and compound leaves.

Comments on Tables 16 and 17

- All the distributions of *L/l* values *leaf by leaf* (Table 16) have a wide range for any leaf form or type. This variability could probably be reduced through a better definition and more categories of leaf forms.
- The distributions are plurimodal and consequently mixtures; a typical case is the orbicular-ovate form where *Coccoloba* with a mean *L/l* of nearly 1.0 is obviously out of place and exceptional in all respects.
- The irregularity of the distributions can be reduced with a larger class interval, e.g. 0.2 instead of 0.1. See Table 19. Modes have been printed in **bold type** so that irregularities become more obvious.

All the distributions show some amount of skewness.

The distribution of the lanceolate leaf form is quite distinct from the so-called elliptic-lanceolate form. Their respective modes are 3.20 and 2.10 and respective means 3.10 and 2.32 (Table 18).

Although pinnate leaves are put together and belong to various populations, their overall distribution is unimodal with a wider scatter than palmate leaves.

Table 16. Distribution of individual values of *L/l* by leaf forms and types (class interval 0.1).

SIMPLE LEAVES			1										2										3										4										5							
	.8	.9	.0	.1	.2	.3	.4	.5	.6	.7	.8	.9	.0	.1	.2	.3	.4	.5	.6	.7	.8	.9	.0	.1	.2	.3	.4	.5	.6	.7	.8	.9	.0	.1	.2	.3	.4	.5	.6	.7	.8	.9	.0	.1	.2	.3	.4	.5	.6	.7
Lanceolate									2	1		4	8	16	28	24	33	30	56	46	**61**	59	**69**	54	**72**	51	41	32	28	24	28	12	13	10	3	6	3	3	3	2		2		2	1		1			
Ovate		2	16	11	1							1	2		4	9	6	6	2																															
Elliptic-lanc.						5	13	29	47	81	119	213	261	**268**	228	**232**	220	166	142	117	88	78	69	28	21	10	8	3	5	1																				
Oblong										1	9	12	6	6	7	10	7	11	13	16	18	14	16	9	16	17	7	3	8																					
Spatulate							1	1	10	8	12	9	9	10	2	5	5	5	9	9	7	3	12	6	5	3	5	1	3	1	1	1						1												
Obovate						1	2	7	12	8	9	15	17	19	21	9	8	4	5	8	3	1		1																										
TOTAL		2	16	11	1	6	16	37	71	99	149	254	303	**319**	290	289	279	222	227	196	177	155	166	98	114	81	61	39	44	26	29	13	13	10	3	6	3	4	3	2		2		2	1		1			
COMPOUND LEAVES																																																		
Pinnate					2	8	7	12	22	45	60	65	115	95	102	**117**	93	108	104	89	74	74	64	35	30	26	14	17	10	11	4	2	3	5	1	2	2	1	2		2			1				1		1
Palmate											2		1	9	7	11	**14**	12	10	10	11	7	7	4	3	3	1	2	3	1																				
TOTAL					2	8	7	12	22	45	62	65	116	104	109	**128**	107	120	114	99	85	81	71	39	33	29	15	19	13	12	4	2	3	5	1	2	3	1	2		2			1				1		1
GRAND TOTAL		2	16	11	3	14	23	49	93	144	211	319	419	423	399	**417**	386	342	341	295	262	236	237	137	147	110	76	58	57	38	33	15	16	15	4	8	6	5	5	2	2	2		3	1		1	1		1

Table 17. Distribution of mean values by species of *L/l*, by leaf forms and types (Class interval 0.1).

SIMPLE LEAVES	1										2										3										4										5							
	.0	.1	.2	.3	.4	.5	.6	.7	.8	.9	.0	.1	.2	.3	.4	.5	.6	.7	.8	.9	.0	.1	.2	.3	.4	.5	.6	.7	.8	.9	.0	.1	.2	.3	.4	.5	.6	.7	.8	.9	.0	.1	.2	.3	.4	.5	.6	.7
Lanceolate													1	2			2	1	4	1	3	3	3	2	2		1		1					1														
Ovate														1																																		
Elliptic-lanc.							1		4	6	12	15	8	3	11	5	4	8	2	2	1	1																										
Oblong										1						1	1			1	2		1																									
Spatulate								1			1					1					1		1																									
Obovate								1	1				2				1																															
TOTAL							1	2	5	7	13	15	11	6	11	7	8	9	6	4	7	4	5	2	2		1		1					1														
COMPOUND LEAVES																																																
Pinnate							1		2	1	3	7	4	4	5	2	5	5	2	2	2	1	1	1						1																		
Palmate															1	1			2																													
TOTAL							1		2	1	3	7	4	4	6	3	5	5	4	2	2	1	1	1						1																		
GRAND TOTAL							2	2	7	8	16	22	15	10	17	10	13	14	10	6	9	5	6	3	2		1		1	1				1														

Table 18. General means of L/l by leaf forms and types and numbers of leaves.

Leaf forms and types	Mean L/l	Number of leaves
Simple lanceolate	3.10	828
— Orbicular-ovate	1.05 2.35	30 30
— Elliptic lanceolate	2.32	2454
— Oblong	2.77	206
— Spatulate-obovate	2.52	144
— Ovate	2.12	150
Compound pinnate	2.51	1426
Palmate	2.69	117
TOTAL	2.51	5385

Form factors

Another way to characterize a leaf form is to measure the ratio K of the blade area to the rectangle $L \times l$ that circumscribes the blade.

As for the ratio L/l, K is computed for each leaf and the distributions of K by leaf forms and types are established. Means and ranges are given in Appendix 3.

Influence of the size of the tree on the form factor of the leaf

This influence cannot be known with accuracy because the available sample is too small; many species were collected without adequately measuring the height of the branches above the ground (Table 20).

The differences observed on the means of K are not significant between height classes < 10 m and 10 to 19 m, and between trees 20 to 29 m and above 30 m but the difference is significant between height classes 10 to 19 m and 20 to 29 m. If all available trees are separated into two groups — more than 20 m and less than 20 m — their difference of 0.681—0.642 is very significant.

Thus we may suspect that the undergrowth trees show smaller leaf form factors than codominant and emergent trees. This result is in agreement with the tendency of L/l ratios to decrease from the undergrowth to canopy trees, leaves becoming less elongated and less lanceolate.

Distributions of K values

The distributions of $K \times 100$ values (originally established with a class interval of 1) are presented in Tables 21 and 22 with a class interval of 2.

In a similar way to L/l ratios K values show a considerable scatter in their distribution, sometimes several modes and some skewness. Modes are indicated in **bold type**.

In Table 21, the column of K averages indicates that the various leaf forms are fairly well distinguishable in spite of their wide variability.

The general average value of K is 0.676 for simple leaves, and 0.682 for the leaflets of pinnate and palmate leaves.

These figures are in agreement with the findings of Cain and Castro (1959). In a rainforest of Brazilian Amazonia near Belém Cooper (1960) indicated that a ratio of two-thirds is fairly satisfactory for a quick measurement of leaf areas and does not yield biased results in the frequencies of the Raunkiaer leaf spectrum for an oak-hickory woodland near Ann Arbor, Michigan, U.S.A. However the value of $K = 2/3$ indicated by Cain — an easy rule of thumb — seems to be a little too low.

The means of K given in Table 22 by species (30 leaves each) are themselves very variable: from 0.453 to 0.855 for simple leaves (136 species) and from 0.505 to 0.805 for compound leaves (54 species). The lowest, obviously isolated value of 0.453 refers to an undetermined species

Table 20. Means of form factors K in function of the size of the tree.

Total height m	< 10	10—19	20—29	⩾ 30	< 20	⩾ 20
Number of trees	9	22	34	13	31	47
Means of K	0.627	0.648	0.685	0.670	0.642	0.681
Standard deviation	0.0377	0.0569	0.0468	0.0448	0.0523	0.0462

Table 19. Distribution of the ratio *L/l* leaf by leaf by leaf types and forms (Class interval 0.2) 0.8 means ⩾0.8 and < 1.0.

Leaf forms and types	0.8	1.0	1.2	1.4	1.6	1.8	2.0	2.2	2.4	2.6	2.8	3.0	3.2	3.4	3.6	3.8	4.0	4.2	4.4	4.6	4.8	5.0	5.2	5.4
Simple lanceolate					3	4	24	52	63	102	120	**123**	**123**	73	52	40	23	9	6	5	2	2	1	1
— Orbicular-ovate	2	27	1			1	2	**13**	12	2														
— Elliptic lanceolate			5	42	128	332	**529**	460	386	261	166	97	31	11	6									
— Oblong					1	21	12	17	18	29	32	25	33	10	8									
— Spatulate-obovate				2	18	21	19	7	10	18	10	18	8	6	4	2			1					
— Ovate			1	9	20	24	**36**	30	12	13	4	1												
Compound																								
— Pinnate			10	19	67	125	210	219	201	193	148	99	56	31	21	6	8	3	3	2	2	1		1
— Palmate						1	10	18	26	20	18	11	6	3	4									

Table 21. Distribution of the ratios 100 K = 100 *S/Ll* leaf by leaf for the various leaf forms and types (Class interval = 2) 38 means ⩾38 and <40.

Simple leaves	38	40	42	44	46	48	50	52	54	56	58	60	62	64	66	68	70	72	74	76	78	80	82	84	86	88	90	92	94	Means of K	Number of leaves
Lanceolate	3	4	5	6	7	6	8	18	38	56	90	84	105	**128**	89	88	59	57	16	15	5	4	2							0.638	893
Ovate						1						2	3	7	5	1	4	2	4	3	2	6	6	5	1	5	3			0.751	60
Elliptic-lanceolate						2	1	6	17	31	52	142	231	322	**387**	341	303	249	182	92	39	17	4	6						0.683	2424
Oblong										1	1		1	4	14	18	29	24	30	17	15	15	9	13	3	4	3	3		0.755	204
Spatulate									1	5	10	30	27	22	24	18	5	2	1	1										0.643	146
Obovate											1	8	17	26	34	30	20	6	6	1	1									0.675	150
TOTAL	3	4	5	6	7	9	9	24	56	93	154	266	384	509	**553**	496	420	336	239	129	62	42	21	24	4	9	6	3		0.676	3877
Compound																															
Pinnate					1		3	6	12	26	54	95	117	178	**203**	153	153	149	115	64	30	22	7	3	2	1		1		0.683	1395
Palmate								3	8	8	16	18	11	13	17	13	8	9	3	3	5	2	3	3	1	3	1		2	0.670	150
TOTAL					1		3	9	20	34	70	113	128	191	**220**	166	161	158	118	67	35	24	10	6	3	4	1	1	2	0.682	1545

of *Myrtaceae* (v. n. Guayabillo negro) the leaves of which are provided with an extremely long drip-tip.

Nothing can be said about the distribution of palmate leaves, since they were too few in number to be of significance (4 species).

The overall distribution of K means for pinnate leaves is unimodal and obscures the presence of several populations, if there are any. On the other hand the overall distribution of K means for simple leaved species shows at least two modes and appears clearly as a mixture of populations: lanceolate leaves are perfectly distinct from elliptic-lanceolate leaves.

The distribution of K for the lanceolate form (leaf by leaf) is bimodal whereas the distribution for K means is nearly unimodal, a sufficient reason to believe that the lanceolate group could be split into two subgroups, the higher K values being possibly allocated to the elliptic-lanceolate form.

The opposite happens with the elliptic-lanceolate form: the means distribution shows a mixture whereas the unimodal distribution leaf of Table 21 obscures this mixture.

A better leaf form characterization could possibly lead to a discussion of the ecological meaning in the light of a comparison between the main great tropical forest formations. Sometimes L and l were measured but not S; this explained the observation that the number of leaves in each leaf form is not always the same for L/l and S.

Another way to compute the leaf form factor

The figures given in Tables 20—22 for form factors K are computed using the total length of the leaf.

The main drawback of this procedure appears when non-lanceolate leaves are provided with a long narrow drip-tip, resulting in small K values, as if the leaves were lanceolate.

A second method of computation consists in dropping the length of the drip-tip and considering the length L' without the apex. A regression $S = a + b\,L' \times l$ is computed for each species. Values of coefficients a and b and characterics of the regressions are found in Appendix 2.

When a is small b is nearly equal to $S/\Sigma L' \times l$ and close to the means of K. The coefficient of correlation between S and $L' \times l$ is always very high (Appendix 2, column $r \times 10^4$) and rarely inferior to 0.98 (19 cases out of 126 for simple leaves and 8 cases out of 49 for compound leaves).

Similarly to K means, the coefficients b vary in a wide interval according to the species, between 0.48 and 0.88. Table 23 gives the distribution of b and the means as a function of leaf forms and types.

The so-called elliptic-lanceolate form is by far the most frequent; 79 species with simple leaves, 36 species with compound leaves, i.e. 115 species out of a total of 175. The oblong leaf form, the most rectangular of all forms yields for K a general means of 0.789, a figure quite near to the theoretical value of the geometric ellipse = 0.7854. The drawback of the term elliptic in leaf morphology is that a shape is still elliptic, geometrically speaking, with L/l varying from 1 (circle) to infinite (linear) when l becomes very small.

The term elliptic is therefore misleading and inconvenient for practical reasons.

In Table 23 K means are displayed in increasing order.

A perfect rhombic form would give $b = 0.500$; a rectangular form $b = 1$; a circular or elliptic form $b = \pi/4 = 0.7854$. Less frequent shapes, such as the hastate form, would yield b values inferior to 0.500 if total length is measured.

Generally speaking b values are slightly larger than K values. For simple leaves $b = 0.697$ $K = 0.676$; for compound leaves $b = 0.709$, $K = 0.682$.

Relationships between L/l K and S. (Figures 1—4)

Mean values of L/l and K are given in Table 24 in decreasing order of L/l and for simple leaves only. The two ratios vary in an opposite way for the two first leaf forms but otherwise they vary rather irregularly.

The graphs of L/l as a function of K means by species show that the cloud of points is fairly

Table 22. Distribution of the means by species of the ratios 100 $K = 100\ S/Ll$ for the various leaf forms and types (Class interval = 2) 38 means ⩾38 and <40.

Simple leaves	38	40	42	44	46	48	50	52	54	56	58	60	62	64	66	68	70	72	74	76	78	80	82	84	86	88	90	92
Lanceolate				1						2	3	4	4	5	3	2	3	2										
Ovate																1						1						
Elliptic-lanceolate										1		2	11	14	**17**	10	9	13	5	1								
Oblong																	2	2	1			1		1				
Spatulate												1	2		1	1												
Obovate													1		3	1												
TOTAL				1						3	3	7	18	19	**24**	15	14	17	6	1		2		1				
Compound																												
Pinnate											1	3	5	5	7	9	10	4	2	2	1							
Palmate											1	1		1		1						1						
TOTAL											2	4	5	6	7	**10**	**10**	4	2	2	1	1						

Table 23. Distribution of *b* coefficients by leaf forms.

b	(1)	(2)	(3)	(4)	(5)	TOTAL	(1)	(3)	(4)	TOTAL	GRAND TOTAL
0.45—0.49	1				1	2					2
0.50—											
0.55—	5	1				6	1	1		2	8
0.60—	10	3	6			19	3	4		7	26
0.65—	5	6	23			34	2	8		10	44
0.70—	4	1	30	2		37	1	14	2	17	54
0.75—	2		19	3		24	1	9	1	11	35
0.80—			1	1		2		2		2	4
0.85—				1	1	2					2
Number of species	27	11	79	7	2	126	8	36	3	49	175
b Means (unweighted)	0.646	0.660	0.717	0.789		0.695	0.665	0.715		0.709	0.698

(1) — Lanceolate
(2) — Ovate-obovate
(3) — Elliptic lanceolate
(4) — Oblong
(5) — Others.

distinct for lanceolate and elliptic lanceolate leaf forms (Figures 1 and 2).

Both clouds demonstrate a weak correlation between *L/l* and *K*. As expected the variability is even greater for compound leaves (all leaf forms mixed), see Figure 3.

In fact the ratios *L/l* and *K* are not independent variables: for a given *S* and *L/l* *K* varies as the inverse of l^2. A graph of *S* over *L/l* might be more meaningful; in other words the elongation of a leaf may have something to do with its area. Figure 4 gives a picture of the logarithm of *S* over *L/l* ratios, with different signs for the main leaf forms; each point is a species.

The more numerous species of elliptic-lanceolate form are shown with points on Figure 4; *L/l* ratios vary roughly from 1.8 to 2.8 and *S* from 9 to 250 cm^2. Next in abundance are lanceolate leaved species shown with small *x*; *L/l* ratios vary from 2.2 to 4.3 and *S* from 6 to 360 cm^2.

For oblong leaves, shown with little squares, *L/l* ratios vary from 2.0 to 3.2 and *S* from 19 to 180 cm^2.

Finally for ovate-obovate leaves *L/l* ratios vary from 1.7 to 3.3 and *S* from 8 to 190 cm^2.

Each cloud of points is individually potato-shaped and it can be fairly safely stated that there is a *quasi-independence between form and size of leaves.*

Table 24. Mean values of *L/l* and *K* by leaf forms (Simple leaves only; decreasing order of *L/l*).

FORMS	*L/l*	*K*
Lanceolate	3.10	0.638
Oblong	2.77	0.755
Elliptic lanceolate	2.32	0.683
Ovate-obovate	2.31	0.662

Drip-tips

Stahl (1893) and Gessner (1956) gave detailed interpretations of drip-tips in relation to the environment.

The importance of drip-tips is very different among the species. Many species, *Burseraceae*, some *leguminosae* and *Myrtaceae*, are provided with conspicuously long narrow apices.

Drip-tips are supposed to accelerate the drying up of leaves; but they do not necessarily hinder the development of epiphylls (observations at Rio Grande, Venezuela in the rainforest).

Mouton (1966) characterizes the apices quanti-

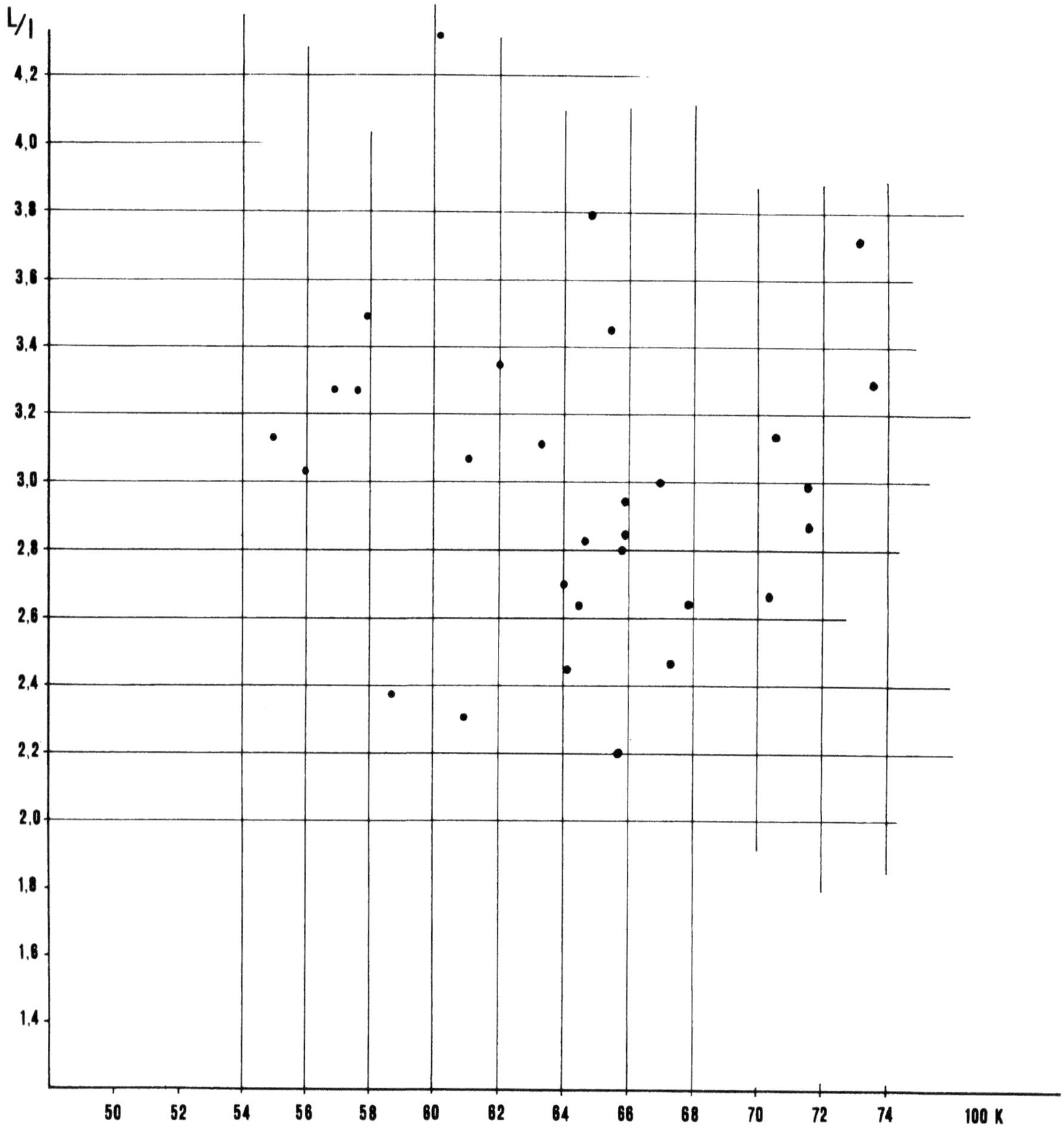

Fig. 1. Ratio *L/l* as a FUNCTION OF LEAF FORM FACTOR *K*.

tatively by means of 3 measurements: length, width and apex angle (p. 495, Figure 3). The apex angle is defined by two bitangential lines to the margin drawn at the point of inversion of curvature on each side of the principal nerve. Width and length are readily measured from the position of these two points.

However the points of inversion of curvature are often ill-defined and consequently so also are the width and length of the apex. Besides it is difficult to speak of apices in bipinnate leaves where the leaflets are rather mucronate.

In Appendix 2 (column 9) driptips are scored in the following way:

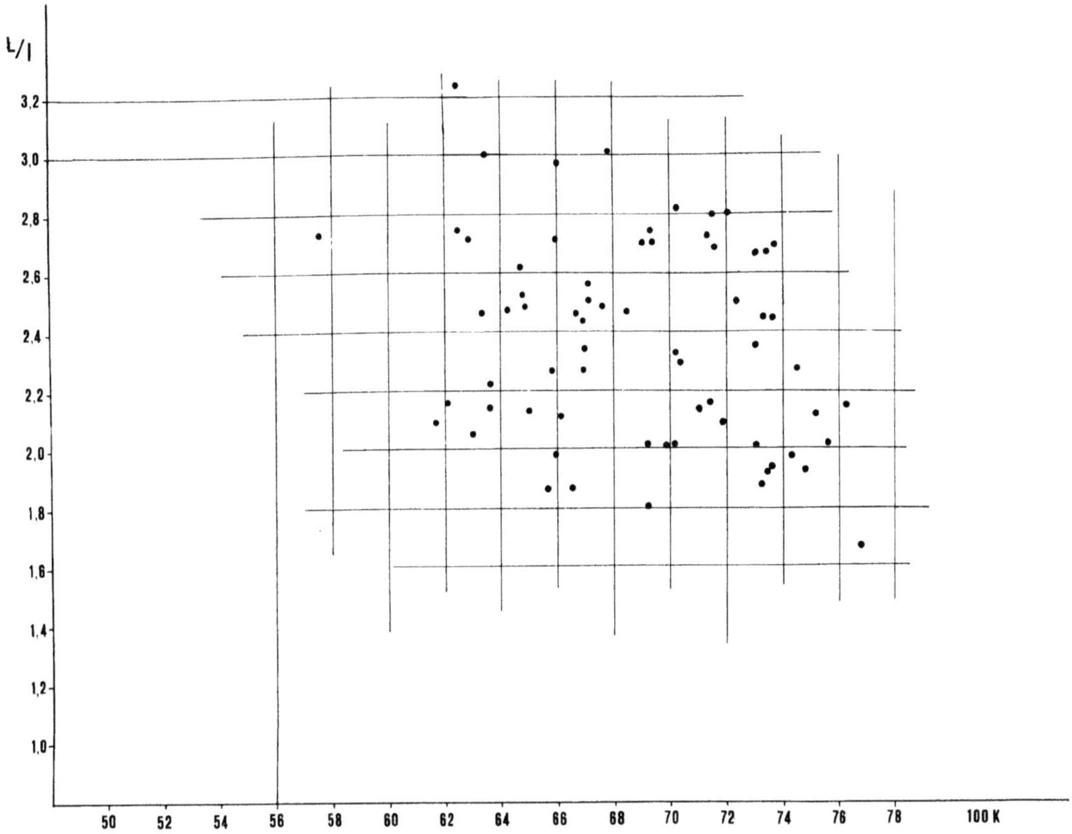

Fig. 2. Ratio *L/l* as a FUNCTION OF LEAF FORM FACTOR *K*. Simple elliptic lanceolate Leaves. Code 3.

— very conspicuous, narrow and long xx
— well marked but short x
— present but weak ε
— -id- sometimes lacking in significant proportion ε-
— always absent 0

A statistical description of drip-tips can be drawn from column 9 (Appendix 2). See Table 25. About half of the observed species do not show any evidence of drip-tips. One fourth have well marked drip-tips, less than 10% is provided with very obvious drip-tips.

Are pinnate leaves more frequently provided with marked driptips than simple leaves? The 2 × 2 chi-square test is used in three cases xx species versus others; x and xx species versus others and finally ε, x and xx species versus others.

The first test corrected for low frequencies gives 1.029 i.e. the difference in proportions is not significant; the second is very significant (χ^2 = 14.5); the third is not; $\chi^2 = 0.02$:

	Simple	Pinnate	TOTAL
Presence of xx	7	6	13
Absence of xx	119	43	162
TOTAL	126	49	175

$\chi^2 = 1.029$

	Simple	Pinnate	TOTAL
Presence of x and xx	34	29	63
Absence of x and xx	92	20	112
TOTAL	126	49	175

$\chi^2 = 14.5$

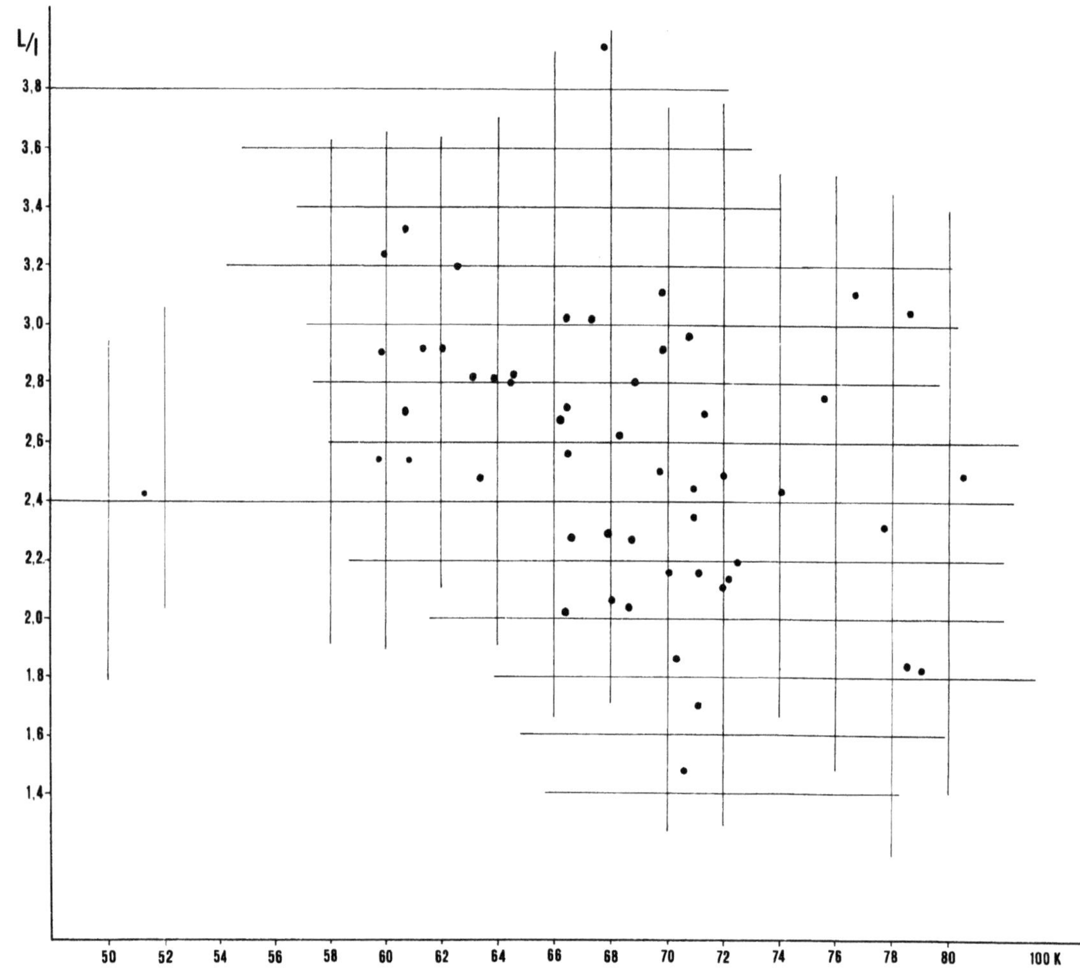

Fig. 3. Ratio *L/l* as a FUNCTION OF LEAF FORM FACTOR *K*. Compound Leaves.

	Simple	Pinnate	TOTAL
Presence ε x xx	73	27	100
Absence ε x xx	53	75	75
TOTAL	126	49	175

$\chi^2 = 0.02$

This result shows that a clearcut conclusion cannot be drawn and that it depends on the definition of the drip-tip. Conspicuous drip-tips are not more frequent in pinnate leaves. In a broader definition of the drip-tip, pinnate leaves are significantly more frequently provided with drip-tips than simple leaves. However, in the first test, the proportion 6/43 = 1/7 might turn out significantly higher than 7/119 = 1/17 if the sample were larger. From this discussion it can be said that *pinnate leaves have a tendency to be more frequently provided with drip-tips than simple leaves.*

Table 25. Number of species with apices and drip-tips in simple and pinnate leaves (Imataca).

	Simple	Pinnate	TOTAL
0	62	25	87
ε−	11	2	13
ε	19	3	22
X	27	13	40
XX	7	6	13
TOTAL	126	49	175

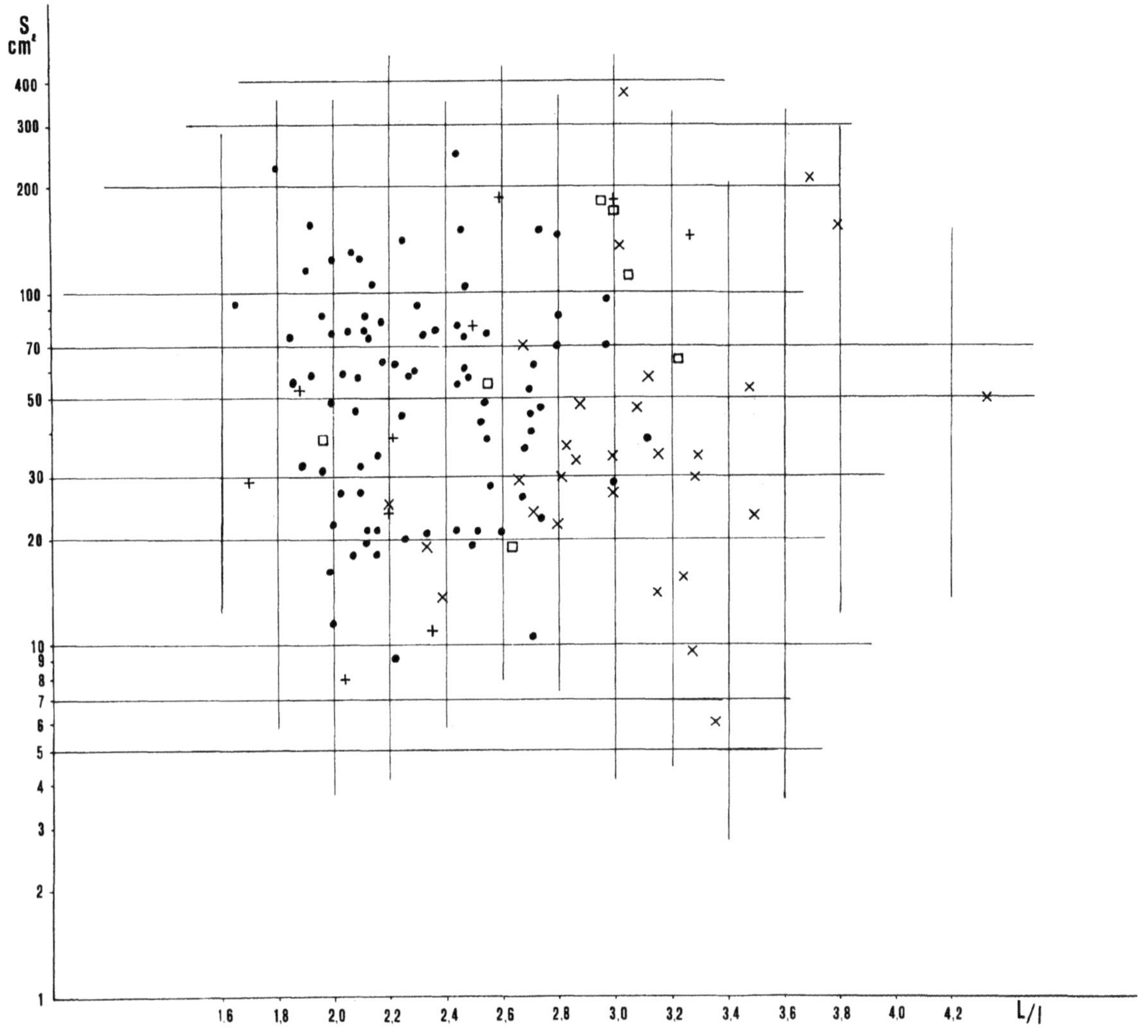

Fig. 4. LEAF AREA as a FUNCTION OF RATIO *L/l*. Simple Leaves •elliptic lanceolate, + ovate-obovate, x lanceolate, □ oblong.

Total percentage of driptips is 7.4% in the first case, 3.60% in the second. It is interesting to compare these figures with the data of Baker (1938) for Ceylon and Gessner (1956) for Guatemala and Venezuela.

After Baker, most of the dominant species are provided with drip-tips in the evergreen rainforest of Ceylon. Unfortunately the result relies only on the 41 most abundant species (of which only 4 have no drip-tip).

Gessner investigated a rainforest of Guatemala where he found that 80% of the species were provided with fairly-well to well marked drip-tips. This proportion reached 85% in a montane rainforest South of Caracas (61% well marked, 24% more or less marked) and only 26% in a dry deciduous area of the coastal forest of Venezuela (13% well marked, 13% more or less marked).

These figures are far different from ours and obviously different methods of work are involved. Therefore we think a considerable amount of caution is required to compare the results. Not only the method of measurement is involved but also the size of the sample under consideration.

Influence of the size of the tree on the length of the drip-tips

Table 26 gives the mean length of drip-tips in mm by species (one line by species) in five levels of height for comparison: <1 m; 1—4.9 m; 5—9.9 m; 10—19.9 m and ⩾20 m.

A complementary list of species, available only for one size and, where no comparisons are possible, inside a species (Table 27) will be pooled with the data of Table 27.

Table 26 is summarized in Table 28. Table 27 and the pooling of Tables 26—27 are summarized in Table 29.

Presence and absence of drip-tips is considered first of all. The proportion of drip-tips among individuals in height class 1—4.9 m is not higher than in height class 5—9.9 m ($\chi^2 = 1.54$), nor than in height class 10—19.9 m ($\chi^2 = 0.41$). On the contrary it is significantly higher in the subgroup <20 m than in subgroup ⩾20 m ($\chi^2 = 12.56$). Scarcely half of the trees ⩾ 20 m have a drip-tip (89 out of 181) whereas three-quarters of the trees <20 m have drip-tips (82 trees out of 116). See Table 30.

Mean length of drip-tips

If we look now at the mean length of the drip-tips in the various height classes (Tables 28 and 29), it can be said that it roughly decreases when total height increases. In particular, trees ⩾20 m show a sudden drop in drip-tip length. The number of individuals <1 m is too small and the mean 10.0 mm has probably no meaning.

In summary drip-tips have a mean length of 10.4 mm for height-class 1—4.9 m based on 66 saplings; 7.7 mm for height-class 5 to 20 m (40 trees) and 3.8 mm for height-class ⩾20 m (181 trees).

The proportions of trees *without drip-tip* in these height classes are 20 out of 82 (2.5/10), 19 out of 50 (4/10), and 92 out of 181 (5/10), respectively.

The proportion is constantly growing from the undergrowth to the top of the emergents. Whether

Table 26. Mean length of drip-tips (mm) by height classes (m) and by species Abnormal figures in **bold** type.

<1	1—4.9	5—9.9	10—19.9	⩾20	<1	1—4.9	5—9.9	10—19.9	⩾20	<1	1—4.9	5—9.9	10—19.9	⩾20
10				7.9		15			8				17	6
12.3				5		10			6		20			
0	0			0			10		4		15/27			0
	15(10 20)			7.5			5		0		20	**8**		9
	10			0	**8**				8		0			0
			0	0			**0**	15	0		0			0
	15			5		20		20	ε		15			0
	20			7		15			6		5	ε		
	13		8			12			6	10				8
	10			4	25				0			**20**	22	
	17			5	12				7	0				0
	0			0		ε			ε	30				7
ε				0		10			8				17	13
20				0			0		0		3			0
	0	ε		5		15			15			0		
		0		0		15			4		10			
	ε			ε		**10**	13	15	14		10			8
	12			10			15		7			ε	ε	
				0		20		15						
		26		11		10			0					
		10		0		7			11					
	13	10		4		10			3					

Table 27. Mean length of drip-tips (mm) by height classes (m). One figure is one individual.

<1	1–4.9	5–9.9	10–19.9	⩾20	⩾20	⩾20	⩾20	⩾20	⩾20
10	5 0	15	12	0	14	8	0	4	0
10	0 12	ε	0	7	0	0	7	ε	0
12	ε 0	18	12	8.5	0	9	4	11	9
0	26 10	0	7	0	0	6	12	0	0
	10 20	15	8	0	0	0	18	ε	0
	5 0	6	6	0	4	0	0	0	0
	10	5	ε	16.5	0	0	0	ε	9
	12	0	ε	23	7	4	0	15	9
	18		10	7	9	0	9	0	0
	0		0	2	6	0	5	0	
	10		ε	0	0	0	ε	12	
	20		ε	15	0	13	0	0	
	ε		7	4	0	0	0	6	
	5		12	7	7	2	7	9	
	15		6	ε	8	ε	0	0	
	25		0	5	0	ε	2	0	
	15			0	16	ε	10	6	
	ε			6	11	0	2	5	
	10			0	9	0	ε	0	
	15			12	0	0	ε	ε	
	20			4	0	0	4	15	
	20			0	ε	7	0	0	
	10			0	12	ε	5	0	

long drip-tips are useful to undergrowth plants is not quite clear: it certainly improves transpiration and consequently mineral nutrition.

6. OTHER LEAF FEATURES

Margin of leaves

Among the 183 species studied 169 are provided with an entire margin: 114 out of 126 species with simple leaves, 47 out of 49 species with pinnate leaves, and the 8 bipinnate species, i.e. 92% of all leaves. If the number of trees were considered instead of the number of species, this percentage would be still higher. Therefore the non-entire margin character is highly diagnostic for 14 species.

Vesture

Unfortunately vesture types have not been adequately quantified and should have taken account of the tentative terminology of Lawrence (1951,

Table 28. Frequencies of drip-tips by height classes. Summary of Table 26.

Total height classes m	<1	1–4.9	5–9.9	10–19.9	⩾20
Number of species without drip-tip	2	5	4	1	20
〃 〃 〃 with drip-tip	2	2	3	1	5
〃 〃 〃 with conspicuous drip-tip	8	30	9	8	32
Total number of species	12	37	16	10	57
Mean length of drip-tip mm	10.6	10.6	7.3	12.9	4.0
Standard deviation mm	10.15	6.55	8.18	7.72	4.31

Table 29. Frequency of drip-tips by height classes.

Total height classes m	Summary of Table 27					Summary of Tables 26 and 27				
	<1	1–4.9	5–	10–	⩾20	<1	1–	5–	10–	⩾20
Number of species without drip-tip	1	5	2	3	56	3	10	6	4	76
〃 〃 〃 with 〃		3	1	4	11	2	5	4	5	16
〃 〃 〃 with conspicuous 〃	3	21	5	9	57	11	51	14	17	89
Total number of species	4	29	8	16	124	16	66	24	26	181
Mean length of drip-tip mm	8.0	10.1	7.5	5.0	3.7	10.0	10.4	7.3	8.0	3.8
Standard deviation mm	5.41	8.24	7.56	4.94	5.14	9.10	7.28	7.81	7.17	4.88

pp. 746—748) who mentions 25 characters, or the systematization of Payne (1978) who deals with 27 qualitative characters and 3 quantitative characters for hair length. Roth presented a paragraph on *hair density*.

A wide majority of species has glabrous leaves in the rainforest. Some are tomentose such as *Calycorectes* (*Myrtaceae*) with an evocative vernacular name "Terciopelo" (literally, velvet); *Hirtella americana* (*Chrysobalanaceae*) v. n. "Terciopelo blanco"; *Inga rubiginosa* (*Mimosaceae*) v. n. "Guamo terciopelo" a secondary forest species found uncommonly in the rainforest. Pilose leaves are found in *Duroia* (*Rubiaceae*) v. n. Conserva; scabrous leaves are common in *Boraginaceae*, especially in the genus *Cordia*: *Cordia alliodora* (v. n. Pardillo), *Cordia viridis* (v. n. Caujaro) and a treelet *Cordia nodosa* (v. n. Alatrique peludo).

Leaf dimorphism

Several characters are involved in leaf dimorphism: size of leaf, form and variation in *L/l* ratio during its lifetime, changes in number of lobes or of leaf type (simple or compound).

Size of leaves

Fully developed leaves of juvenile or young stages are generally larger than adult leaves (see below and Appendix 3), but there are exceptions. The most outstanding is the *Cecropia* spp. where adult leaves are huge compared with young stages. Other exceptions are *Palicourea* (*Rubiaceae*, v. n. Dantera), *Erythrochiton brasiliensis* (*Rutaceae*, v. n. Cola de Pava), *Alexa imperatricis* (*Papilionaceae*, v. n. Leche de cochino), *Cordia exaltata* (*Boraginaceae*, v. n. Alatrique blanco), *Aparisthmium cordatum* (*Euphorbiaceae*, v. n. Tabali morado).

Sometimes the largest size of the leaves occur in intermediate stages (treelets), as in *Parinari rodolphii* (*Chrysobalanaceae*, v. n. Merecure montañero), *Sterculia pruriens* (*Sterculiaceae*, v. n. Majagua), *Didymopanax morototonii* (*Araliaceae*, v. n. Sunsun), *Terminalia amazonia* (*Combretaceae*, v. n. Pata de Danto).

During its lifetime the form of a leaf varies, especially the *L/l* ratio which has been already studied e.g. *Rollinia multiflora* (*Annonaceae*, v. n. Anoncillo) has long narrow leaves when young, rather oblong when adult.

Table 30. Frequencies of drip-tips by height classes.

Height classes m	<1	1–	5–	10–19	⩾20	<20	TOTAL
Without drip-tip	5	15	10	9	92	34	126
With drip-tip	11	51	14	17	89	82	171
TOTAL	16	66	24	26	181	116	297

In general *L/l* is higher on juvenile leaves but there are exceptions: see Appendix 3.

Leaf type changes

Changes of leaf type (simple to compound leaves) are not uncommon during the course of the life of certain species.

The phenomenon is very common for seedlings among species with compound leaves, e.g. *Pterocarpus officinalis* (*Papilionaceae*, v. n. *Cacú*), *Lonchocarpus sericeus* (*Papilionaceae*, v. n. Jebe) start with simple leaves.

Inga sp. (*Mimosaceae*, v. n. Guamo verde) has bifoliolate leaves on seedlings and three pairs of leaflets on adult leaves.

Didymopanax morototonii (*Araliaceae*, v. n. Sun-Sun) starts with simple leaves on seedlings, then shows palmate leaves with 3—5 leaflets on saplings and in general 10 leaflets on the adult palmate leaves.

Alternatively some adult compound leaves have fewer leaflets on adult leaves; e.g. 3 in adult *Protium decandrum* (*Burseraceae* v. n. Azucarito) and 3—5—7 leaflets on juvenile leaves; *Trichilia schomburgkii* (*Meliaceae*, v. n. Suipo) has about 5 leaflets on trees, 7 on seedlings.

The changing in number of lobes and a return to the juvenile form at the adult stage in not uncommon. *Sterculia pruriens* shows simple entire leaves up to the the 5th—6th leaf, 2—3—5 lobed leaves from the 8th—9th leaf onwards, again simple entire leaves on adult trees. *Catostemma commune* (*Bombacaceae*, v. n. Baramán) has trifoliate seedlings (sometimes 1—2 or 4 leaflets); the compound leaf type is kept on small trees e.g. 10 cm to 30 cm DBH but old trees have smaller simple leaves.

Cecropia spp. (*Moraceae*, v. n. Yagrumo blanco) start with entire simple leaves; later on, leaves show 2—3—5—7 lobes. *Cecropia sciadophylla* (v. n. Yagrumo montañero) has its first leaves simple entire and peltate, later on palmatifid, finally palmate-compound leaves.

Leaves on shoots and reiteration occurring after accidental uprooting, or broken branches by windthrow show sometimes exceptionally large size, e.g. *Simarouba amara* (*Simaroubaceae*, v. n. Cedro blanco), *Clathrotropis* spp. (*Papilionaceae*, v. n. Caicareños), *Chaetocarpus schomburgkianus* (*Euphorbiaceae*, v. n. Cacho).

Colour of leaves

Changes of colour in leaves are generally a very attractive phenomenon in tropical deciduous forests after leaf fall. They are very aperiodic and may cover large tracts of the hill forests e.g. in Malaya according to Burgess.

In lowland tropical rain forests, even in the so-called seasonal rain forests, there are several possible behaviours: 50 to 60 species are deciduous in Imataca but their impact is negligible on the evergreenness of the forest either because these species are present as scattered trees, or because the deciduousness occur at different times in the course of the year.

For the so-called evergreen species, some show conspicuous rhythmic flushes of leaves, affecting sometimes only parts of the crowns, with various shades of yellow, orange, crimson or bronze.

A more subtle phenomenon is the colour of the young leaves in the undergrowth with delicate shades ranging from white, pink, mauve, pale green or grey, or carmine to deep purple. For example

White:	*Protium neglectum* (*Burseraceae*, v. n. Caraño), *Tetragastris panamensis* (*Burseraceae*, v. n. Caraño negro); *Pithecellobium basijugum* (*Mimosaceae*, v. n. Curarina chiquita); *Licania densiflora* (*Chrysobalanaceae*, v. n. Hierrito); *Licania hypoleuca* (v. n. Hierrito blanco); *Anaxogorea* sp. (*Annonaceae*, v. n. Yarayara amarilla).
Pale green:	*Brownea latifolia* (*Caesalpiniaceae*, v. n. Rosa de montaña); *Crudia glaberrima* (*Caesalpiniaceae*, v. n. Arepito rebalsero); *Guarea schomburgkii* (*Meliaceae*, v. n. Carapillo).

Pink:	*Catostemma commune* (*Bombacaceae*, v. n. Baramán, on seedlings); *Mouriri sideroxylon* (*Melastomaceae*, v. n. Guaratarillo).
Carmine:	*Carapa guianensis* (*Meliaceae*, v. n. Carapa); *Inga alba* (*Mimosaceae*, v. n. Guamo colorado); *Licania alba* (*Chrysobalanaceae*, v. n. Hierro), *Tetragastris* sp. (*Burseraceae* v. n. Aracho).
Mauve or light purple:	Many *Melastomaceae, Licania alba*; *Cecropia sciadophylla* (*Moraceae*, v. n. Yagrumo montañero).
Deep purple:	*Eugenia florida* (*Myrtaceae*, v. n. Guayabito morado).
N. B.:	Many adult leaves have a purple colour underneath: *Palicourea*, (*Rubiaceae*, v. n. Dantera), *Miconia amacurensis* (*Melatomaceae*, v. n. Saquiyak rojo).

Fallen leaves of some species have conspicuous colours.

Black:	*Didymopanax morototonii* (upper surface; the lower is golden); *Cordia alliodora*; *Aspidosperma* spp.; *Jacaranda obtusifolia* (v. n. San José); *Mabea piriri* (v. n. Pata de pauji); *Aspidosperma* spp. (v. n. Canjilón amarillo, Canjilón negro, Hielillo blanco, Hielillo negro); *Torrubia* spp. (v. n. Casabe); *Genipa americana* (v. n. Caruto montañero); *Sacoglottis cydonioides* (v. n. Ponsigue montañero).
Carmine:	*Couratari multiflora* (v. n. Tampipio); *Manilkara bidentata* (v. n. Purguo); Manilkara sp. (v. n. Pendare).
Bright orange:	*Croton gossypifolia* (v. n. Sangre de drago); *Croton xanthochloros* (v. n. Canelón).

7. SIZE OF LEAVES

Comments on Raunkiaer's classification

Raunkiaer proposed a classification of leaves by area classes in various publications (1916; 1918; 1934) which is in current use by plant ecologists.

Leptophyll leaves		< 25 mm^2		
Nano	"	25 to 25×9 mm^2	or	25 to 225 mm^2
Micro	"	25×9 to 25×9^2 mm^2	or	225 to 2025 mm^2
Meso	"	25×9^2 to 25×9^3 mm^2	or	20.25 to 182.25 cm^2
Macro	"	25×9^3 to 25×9^4 mm^2	or	182.25 to 1640.25 cm^2
Mega	"	$\geqslant 25 \times 9^4$ mm^2	or	$\geqslant 1640.25$ cm^2

Webb (1959) suggested a subdivision of the mesophyll class, by far the most frequent into two subclasses.

Notophyll leaves	10.25 to 45 cm^2
Mesophyll	45 to 182.25 cm^2

and claimed its interest in the classification of the Australian vegetation.

Originally Raunkiaer had in mind a decimal logarithmic scale, multiplying successively a base number 25 mm^2 by powers of 10, but he finally adopted a constant ratio of 9 instead of 10, just because he wanted to retain the possibility of splitting each class into 3 parts (small, medium, large).

He had questioned the treatment of compound leaves, whether they should be considered as a whole, or if the unit should be the leaflet. He adopted the second point of view arguing that in Europe and in the West Indies compound leaves are generally much larger than simple leaves and that their leaflets are much nearer in size to simple leaves. Although many ecologists use the Raunkiaer system as it is (except for the minor improvement by Webb), it is not unreasonable to use both approaches for the sake of comparison.

Obviously Raunkiaer did not process much tropical material; had he obtained most of his observations from the tropics, he could not have overlooked the fact that the areas of simple leaves, leaflets of pinnate and bipinnate leaves show a higher variability within a species and that the range of the areas of simple leaves and leaflets of compound leaves is higher than the range of the area of simple leaves, pinnate and bipinnate leaves considered as units which invalidate to some extent the reason for the use of leaflets as units.

Raunkiaer did not actually measure the areas of the leaves except to check dubious sizes at the limits of his classes; with some experience most of the leaves can be readily graded into the six classes of Raunkiaer.

The wide range of the areas of leaflets and simple leaves 5 mm^2 to 1600 cm^2 and more, i.e. in a proportion of 1 to 3×10^5 invites us to process the areas using a logarithmic scale.

We show below that there are no discontinuities between size classes and no modal values which would betray a mixture of populations but, on the contrary, a continuous variation of the frequencies of areas, so that breaking the whole range into six classes necessarily looks somewhat arbitrary.

This continuity is due to the wide variability of leaf area for each species resulting in many overlaps of the various species distributions.

Thirty leaves were measured for each species. Mean areas by species and corresponding coefficients of variation are given in Appendix 2.

Distributions of leaf areas

Several methods will be used to establish these distributions (without and with weighting).

(1) – *Without weighting*:
- Use of an arithmetical scale for areas with class interval 10 cm^2 and 20 cm^2.
- Use of a logarithmic scale for areas.
- Distribution of area *means* by species in two ways: leaflets as units or whole compound leaves as units.
- General distribution of areas *leaf by leaf, all species together.*

(2) – A weighting takes into account the relative importance of the species in the forest.

Distribution of leaf area means by species (183 species) Table 31

Table 31 gives the *distributions of means* without weighting.

The adoption of a class interval of 10 cm^2 results in a considerable scatter of the frequencies which become erratic above 150 cm^2.

Each species by itself shows a considerable variability around its means; coefficients of variation are given in column 8 of Appendix 3. They are generally near to 35% (above 25 and 50%).

The distribution (2) of leaflets in Table 31 is similar to distribution (1) for simple leaves, very skewed with a long tail for large leaf areas. All the leaflets of bipinnate leaves distribution (3) are concentrated in leaf area class 0 to 9.9 cm^2; class interval 10 cm^2 is not fit for this type of leaflet.

One species with a simple leaf is very abnormal, nearly megaphyll with a mean of 1622 cm^2: it refers to *Coccoloba* sp. (v. n. Uvero), a tree of the undergrowth rarely above 30 cm DBH with less than 0.05% of the total number of trees $\geqslant 10$ cm DBH; its leaves are much larger than any other species with simple leaves; this *Coccoloba* cannot be mistaken in the undergrowth whenever it is present.

Palm tress have not been considered here; although they play a minor role with less than 0.6% of the total number of individuals $\geqslant 10$ cm DBH they are very striking in the undergrowth landscape, because of their huge leaves: each pinna is megaphyll in many species so that the total leaf area would need an extension of Raunkiaer's classification; they could be called gigaphyll. The prefixes of Raunkiaer are not always adequate and are confusing with respect to the decimal system. 'Macro' means long in morphological terminology. Let us recall the scale used in physics: pico 10^{-12}; nano 10^{-9}; micro 10^{-6}; mega 10^6; giga 10^9.

With class interval of 20 cm^2, the distribution (1) + (2) + (3), i.e. of leaves and leaflets, is less erratic for large areas but very asymmetrical and truncated for small areas.

If *total areas of leaves* (simple or compound) are considered with class interval 20 cm^2, we obtain erratic distributions (4) and (5) for compound leaves, the means of which are well above

Table 31. Number of species by leaf area classes. Total 183 species.

	CLASS INTERVAL 10 cm²				CLASS INTERVAL 20 cm²						
Leaf area	(1)	(2)	(3)	(1) + (2) + (3)	(1)	(2)	(3)	(1)	(4)	(5)	(1) + (4) + (5)
0—9.9	3	2	8	13	0—19.9		35	17			17
10—	14	8		22	20—		**55***	**36***	2		**38***
20—	**20***	**12***		**32***	40—		34	25	1	1	27
30—	16	7		23	60—		22	16	2		18
40—	11	7		18	80—		12	10	2		12
50—	14	2		16	100—		5	4	2		6
60—	4	4		8	120—		5	5	5		10
70—	12	2		14	140—		6	5	6	1	**12***
80—	8	2		10	160—		2	2	2		4
90—	2			2	180—		2	2	1	1	4
100—	3	1		4	200—		1	1	3		4
110—	1			1	220—		2	1	1	2	4
120—	2			2	240—		1	1	1		2
130—	3			3	260—				4		4
140—	1			1	280—				1		1
150—	4	1		5	300—						
160—					320—						
170—	2			2	340—				2		2
180—	2			2	360—				1		1
190—					380—				1	2	**3***
200—					400—				1		1
210—	1			1	420—				1		1
220—	1	1		2	440—				3		3
230—					460—						
240—					480—						
250—	1			1	500—				1		1
					520—				1		1
					540—						
					560—				1		1
					″						
					800—899				1	1	
					1300—1399				1		
					1400—1499				1		
					1600—		1	1	1		

* Modes in **bold** type
(1) — Simple leaves.
(2) — Leaflets of pinnate and palmate leaves.
(3) — Leaflets of bipinnate leaves.
(4) — Pinnate and palmate leaves (total leaf area).
(5) — Bipinnate leaves (total area).

the means of simple leaves; only 8 species with bipinnate leaves were studied out of the 16 species occurring in the 155 ha inventory.

When the three distributions (1) (4) (5) are pooled, we obtain a distribution with three modes, respectively for area class midpoints 30, 150 and 390 cm², the latter not very well defined.

Obviously the interval of variation is very high for compound leaves areas and subpopulations are poorly individualized with an arithmetical scale.

Distribution of area means according to Raunkiaer's classification (183 species)

Five distributions are given in Table 32 according to the area classes of Raunkiaer. The distributions for simple leaves (1) and pinnate leaflets (2) are

Table 32. Distribution of area means for 183 species according to Raunkiaer's classification.

Type of leaves	Area in cm^2	Number of species							
		(1)	(2)	(3)	(1+2+3)	(1)	(4)	(5)	(1+4+5)
Leptophyll	<0.25			3	3				
Nano	0.25 to 2.24			4	4				
Micro	2.25 to 20.24	17	10	1	28	17			17
Meso	20.25 to 182.24	104	38		142	104	22	2	128
Macro	182.25 to 1640.24	5	1		6	5	26	6	37
Mega	⩾1640.25						1		1
TOTAL		126	49	8	183	126	49	8	183

(1) — Simple leaves.
(2) — Leaflets of pinnate leaves.
(3) — Leaflets of bipinnate leaves.
(4) — Pinnate leaves (total area).
(5) — Bipinnate leaves (total area).

quite similar whereas the distribution for bipinnate leaves (3) is almost completely outside the range of (1) and (2).

The area of *simple leaves* varies from 6.1 cm^2 (undetermined *Myrtaceae*, v. n. Guayabito piedrero chiquito) to 1623 cm^2, a *Coccoloba* of abnormal area already mentioned. The largest leaf area outside *Coccoloba* is 250 cm^2 for *Rheedia* sp. (v. n. Cozoiba picuda).

The leaflet areas of pinnate leaves vary from 4.8 cm^2 (*Hymenolobium* sp., v. n. Alcornoque montañero) to 227 cm^2 (*Carapa guianensis*, v. n. Carapa).

The leaflet areas of bipinnate leaves vary from 5.6 mm^2 (*Piptadenia psilostachya*, v. n. Yigüire) to 476 mm^2 (*Pithecellobium jupunba*, v. n. Saman montañero).

In the three cases leaf areas vary in a proportion of 1 to 50 (simple leaves and leaflets of pinnate leaves) and 1 to 100 (leaflets of bipinnate leaves). The latter are at least as variable, if not more so than the former and are a hundred times smaller on the whole.

If the total area of the leaf is considered, the interval of variation is shorter (Table 32, column on the extreme right), all leaves falling only into three classes: macro — meso — and microphylls.

The adoption of leaves and leaflets as statistical entities rather than leaves in all cases is questionable concerning bipinnate leaves: compare distributions (1 + 2 + 3) and (1 + 4 + 5) of Table 32. On one hand the interval of variation is in the proportion of 1 to 3 × 10^5 whereas it is in the proportion of 1 to 300 i.e. 1000 times less when complete leaves are considered: 6.1 cm^2 for the unknown *Myrtaceae* (Guayabillo piedrero chiquito) and 1660 cm^2 for *Toulicia guianensis* (v. n. Carapo blanco).

Distribution of leaf areas leaf by leaf in case of simple leaves

Arithmetical scale

A distribution of leaf area with class interval 10 cm^2 is established for all *individual* simple leaves available, about 3600 distributed in 119 species. See Figure 5. The distribution is truncated on the left, strongly skewed and with a long tail on the right; the mode is 25 cm^2; a second mode of 155 cm^2 is poorly defined. This distribution gives a more satisfactory view of the phenomenon than column 1 of Table 32. Compare also with the distribution of the means by species, column (1) of Table 31, the irregularities of which are buffered on Figure 5. Table 33 gives the actual number of leaves by 10 cm^2 classes corresponding to Figure 5.

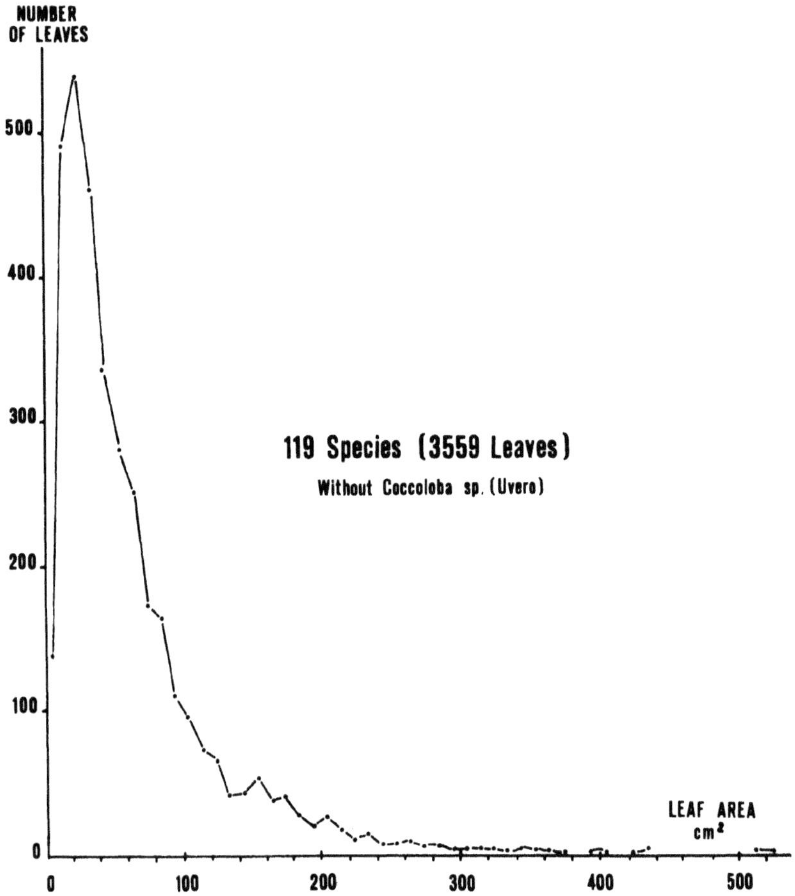

Fig. 5. DISTRIBUTION OF FREQUENCIES OF SIMPLE LEAVES as a FUNCTION OF LEAF AREA. 119 Species 3559 Leaves. *Coccoloba* sp. (v.n. Uvero) excluded.

Table 33. Number of simple leaves (F) in 10 cm^2 classes. (3559 simple leaves distributed in 119 species.)

cm^2	F	cm^2	F	cm^2	F	cm^2	F
0—9.9	138	120—	65	240—	7	360—	2
10—	492	130—	40	250—	8	370—	1
20—	**541**	140—	43	260—	7	380—	
30—	460	150—	**52**	270—	5	390—	1
40—	336	160—	38	280—	5	400—	1
50—	279	170—	39	290—	4	410—	
60—	250	180—	26	300—	2	420—	1
70—	174	190—	21	310—	2	430—	2
80—	164	200—	24	320—	2		
90—	110	210—	17	330—	1	510—	1
100—	95	220—	10	340—	3	520—	1
110—	73	230—	14	350—	2		

Table 34 gives the number of species (simple leaves) by area classes of 10 cm^2 and the percentage of leaves in the same classes after weighing with the 155 ha inventory. Only 112 species are used because 7 species of the 119 studied species do not occur on the 155 ha.

Modes 25 and 155 existed already in the distribution leaf by leaf: in Table 34 another mode appears in the class 70—79.9 cm^2 which seems to indicate that the interval of Raunkiaer for mesophyll leaves (20.25 cm^2 to 182.25 cm^2) could be split. Webb kept these interval limits but introduced a new limit 45 cm^2.

In the case of Venezuela Guiana, the limit 20.25 cm^2 too close to the mode 25 cm^2 cannot be retained, but the limit 45 cm^2, intermediate

Table 34. Number of simple leaved species in 10 cm^2 classes and percentage of leaves in the same classes after weighing by D^2 using the 155 Ha inventory.

Area in cm^2	0—9.9	10—19.9	20—29.9	30—39.9	40—49.9	50—59.9	60—69.9	70—79.9	80—89.9
Number of species	0	15	17	16	10	11	4	12	7
Importance %	0	7.67	**17.43**	15.78	13.41	13.78	1.37	**13.84**	11.51
Area in cm^2	90—99.9	100—109.9	110—119.9	120—129.9	130—139.9	140—149.9	150—159.9	160—169.9	170—179.9
Number of species	2	3	1	1	3	1	4		1
Importance %	0.09	0.54	0.01	0.07	0.22	0.17	**3.76**		0.07
Area in cm^2	180—189.9	190—199.9	200—209.9	210—219.9	220—229.9	230—239.9	240—249.9	250—259.9	
Number of species	1			1	1			1	
Importance %	0.02			0.19	0.02			0.02	

between the two 25 and 75 cm^2 modes, could be adopted if the objective is to deal with area means by species.

But the distribution of areas leaf by leaf indicates that *the subdivision with limits 20 and 182 cm^2 is arbitrary for simple leaves.*

If a representative sample of leaves could be drawn for each species occurring in a large forest inventory (155 ha in our case), the overall sample means could be improved by weighing the mean leaf area of each species. It could not be weighed by the number of leaves occurring in each class – an impossible task – but by the number of trees of this species or by a coefficient highly related to this number of leaves. I have already indicated a possibility of weighing with basal area and it would be necessary to adopt another scale for areas to obviate the skewness of the curve.

Logarithmic scale

The processing of the 119 species with simple leaves leaf by leaf with a logarithmic scale gives the following cumulated frequencies using inferior limits in geometric progression (Table 35).

An almost perfect sigmoid curve is obtained (Figure 6), though it is truncated on the left. Using log-normal coordinates, the sigmoid curve is transformed into a line called the line of Henri (Figure 7).

Actually it could be guessed from the shape of Figure 5 that the distribution had something to do with a log-normal law.

It would be interesting to check whether this property would apply when total leaf area is used including the 30 leaves of *Coccoloba* which had been discarded in Table 35, and all the compound leaves with large area such as *Carapa guianensis* 1400 cm^2, *Toulicia guianensis* 1660 cm^2, *Alexa imperatricis* 1340 cm^2, *Didymopanax morototonii* 900 cm^2, *Cecropia* sp. and *Cecropia sciadophylla*, (unmeasured) and other species with large area leaves that are present in the forest but absent in the 155 ha.

Unfortunately only means were estimated for such large compound leaves, the very lengthy measurement of areas or at least the L and l of *all* the leaflets for all the leaves of the sample were not carried out and the indirect measurement of area by weight was not done, so that the last points of the curve corresponding to the limits $\geqslant 640$ cm^2 $\geqslant 1280$ cm^2 and $\geqslant 2560$ cm^2 are missing.

Adding only the 30 leaves of *Coccoloba* just introduces an abnormality in the curve (Figure 7 and Table 36).

There is sufficient evidence to believe that *the total leaf areas in the seasonal rainforest of Imataca show a tendency to be log-normally distributed.*

Use of Raunkiaer's classification of leaf areas in the tropics

A quick review of the literature on this subject must include the work by Webb (1959) for Australia, Ashton (1964) in Brunei (Borneo), Duvigneaud *et al.* (1951) in Zaire, Cain *et al.* (1956) in Brazil, Stehlé (1945) in the Lesser Antilles, Tasaico (1959) and Petit Betancourt (1964) in Costa Rica, Loveless and Asprey (1955) in Jamaica. Some results are also mentioned by Richards (1952) for Nigeria and the Philippines. The world-wide evaluation by Vareschi (1980) on the interest of leaf forms and size for comparing the tropical vegetation formations is worth mentioning.

Influence of the height of the trees on leaf size

Appendix 3 contains mean leaf areas by species and will be used to answer the two questions:

Table 35. Distribution of $N = 3559$ leaf areas arranged in cumulated frequencies n/N.

Inferior limits cm^2	$\geqslant 1.25$	$\geqslant 2.5$	$\geqslant 5$	$\geqslant 10$	$\geqslant 20$	$\geqslant 40$	$\geqslant 80$	$\geqslant 160$	$\geqslant 320$
Number of leaves n	3559	3558	3547	3421	2939	1928	889	247	18
n/N	1.0000	0.9997	0.9996	0.9612	0.8258	0.5417	0.2498	0.0694	0.0051

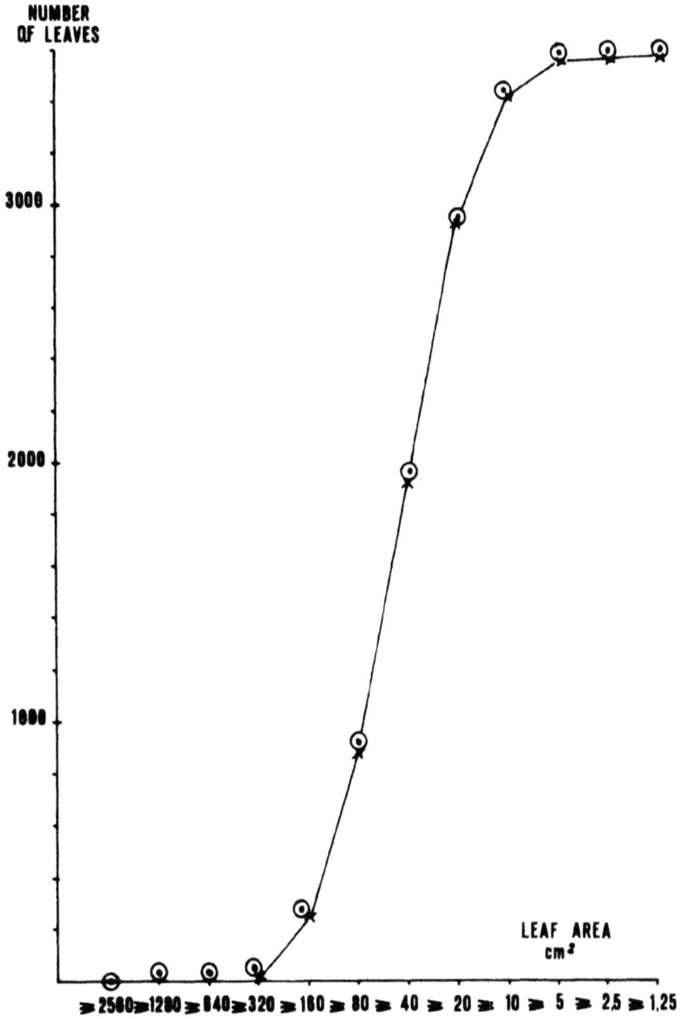

Fig. 6. DISTRIBUTION OF CUMULATED FREQUENCIES OF SIMPLE LEAVES as a FUNCTION OF AN INFERIOR LIMIT OF LEAF AREA. With and without *Coccoloba* (see Figure 7).

- Are leaves belonging to the shrub or undergrowth layers larger on the whole than leaves in codominant and emergent trees?
- Are juvenile leaves larger than adult leaves of the same species?

The leaf areas of individuals with measured total height are arranged in four height classes < 9 m; 10—19 m; 20—29 m; ≥30 m. Means and standard deviations are computed for each height class in two cases: simple leaves plus leaflets of compound leaves; total leaf area for all leaves (Table 37). The variation of leaf area means with total height is irregular in both cases. The differences of area means between height classes are not significant because of the small number of observations and high standard deviations; e.g. means for leaves and leaflets are 76.7 and 47.2 cm^2 for height-classes 10—19 m and 20—29 m respectively. Even if all individuals are grouped in only two categories ≥ 20 and < 20 m total height, the difference is not significant in spite of a fairly large difference in the case of leaves and leaflets (Table 38).

If only simple leaves are considered their mean leaf area in the four height classes are 58.2 85.8

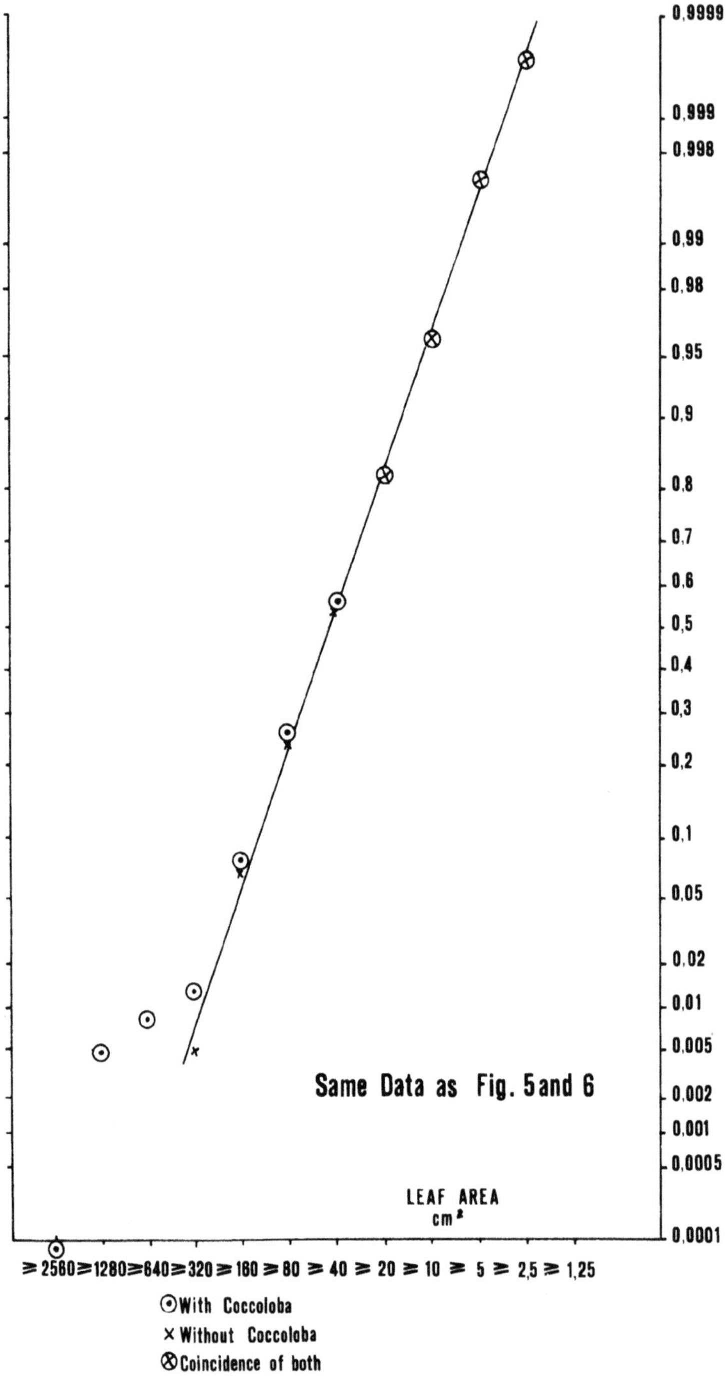

Fig. 7. DISTRIBUTION OF CUMULATED FREQUENCIES OF SIMPLE LEAVES as a FUNCTION OF AN INFERIOR LIMIT OF LEAF AREA. Adjustment to a lognormal distribution (with and without *Coccoloba*).

Table 36. Distribution of $N = 3589$ leaf areas (including *Coccoloba*) arranged in cumulated frequencies n/N.

Inferior limits cm^2	⩾1.25	⩾2.5	⩾5	⩾10	⩾20	⩾40	⩾80	⩾160	⩾320	⩾640	⩾1280	⩾2560
Number of leaves n	3589	3588	3577	3451	2959	1958	919	277	48	30	19	3
n/N	1.0000	0.9996	0.9966	0.9614	0.8244	0.5455	0.2560	0.0772	0.0134	0.0084	0.0053	0.0008

Table 37. Mean leaf size in cm^2 by height classes (m).

	<10 m	10—19 m		20—29 m		⩾30 m
	108	76.3	145	425* (42.2)	54.2	52.3
	202* (67.5)	250	130	11.8	211	22.9
	47.1	34.1	240* (21.8)	34.2	42.8	587* (49.6)
	31.1	57.4	205* (23.5)	17.7	38.4	89
	272* (57.4)	21.2	19.7	19.5	58.6	52.7
	27.2	47.1	401* (71.6)	30.5	24.9	16.2
	105.6	59.5	63.6	21.2	32.2	45.2
	30.0	213* (1.51)	95.1	59.6	10.3	132* (65.9)
		78.7	77.2	86.5	19.0	119
		46.4	139	51.7	344* (36.9)	71
		54.5	224	70.3	142* (4.76)	222* (11.1)
		14.0		186.0	132* (18.9)	148* (32.3)
				262* (40.2)	76.6* (4.79)	205* (21.5)
				242* (23.5)	150	571* (59.5)
				123* (14.4)	137* (20.5)	
				71.0	37.3* (7.46)	
Number of species	8	23		32		14
Leaves and leaflets						
Mean area cm^2	59.2	76.1		47.2		50.4
Standard deviation cm^2	32.5	64.1		49.3		30.4
TOTAL LEAF AREA						
Mean area cm^2	102.7	117.7		100.6		130.4
Standard deviation cm^2	90.9	98.2		101.8		162.0

* Total area of compound leaves (mean area of leaflet between brackets).

Table 38. Significance of the differences of leaf area between two height classes.

TOTAL HEIGHT	LEAVES AND LEAFLETS		TOTAL LEAF AREAS	
	<20 m	⩾20 m	<20 m	⩾20 m
Number of species	31	46	31	46
Mean area cm^2	71.7	48.2	113.6	109.6
Standard deviation cm^2	57.6	44.0	95.1	121.2
Ratio of difference in area to standard deviation of the difference	23.5/12.2 Nearly significant		4.0/23.5 Not significant	

59.0 58.5 respectively (general means 68.1 cm^2). Their differences are not significant, even between the two ⩾20 m and <20 m subgroups (respective means 79.2 and 58.9). The conclusion is the same for the total areas of compound leaves; the respective total area means are 258.6 cm^2 for trees <20 m and 211.6 for trees ⩾20 m.

In short, the conclusion to the first question is not clear-cut; though negative on the whole, especially when total leaf areas are concerned, *there is a tendency to some difference in leaf area between undergrowth and canopy* when simple leaves and leaflets of compound leaves are considered. Large standard deviations indicate that the distributions of leaf areas are far from normal and are responsible for the dubious answer.

Although Cain *et al.* (1956) did not draw statistical conclusions from their data on Brazilian Amazonia, it can be shown from their means and standard deviations that the differences in leaf areas between the various strata >30 m 8—30 m and <2 m are generally not significant (nearly significant in one case only) just because of the high variability and too few observations. However the tendency is towards an increase of mean leaf area (simple leaves) or mean leaflets area (compound leaves) from canopy trees to the undergrowth and shrub layer.

Leaf areas of juvenile and adult leaves of the same species

The data are presented in Appendix 3. Out of 83 comparisons at all heights, 59 cases showed definitely larger juvenile leaves (48 very significant differences, and 11 significant); in 14 cases juvenile leaves were not significantly larger than adult leaves; in 10 cases they were smaller (8 very significant, 2 significant cases).

More precisely for height class 1 to 4.9 m, 32 juvenile leaves (out of 42) are larger than adult leaves (they have the same size in 5 cases and are smaller in 5 cases). In height class 5 to 9.9 m juvenile leaves are larger in 12 cases out of 14 (smaller in 2 cases).

The answer to the second question is therefore that *in general juvenile leaves are larger than adult leaves*, a result obtained through the comparison of paired observations by species, whereas the gross comparison of the height class means in Table 37 does not lead to a convincing interpretation.

8. THICKNESS OF LEAVES

The measurements were supplied by Roth. There is a special difficulty in measuring the thickness of the lamina especially when provided with prominent nerves. In some cases (e.g. *Chrysobalanaceae*) the nerves are twice as thick as the lamina.

Thickness varies obviously with the age of the tree. It is roughly twice thicker in the adult than in the juvenile stage. The distribution with 50 microns class interval is given in various height strata 1—4.9 m; 5 to 9.9 m; 10 to 19 m; ⩾20 m. See Table 39.

An analysis of variance shows that the means are very significantly different, although the difference is not significant between strata 1—4.9 and

Table 39. Distribution of leaf thickness by height-classes (m). Class interval 50 μ.

Microns	1—4.9 m	5—9.9 m	10—19.9 m	⩾ 20	TOTAL
0—49					
50—	5	7	4	3	19
100—	9	16	5	18	48
150—	9	14	7	28	58
200—	1		6	40	47
250—	1	4	5	19	29
300—			5	13	18
350—				6	6
400—				2	2
450—				3	3
500—				1	1
550—				1	1
Number of species	25	41	32	134	232
Mean thickness microns	143	148	205	235	205

5—9.9 m. The mean thickness is 144 microns between 1 and 9.9 m; 205 microns between 10 and 19 m and 235 microns above 20 m.

Among 134 trees higher than 20 m, *compound leaves are significantly thinner* (mean thickness of 214 microns on 52 trees) *than simple leaves* (mean thickness 248 microns on 82 trees).

The distributions are slightly asymmetrical in each height class. The mode of the general distribution is 175 microns and the means 205.2 microns.

Leaf thickness will be used for the computation of the corrected densities d_i.

9. LEAF DENSITY

All the leaves were weighed individually air dry (12% humidity) with a precision of 1 cg.

The densities were expressed in two ways: d ratio of weight in cg. to area in cm^2; and the corrected value d_i expressed in g/cm^3, which is d divided by thickness. Since leaf thickness was obtained on material saturated with water, d_i has something in common with the specific gravity used in wood technology (ratio of oven-dry weight to saturated volume). However in our case weights are measured with 12% humidity and air dry leaf areas lose about 12 to 20% by retractibility. The average thickness of the lamina tabulated by Roth was used to obtain a rough estimate of the volume of the leaf. The question is whether d_i is ecologically more meaningful than d, in relation to the various strata of the forest and leaf consistency. On the other hand d seems more useful in biomass studies and leaf area index estimation.

Densities d cg/cm^2

The column subtotals and total in Table 40 give the distribution of densities of simple leaves; compound leaves and the grand total by species. The grand total distribution is fairly symmetrical with a general mean of about 1.30 and a rather high variability; the highest d is 2.665 the lowest 0.428, i.e. a variation in the proportion of 6 to 1.

As for leaf size some weighing is necessary to avoid bias in the estimation of total leaf area from leaf weight. Since the various species have different leaf densities and are represented in widely different proportions, it seems to be difficult to estimate the leaf area of any leaf biomass without some caution and experimentation.

There is no evidence a priori that the most frequent medium densities will correspond to the species with the highest numbers of trees.

Table 40. Distribution of densities d cg/cm² by species and by consistency categories.

Density cg/cm²	CONSISTENCY CODES (see p. 46) Simple leaves								Compound leaves						Grand TOTAL
	4	5	6	1	7	3	2	TOTAL	4	5	6	1	3	TOTAL	
0.40—0.59	1		2					3			1			1	4
0.60—	1	2	2					5		2	1			3	8
0.80—	3	3	8		1			15		6	2			8	23
1.00—	3	5	2	1		1	1	13	2	9	1			12	25
1.20—	6	15	2					23	1	2			1	4	29
1.40—	6	7		2		1		16	6	1				7	23
1.60—	6	3		2		2		13	2	1			1	4	17
1.80—	6	1		1		4		12	1				1	2	14
2.00—	3							3	1			1		2	5
2.20—	3			1				4							4
2.40—															
2.60—2.79				1		2		3							3
TOTAL number of species	48	36	16	8	1	10	1	120	13	21	5	1	3	43	163

The weighing by basal area gives proportions in the various density classes quite different from the mere proportion of species (without weighing) in these same density classes. This alteration in the proportions is shown in Table 41 and is due mainly to the abundance in the stands of *Pentaclethra macroloba* (v. n. Clavellino, $d = 0.587$), of *Chaetocarpus schomburgkianus* (v. n. Cacho; $d = 2.293$) and of *Eschweilera subglandulosa* (v. n. Majagüillo, $d = 2.202$).

Weighing was carried out on mean densities (30 leaves by species). The irregularities of the distribution of leaf areas leaf by leaf, suggest that at least some modes would disappear in a distribution of density leaf by leaf. But the 155 ha inventory could probably be interpreted in terms of leaf density as a mixture of two populations, one with density 1.4 the other with density 2.2. The latter being the translation of the abundance of the two high leaf density species mentioned above (*Chaetocarpus* and *Eschweilera*). The distribution of leaf densities by species is symmetrical without irregularities (column Total of Table 40) and contradict this conclusion.

Variability of the density within species

The following list gives the weight in g of 30 air

Table 41. Distribution of species according to the mean leaf density of a sample of 30 leaves by species. Proportion without and with weighing by basal area.

Density interval 0.2 cg/cm². Total number of species: 162

Density cg/cm²	0.40 0.59	0.60 0.79	0.80 0.99	1.00 1.19	1.20 1.39	1.40 1.59
Number of species	5	8	22	26	28	33
Total %	3.09	4.94	13.58	16.05	17.28	20.37
-id- weighed by ΣD^2	3.42	0.93	4.25	15.78	13.36	24.50
Density cg/cm²	**1.60 1.79**	**1.80 1.99**	**2.00 2.19**	**2.20 2.39**	**2.40 2.59**	**2.60 2.79**
Number of species	17	11	5	4	0	3
Total %	10.49	6.79	3.09	2.47	0	1.85
-id- weighed by ΣD^2	14.74	4.20	5.28	11.99	0	1.53

dry leaves (or leaflets) by species with two or three replicates on different adult trees e.g.: *Aspidosperma excelsum* with 35.1 and 36.4 g.

Aspidosperma excelsum 35.1/36.4
Carapa guianensis 49.2/74.1
Catostemma commune 16.2/16.6
Clarisia racemosa 11.8/8.0/13.9
Didymopanax morototoni 50.1/22.5
Eschweilera chartacea 12.2/8.4
Eschweilera grata 7.8/9.6
Eschweilera odora 11.1/19.8
Eugenia pseudopsidium 6.8/15.2
Himatanthus articulata 25.3/40.2
Inga (N.V. Guamo verde) 20.7/18.6
Laetia procera 24.3/21.2
Lecointea amazonica 4.6/8.1
Lecythis davisii 4.7/10.5
Manilkara bidentata 32.7/41.2
Manilkara (N.V. Pendare) 38.3/42.9/49.8
Ouratea sagotii 38.2/32.4
Parinari excelsa 5.4/6.7
Pera schomburgkiana 7.2/6.8
Picramnia macrostachya 6.3/6.5
Piranhea longepedunculata 6.7/7.5
Protium neglectum 24.2/25.3
Protium (N.V. Caraño) 9.4/5.5
Protium (N.V. Sipuede) 6.1/9.3
Sacoglottis cydonioides 21.9/13.3
Sclerolobium paniculatum 18.7/13.0/13.4
Simaba multiflora 4.0/7.2
Tapura guianensis 15.8/17.6
Tetragastris panamensis (N.V. Aracho) 11.7
Vismia (N.V. Lacre amarillo) 5.1/5.4
Sapotaceae (N.V. Purğuillo) 10.3/8.0

Density in cg/cm^2 of the lamina and leaf area classes

On mechanical grounds one would think that leaf density increases with leaf-area; Figure 8 shows no such evidence, neither for simple nor for pinnate leaves; there is a wide variability of densities for any given area for both types of leaves, around of means of about 1.4; i.e. small leaves show small as well as high leaf densities, and the same is true for pinnate leaves, although few observations are available for large areas.

Therefore there is a *quasi-independence of leaf density and leaf area*: this observation has an important practical consequence. Since leaf densities are rather symmetrically distributed (Table 40), an average density can be calculated from the large sample available for 180 species with one individual and about 30 leaves by species. The weight of this sample is $P = 387\,954$ cg and its area $S = 289\,723$ cm^2 and the ratio $S/P = 0.7468$ is the inverse of the average density. This ratio, multiplied by the air dry weight of any large amount of leaves (all species mixed), gives a good prediction of the leaf area. Hence a rule of thumb for leaf area index estimations: the leaf area in cm^2 is close to three-fourths of the air dry weight of the leaves in cg, i.e. a hectare for every ton of air dry leaf batch. An allowance can easily be made to convert air dry weight into green weight by subsampling.

In fact the ratio should be corrected because simple leaves and leaflets of pinnate leaves were weighed with their petioles but the rachis of pinnate and bipinnate leaves were not included. For the same leaf area, P would be larger and the coefficient smaller; the value of 0.7468 would overestimate the area.

Furthermore the results of weighing densities by basal area indicate that there is probably a mixture of at least two populations of densities; some stratification would improve the accuracy of the sampling, at least by separately processing the few species with high leaf density and large populations of trees. It should be remembered that in any large collection of leaves from the forest biomass (all species mixed), densities in cg/cm^2 on the whole vary in the proportion of 1 to 6 with possible local variations.

The six most abundant species in number of trees above 10 cm DBH in the 155 ha inventory of Imataca are given in Table 42 in the decreasing order of number of trees: altogether 26.5% of the grand total of trees (17961/67777) or 31.7% of the total basal area (relative "importance" in the stands). This example is given to stress the wide differences in leaf density of 0.587 to 2.293 (proportion 1 to 4) among the six most common species (mean density 1.594, coefficient of variation 39.9%).

Another way to look at the relationship between leaf area and weight is to calculate a regression S (cm^2) $= a + b \times P$ (cg). With the available 180 species mentioned above, $a = 322.6$; $b = 0.597$; coefficient of correlation $r = 0.889$. The correlation would be better without species either with high or low leaf density which give points that fall somewhat outside the main cloud and have an incidence on the regression according to local peculiarities of floristic composition.

Since the value of a is rather high it is more convenient to use the ratio of S/P.

Leaf density d and total height of trees

The comparison of leaf density means in the three height classes 10 to 19 m, 20 to 29 m and above

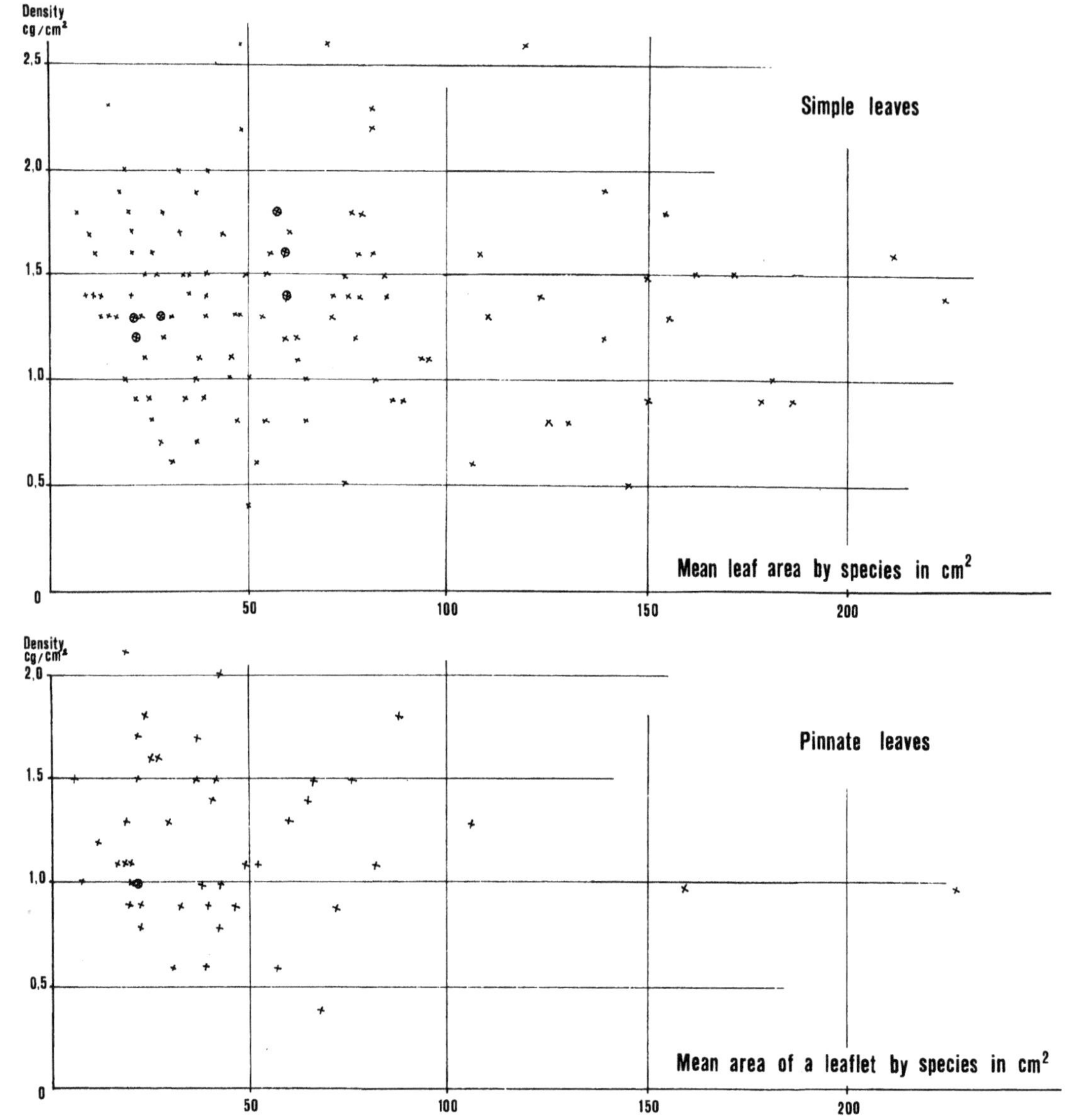

Fig. 8. DENSITY *d* cg/cm² OF LEAF LAMINA as a FUNCTION OF LEAF AREA. One point is one species (mean density of 30 leaves versus mean area) Simple leaves. Pinnate Leaves.

Table 42. Leaf density of the six most abundant species of the 155 Ha inventory (Imataca).

Scientific name	v.n.	Number of trees ⩾ 10 cm DBH	Importance of the species weighed %*	density cg/cm²
Chaetocarpus schomburgkianus	Cacho	3679	6.82	2.293
Eschweilera grata	Cacaito	3446	5.06	1.211
Licania densiflora	Hierrito	3258	6.81	1.598
Protium decandrum	Azucarito	2669	4.65	1.676
Pentaclethra macroloba	Clavellino	2535	3.23	0.587
Eschweilera subglandulosa	Majagüillo	2374	5.14	2.202

* Relative importance of the species in the stand ⩾ 10 cm DBH, weighed in basal area.

30 m, giving respectively 1.283 1.337 and 1.309, shows that they are not significantly different, whereas the mean leaf density of height class below 10 m (0.832) is very significantly lower than leaf densities of trees higher than 10 m (1.313) (Table 43).

Table 43. Leaf density d and total height of trees.

Total height m	< 10	10—19	20—29	⩾ 30
Number of trees	15	23	32	11
Mean density cg/cm²	0.832	1.283	1.337	1.309
Standard deviation cg/cm²	0.265	0.350	0.444	0.517

Density of compound and simple leaves in codominant and emergent trees

Thirty compound-leaved species have a mean density of 1.285 with a standard deviation 0.377. Sixty-five simple-leaved species have a mean density of 1.476 with a standard deviation 0.375. The difference between the two density means 0.191 is significant ($t = 2.30$). *Simple leaves are heavier than compound leaves in the canopy of the forest.*

Corrected densities d_i g/cm³

The same steps of research will be followed for densities d_i. The distribution of d_i (all species and heights mixed) is fairly symmetrical with a gap for d_i values of 1.1 and 1.2 (Table 44). The interval of variation is wide, being 0.15 to 1.35 (proportion 1 to 9, i.e. higher than for d) around the mean of 0.632 for the 158 species.

Thirty compound-leaved species and 45 simple-leaved species of the canopy do not show any significant difference between their corrected density means, which are respectively 0.604 and 0.639 (with respective standard deviation 0.150 and 0.195).

Table 45. Corrected leaf density d_i and total height of trees.

Total height m	< 10	10—19	20—29	⩾ 30
Number of trees	11	15	24	9
Mean density g/cm³	0.562	0.661	0.655	0.578
Standard deviation g/cm³	0.262	0.243	0.161	0.173

The variation of d_i with height is quite different from d; d_i shows a maximum for height class 10—29 m. However, differences between height class means are not significant; since the general distribution of d_i is unimodal, it may be accepted that corrected densities are rather the same at any height of the forest profile: this result is a possible consequence of sparse observational data.

Corrected densities in g/cm³ possibly have a lesser ecological meaning than the uncorrected density cg/cm² and leaf thickness interpreted separately.

Relationship between density d and leaf thickness

The linear regression $1000\ d = a + b \times e$, where d is the density in cg/cm² and e is leaf thickness in microns has been studied for various groups of trees, see Table 46. The correlation between the two parameters is rather loose, being almost lacking in the small group of trees below 10 m. Slopes are fairly different in the two tree groups

Table 46. Linear regression between density d cg/cm² and leaf thickness.

Height class	Number of trees	a	Slope b	Coefficient of correlation
Trees < 10 m	11	726.2	0.652	0.1723
10—19 m	18	761.2	2.500	0.5187
20—29 m	24	245.8	5.127	0.7161
⩾ 20 m	33	87.6	5.641	0.7032
SUBTOTAL	62	408.1	3.928	0.5889
Other canopy trees	72	776.5	2.641	0.6285

Table 44. Distribution of corrected densities, d_i g/cm³. Class interval 0.1.

Density d_i g/cm³	0.1—	0.2—	0.3—	0.4—	0.5—	0.6—	0.7—	0.8—	0.9—	1.0—	1.1—	1.2—	1.3—
Number of species	1	4	11	18	35	39	25	14	7	2			2

10—19 m and 20—29 m or ≥ 20 m. Although the two clouds overlap each other a great deal this is a confirmation of the result found earlier that thickness and density *d* are significantly different in these two height classes.

10. LEAF CONSISTENCY

Leaf consistency is very useful, among other characteristics, to identify trees in the field. However an attempt to classify or grade leaf consistency is somewhat subjective because several parameters are involved: thickness, density, anatomical features, touch and hearing. Leaf consistency is evaluated in the field by crumpling fresh (but not wet) leaves. Some leaves are outstanding for their leathery texture, soft or succulent feel (often due to the presence of a hypodermis), with a crackling or metallic sound. Some species fall into small pieces upon crushing or, on the contrary, resume their original aspect as if they had not been crumpled. Some leaves turn whitish at the instant they are crushed.

Although there are many intermediate situations where species will yield low reproducibility in grading tests, it is easy to select typical species for each category of consistency.

1 — very coriaceous: *Parinari rodolphii*; *Chaetocarpus schomburgkianus*.
2 — metallic sound: *Faramea torquata*; *Ouratea* (*sagotii, guianensis*).
3 — leather: *Manilkara bidentata*: *Rheedia* sp. (v. n. Cozoiba picuda).
4 — like leather, slightly breakable: *Tapura guianensis*; *Ecclinusa guianensis*.
5 — papery sound: *Brownea latifolia*; *Pouteria* cf. *trilocularis* (v. n. Rosado).
6 — weaker than paper: *Rollinia multiflora*; *Hirtella davisii*.
7 — soft: *Chimarrhis microcarpa*.
8 — very soft, and similar to the leaves of mangelwurzel: *Torrubia* spp.; many epiphytes.
9 — very fibrous: leaflets of palm trees, Poaceae, Cyperaceae, Marantaceae, Musaceae.
10 — herbaceous: very weak watery structure (generally not trees).
11 — succulent: Zingiberaceae, Orchidaceae, many epiphytes.

N.B.: Codes 5 and 6 are somewhat difficult to separate and may better be pooled.

The consistency codes are given in column 6 of Appendix 2 for the various species.

Consistency, a subjective qualitative character, is obviously linked to density *d* cg/cm^2, an easily quantifiable character. Table 40 shows that the most frequent consistency categories 4—5—6 have decreasing densities, from category 4 with mainly *d* between 1.3 and 1.4, to category 6 with most frequent density between 0.7 and 0.9. This is true for simple as well as for compound leaves (Table 40).

On the bottom line of Table 40, 48 + 13 = 61 species out of 163 have leathery leaves (consistency 4) and 36 + 21 = 57 species have papery leaves.

Consistencies 1—2—3—7 are infrequent among trees; consistencies 8—9—10—11 refer mainly to palms and other monocots, herbs, forbs and epiphytes.

11. SUMMARY AND CONCLUSION

Leaf morphology has been studied qualitatively and quantitatively in a rainforest of Venezuela with some examples from Brazilian Amazonia and different formations such as the montane rain forest and the semideciduous tropical forest.

The main difficulties inherent in the study of leaves are the collection of material in the field and sampling.

Collecting in the canopies is a prolonged and dangerous operation. Damage to leaves by insects, the presence of epiphylls and abnormal or young leaves among adult leaves are only minor difficulties.

On the other hand, representativeness is the central problem for a reasonable study. It involves an advanced knowledge of the tree flora, including juvenile stages, and requires large forest enumerations for an adequate species weighting. Leaf samples must be large enough to analyze the different parameters in the forest profiles. Such studies have acute limitations, mainly because of the time involved in forest surveys and leaf measurements.

The high proportion of compound leaves is a characteristic of the rain forest in spite of much floristic variation in space. There is a tendency towards enrichment in compound leaves from the lower strata of the forest to the canopy. It has been shown that compound leaves are more deciduous and more frequent than simple leaves among light-demanding species in contrast to tolerant species. Compound leaves are no more common in gaps than in the neighbouring forest. The proportion of compound leaves is lower in montane rainforest, higher in semi-deciduous tropical forest. Some lowland rainforests are particularly rich in compound leaves owing to a dominance of Leguminosae, especially *Caesalpiniaceae*.

The proportion of simple leaves in the rain forest of Imataca is close to two thirds, whatever weighting one uses. Only 20% of the total number of species are provided with opposite or decussate leaves.

Concerning leaf form, no phylogenetic or morphogenetic considerations are offered. The influence of the environment and the physiological age have not been studied. Leaf dimorphism and vesture have only been briefly commented on. Venation has been left aside.

The main objective is the description of the leaves of the different species as they occur in the forest profile, and a discussion of the quantitative relationships existing between length and width of the leaves, drip-tips, form factor, size thickness, densities and consistency.

Length—width ratios are very variable within and between species in the profile (1.0 to 5.4) with flattened, bell-shaped or plurimodal distributions. The ratios are definitely higher in the lower strata of the forest.

Form factors (with and without drip-tips) have been discussed. On the whole they are smaller in the undergrowth; they vary from 0.45 to 0.85 for simple leaves (136 species) and from 0.50 to 0.80 for compound leaves (54 species). Their distributions resemble the length—width ratio distribution. The overall average of 2/3 suggested by Cain *et al.* seems to be something of an underestimate (0.676 for simple leaves; 0.682 for leaflets of compound leaves in Imataca).

Drip-tips are more frequent and longer in the undergrowth and decrease progressively in frequency and length from the sapling stratum to the emergent trees. Pinnate leaves are more often provided with drip-tips than simple leaves. The wide differences in appreciation of drip-tips in the literature indicate that some standardization in measurements is necessary.

The size of simple leaves and leaflets of pinnate and bipinnate leaves shows a considerable range of variation. The question is still open whether leaflets of compound leaves should be preferred to total leaves for the comparison of size. Sizes are in the proportion of 3×10^5 to 1; the proportion falls when the total surface of the compound leaves as a whole is considered and becomes 300 to 1. The wide variation of leaflet and leaf sizes justifies the adoption of a logarithmic scale for size; the distribution established with an arithmetic scale has a long and erratic tail. It has been shown that the limits adopted by Raunkiaer are somewhat arbitrary; the distribution of sizes for simple leaves is continuous and nearly log-normal though strongly skewed and truncated on the left. The histogram for size of compound leaves (total area) is multimodal with an arithmetic scale, which speaks in favour of some size stratification.

It is doubtful whether the average area of the complete leaves of the different species occurring at various levels of heights in the profile is bigger in the undergrowth. But leaves are thinner (140 microns) whereas the average thickness in the canopy is 240 microns. Compound leaves are thinner than simple leaves in the canopy.

Leaf density is computed in two ways: d in cg/cm^2 and d_t in g/cm^3. The latter is obtained by dividing the former by leaf thickness and does not show much variation in the profile, whereas d in cg/cm^2 is higher in the canopy than in the undergrowth; d increases with thickness.

An important consequence has been drawn

from the independence of leaf area and leaf density concerning leaf area index estimations.

Consistency is a characteristic that is fairly well correlated with density and one that is of considerable interest for the field determination of species.

To summarize: the quantified parameters *L/l*, *K*, size, thickness and densities of leaves may be said to be very variable, showing a trend in the forest profile from the ground to the top of the trees, increasing for density, *K* and thickness, decreasing for length—width ratio, size and drip-tips.

ILLUSTRATIONS OF ADULT LEAVES. 197 SPECIES IN ALPHABETICAL ORDER OF FAMILIES

ANACARDIACEAE
1 *Tapirira guianensis* Aubl.
2 *Astronium lecointei* Ducke
ANNONACEAE
3 *Unonopsis glaucopetala* R.E.Fr.
4 *Rollinia multiflora* Splitg.
APOCYNACEAE
5 ? (N.V.: Leche de Burra)
6 *Aspidosperma megalocarpon* M.Arg.
7 *Aspidosperma album* (Vahl.) Ben ex Pichon
8 *Aspidosperma marcgravianum* Woodson
9 *Aspidosperma excelsum* Benth.
10 *Himatanthus articulata* (Vahl.) Woodson
ARALIACEAE
11 *Didymopanax morototoni* (Aubl.) Dcne et Planch.
BIGNONIACEAE
12 *Tabebuia serratifolia* (Vahl.) Nicholson
BOMBACACEAE
13 *Catostemma commune* Sandw.
14 *Eriotheca* sp. (N.V.: Cedro dulce)
BORAGINACEAE
15 *Cordia exaltata* Lam.
16 *Cordia alliodora* (R.E.P.) Cham.
17 *Cordia viridis* (Rusby) Johnston
18 *Cordia nodosa* Lam.
19 *Cordia fallax* Johnston
20 *Cordia bicolor* A.DC.
21 *Lepidocordia punctata* Ducke
BURSERACEAE
22 (*Tetragastris panamensis* Engl.)? N.V.: Aracho
23 (*Tetragastris*)? (N.V.: Aracho blanco)
24 *Protium decandrum* (Aubl.) March.*
25 *Protium neglectum* SW.
26 *Protium* aff. *neglectum* (N.V.: Caraño)
27 *Tetragastris* aff. *panamensis* (N.V.: Caraño negro)
28 *Protium* aff. *decandrum* (N.V: Caraño blanco)

* Young leaf on the left.

29 ? (N.V.: Maro)
30 *Protium* sp. (N.V.: Sipuede)
CAPPARIDACEAE
31 *Capparis guaguaensis* Steyerm.
32 *Crataeva tapia* L.
33 *Capparis amplissima* Lam.
CELASTRACEAE
34 *Goupia glabra* Aubl.
COMBRETACEAE
35 *Terminalia amazonia* (Gmel.) Exell
36 *Terminalia guayanensis* Eichler
37 *Buchenavia capitata* (Vahl.) Eichler
DICHAPETALACEAE
38 *Tapura guianensis* Aubl.
EBENACEAE
39 *Diospyros guianensis* (Aubl.)
40 *Diospyros lissocarpioides* Sandw.
41 *Diospyros melinonii* (Hieron.) A.C.Sm.
ELAEOCARPACEAE
42 *Sloanea guianensis* (Aubl.) Benth. ? N.V.: Espina de erizo
43 *Sloanea laurifolia* (Benth.) Standl.
44 *Sloanea guianensis* (Aubl.) Benth.
EUPHORBIACEAE
45 *Pausandra flagellorachis* Lanj.
46 *Chaetocarpus schomburgkianus* (O. Ktze) Pax et Hoffm.
47 *Hieronyma laxiflora* (Tul.) M.Arg.
48 *Conceveiba guyanensis* Aubl.
49 *Mabea piriri* Aubl.
50 *Drypetes variabilis* Uitt
51 *Pera schomburkiana* (Benth.) M.Arg.
52 *Piranhea longepedunculata* Jablonski
53 *Croton matourensis* Aubl.
54 *Sapium* sp. (N.V.: Lechero blanco)
55 *Sapium* sp. (N.V.: Caucho morado)
56 *Amanoa guianensis* Aubl.
FLACOURTIACEAE
57 *Casearia guianensis* Sandw.
58 *Homalium racemosum* Jacq.
59 ? (N.V.: Maspara)
60 *Banara nitida* Spruce
61 *Laetia procera* (P.et E.) Eichl.
GUTTIFERAE
62 *Clusia* sp. (N.V.: Copey)
63 *Rheedia* aff. *spruceana* Engl. (N.V.: Cozoiba)
64 *Rheedia* sp. (N.V.: Cozoiba picuda)
65 *Tovomita brevistamina* Engler
66 *Vismia guianensis* Choisy
67 *Vismia* sp. (N.V.: Lacre amarillo)
68 *Calophyllum brasiliense* Camb.
69 *Symphonia globulifera* L. f. var.? (N.V.: Mangle amarillo)
70 *Symphonia globulifera* L. f.
HUMIRIACEAE
71 *Sacoglottis cydonioides* Cuatr.
LACISTEMACEAE
72 *Lacistema aggregatum* Rusby
LAURACEAE
73 *Ocotea nicaraguensis* Mez
74 *Ocotea* aff. *subalveolata* C. K. Allen (N.V.: Laurel paragüito)
75 *Beilschmiedia curviramea* (Meissn.) Mez
76 *Ocotea martiana* (Nees.) Mez
77 *Aniba riparia* (Nees.) Mez

78 ? (N.V.: Laurel rastrojero)
79 *Nectandra grandis* (Mez) Kosterm.
80 ? (N.V.: Laurel canelo)
81 *Ocotea duotincta* C. K. Allen
82 *Aniba excelsa* Kosterm.
83 *Endlicheria cocuirey* Kosterm.
LECYTHIDACEAE
84 *Eschweilera subglandulosa* (Steud.) Miers.
85 *Eschweilera* sp. (N.V.: Majagüillo erizado)
86 *Eschweilera* aff. *trinitensis* A. C. Smith et Beard (N.V.: Majagüillo negro)
87 *Eschweilera odora* (Poepp.) Miers.
88 *Eschweilera grata* Sandw.
89 *Eschweilera chartacea* (Berg.) Eyma
90 *Eschweilera corrugata* (Poit.) Miers.
91 *Couratari multiflora* (Smith.) Eyma
92 *Couratari pulchra* Sandw.
93 *Gustavia augusta* Alm.
LEGUMINOSAE (CAESALPINIACEAE)
94 *Tachigalia paniculata* Aubl.
95 *Brownea latifolia* Jacq.
96 *Sclerolobium* sp. (N.V.: Guamillo blanco)
97 *Peltogyne* sp.?
98 *Crudia glaberrima* (Steud.) Macbr.
99 *Cassia grandis* L.
100 *Hymenaea courbaril* L.
101 *Peltogyne porphyrocardia* Griseb.
LEGUMINOSAE (MIMOSACEAE)
102 *Inga rubiginosa* (Rich.) D.C.
103 *Inga heterophylla* Miq.
104 *Inga capitata* Desv.
105 *Inga* sp. (N.V.: Guamo verde)
106 *Inga* sp. (N.V.: Guamo caraota)
107 *Inga* sp. (N.V.: Guamo camburito)
108 *Parkia oppositifolia* Spruce ex Benth.
109 *Piptadenia psilostachya* (D.C.) Benth.
110 *Pentaclethra macroloba* (Willd.) Ktze.
111 *Stryphnodendron polystachyum* (Miq.) Kleinh.
LEGUMINOSAE (PAPILIONACEAE)
112 *Centrolobium paraense* Tul.
113 *Lonchocarpus sericeus* H.B.K.
114 *Pterocarpus* sp. (N.V.: Sangrito)
115 *Pterocarpus rohrii* Vahl.
116 *Dipteryx odorata* (Aubl.) Willd.
MELASTOMACEAE
117 ? N.V.: Saquiyak
118 *Mouriria huberi* Cogn.
119 *Mouriria sideroxylon* Sagot ex Triana
MELIACEAE
120 *Trichilia propingua* (Miq.) DC.
121 *Guarea schomburgkii* C.DC.
122 *Trichilia smithii* C.DC.
123 *Cedrela odorata* L.
MORACEAE
124 *Clarisia racemosa* (R. et P.) Berg.
125 *Helicostylis tomentosa* (P. et E.) Rusby
126 ? (N.V.: Charo negro)
127 *Ficus* sp. (N.V.: Matapalo)
128 *Ficus* sp. (N.V.: Higuerón)
129 *Cecropia* sp. (N.V.: Yagrumo morado) *
130 *Cecropia* sp. (N.V.: Yagrumo sabanero) *
131 *Cecropia sciadophylla* Mart.*
MYRISTICACEAE
132 *Virola surinamensis* (Rol) Warb.
133 *Virola sebifera* Aubl.
134 *Iryanthera lancifolia* Ducke
MYRTACEAE
135 ? (N.V.: Guayabito piedrero)
136 ? (N.V.: Guayabito piedrero chiquito)
137 *Myrcia amazonica* DC.
138 ? (N.V.: Guayabillo negro)
139 *Eugenia pseudopsidium* Jacq.
140 ? (N.V.: Guayabito Zaba)
141 *Eugenia anastomosans* DC.
NYCTAGINACEAE
142 *Torrubia cuspidata* (Heim.) Standl.
143 *Torrubia* sp. (N.V.: Casabe blanco)
OCHNACEAE
144 *Ouratea sagotii* (Van Tieg.) Cowan
OLACACEAE
145 *Heisteria iquitensis* Aubl.
OPILIACEAE
146 *Agonandra brasiliensis* Miers.
PIPERACEAE
147 *Piper* sp. (N.V.: Anisillo redondo)
POLYGONACEAE
148 *Coccoloba* sp. (N.V.: Uvero)
149 *Coccoloba* sp. (N.V.: Arahueque)
150 *Coccoloba* sp. (N.V.: Arahuequillo)
151 *Triplaris surinamensls* Cham.
QUIINACEAE
152 *Quiina guianensis* Aubl.
153 *Touroulia guianensis* Aubl.
RHIZOPHORACEAE
154 *Cassipourea guianensis* Aubl.
ROSACEAE
155 *Hirtella davisii* Sandw.
156 *Hirtella racemosa* Lam.
157 *Hirtella americana* L.
158 *Licania densiflora* Kleinh.
159 *Licania alba* (Bernouilli) Cuatr.
160 *Licania rufescens* Klotzsh ex Fritsch
161 *Licania parvifructa* Fanshawe et Maguire
162 *Licania* sp. (N.V.: Pilón nazareno)
163 *Parinari rodolphii* Hub.
164 *Parinari excelsa* Sabine
RUBIACEAE
165 *Genipa americana* L.
166 *Chimarrhis microcarpa* Standl. var *microcarpa*
167 *Amaioua guianensis* Aubl.
168 *Faramea torquata* M.Arg.
169 *Duroia* sp. (N.V.: Conserva)
RUTACEAE
170 *Erythrochiton brasiliensis* Nees. et Mart.
171 ? (N.V.: Erizo)
172 ? (N.V.: Mapurite)
SABIACEAE
173 *Meliosma herbertii* Rolfe
SAPINDACEAE
174 *Talisia reticulata* Radlk.
175 *Toulicia guianensis* Aubl.
SAPOTACEAE

* Young leaf.

176 *Chrysophyllum auratum* Miq.
177 *Chrysophyllum* sp. (N.V.: Chupón)
178 *Ecclinusa guianensis* Eyma
179 ? (N.V.: Chicle Rosado)
180 *Manilkara bidentata* (A. DC.) Chev.
181 ? (N.V.: Pendarito)
182 ? (N.V.: Capurillo negro)
183 *Pouteria* aff. *anibaeofolia* (A. C. Smith) Baehni (N.V.: Caimito blanco)
184 *Pouteria* ? (N.V.: Capure)
185 *Pouteria egregia* Sandw.
186 *Pouteria* aff. *eugeniifolia* (Pierre) Baehni (N.V.: Purgüillo Felix)
187 *Pouteria* aff. *trilocularis* Cronquist (N.V.: Rosado)
SIMAROUBACEAE
188 *Picramnia macrostachys* Kl.
STERCULIACEAE
189 *Sterculia pruriens* (Aubl.) Schum.*
TILIACEAE
190 *Apeiba echinata* Gaertn.
191 *Apeiba* sp. (N.V.: Cabeza de negro)
VERBENACEAE
192 *Vitex stahelii* Moldenke
VIOLACEAE
193 ? (N.V.: Gaspadillo blanco)
194 *Paypayrola longifolia* Tul.
195 *Rinorea riana* (DC.) O. Ktze.
VOCHYSIACEAE
196 *Erisma uncinatum* Warm.
197 *Qualea dinizii* Ducke

Appendix 1. Comparison of the length—width ratios of leaves (*means by species*) by height classes (single, non-coupled observations) in m.

<1 m	1—4.9	5—9.9	10—19	(4) ≥20	(4) ≥20	(4) ≥20	(2) ≥20	(3) ≥20	(3) ≥20
3.81	2.83	2.23	3.00	2.44	1.96	3.30	2.90		
2.01	1.20	2.60	3.13	3.49	1.97	2.87	2.99	2.67 ++	
	2.34	2.47	3.09	2.32	3.16	3.80	2.37	2.15 ++	
	2.53	3.26	1.81	2.37	2.84	2.26	2.68	2.00 ++	
	2.55	2.60	3.03	2.16	2.22	3.09	2.27	1.63 ++	
	3.72	3.16	2.17	2.04	2.48	2.00	3.21	2.88 ++	
	2.78	3.09	2.08	2.82	1.75	2.97	1.46		
	2.94	3.18	2.45	2.29	2.81	2.14	3.21	3.43 ++	
	2.34		2.17	2.25	3.00	2.11	2.46	2.29 ++	
	2.64		2.18	2.29	3.28	1.80	3.12	2.90 ++	
	2.30		2.12	3.91	2.11	2.00	2.18	2.13 −	
	2.94		2.26	1.86	3.36	1.65	2.80	3.10 ++	
	2.19			2.53	1.92	2.39	2.23	1.94 ++	2.43 ++
	3.07			2.20	1.70	2.55	2.48	2.47 −	
	2.68			1.87	2.15	2.10	2.81	2.74 −	
	2.88			2.71	2.39	2.12	2.90	2.64 ++	
	2.43			2.95	2.13	2.51	1.92	1.86 −	
	3.95			2.66	2.15	2.00	2.91		
	2.41			2.50	2.10	3.16	1.97	2.04 −	
	1.95			2.47	2.03	3.03	2.71	2.66 −	
	2.75			2.05	1.82	2.67	1.82	1.97 −	2.01 ++
	2.24			2.20	2.53	2.68	2.22	2.52 ++	
	2.62			3.00	2.33	1.68	2.32	2.17 +	
	2.46			2.61	1.97	2.78	2.00	2.49 ++	2.57 ++
	3.28			1.91	2.72	1.93	2.46		
				2.49	2.04		2.00	2.05 −	
							2.02	2.09 −	2.17 −
							2.37	2.54 +	
							2.26	1.93 ++	
							2.09	2.20 −	
							1.47	1.53 −	
							2.80	2.13 ++	
							2.60	2.71 −	2.74 −

−: non significant difference.
+: significant ($P = 0.05$).
++: very significant ($P = 0.01$).

* The young leaf is palmatifid.

Appendix 2. Characteristics and measurements on leaves of 183 species
Rain Forest of Imataca, Venezuela. (30 leaves by species).

1 — Code of the species.
2 — Type of leaf. See codes in Text.
3 — Leaf arrangement. See codes in Text.
4 — Leaf form. See codes in Text.
5 — Leaf margin. See codes in Text.
6 — Consistency. See codes in Text.
7 — Mean leaf (or leaflet) area.
8 — Coefficient of variation of 7 in %.
9 — Drip-tips. See codes in Text.
10 — Mean number of leaflets in a pinnate leaf, or of leaflets in a pinna (bipinnate leaves).
11 — Coefficient of variation of 10 in %.
12 — Estimation of the mean leaf area of a pinnate leaf.
13 — Mean number of pinnae in a bipinnate leaf.
14 — Coefficient of variation of 13 in %.
15 — Estimation of the mean leaf area of a bipinnate leaf.
a, b. Coefficients of the regression $S = a + b\,(L \times l)$ with L = length of leaf or leaflet; l = width and S = leaf area.
r: coefficient of correlation.
s^2: variance of 7.
s^2 *S/Ll*: variance of deviations from regression.
? no measurement.

Appendix 2

		1	2	3	4	5	6	7	8	a	$b \cdot 10^4$	s^2 S/Ll	s^2	$r \cdot 10^4$	9	10	11	12	13	14	15
Anacard.																					
TAPIRIRA GUIANENSIS	Patillo	444	3	2	3	1	5	49.6	40.4	−0.41	7377	6.87	402	9917	x	7.80	18.6	386.9			
Annon.																					
ANAXAGOREA	Yarayara amarilla	573	1	1	1	1	5	30.5	30.7	−1.17	6835	2.76	88	9847							
ROLLINIA MULTIFLORA	Anoncillo	20	1	1	3	1	6	51.7	41.7	−3.88	7911	8.00	465	9917							
UNONOPSIS GLAUCOPETALA	Yarayara negra	576	1	1	4	1	4	109.7	29.2	4.28	7267	29.21	1025	9856	x						
Apocyn.																					
ASPIDOSPERMA ALBUM	Hielillo negro	292	1	1	4	1	4	55.1	39.9	−3.22	7744	4.50	484	9953							
— EXCELSUM	Canjilón negro	89	1	2	3	1	4	74.4	27.2	−0.23	7042	8.38	410	9901							
— MARCGRAVIANUM	— amarillo	88	1	3	5	1	5	19.0	27.4	−0.10	7262	0.58	27.3	9899							
— MEGALOCARPON	Hielillo blanco	291	1	1	3	1	3	74.6	25.9	8.00	6324	55.98	373	9247							
HIMATANTHUS ARTICULATA	Mapolo	371	1	2	5	1	6	144.7	24.2	10.08	6065	39.85	1229	9842	x						
Aral.																					
DIDYMOPANAX MOROTOTONI	Sunsun	531*	6	2	1	1	3	88.5	47.5	2.29	7266	6.78	1769	9981	x	10.16	9.6	899.2			
Bignon.																					
JACARANDA OBTUSIFOLIA	San José	513	4	4	8	1	5	46.6	57.4	12.70	4469	84.29	714	9413		25.62	26.6		15.37	26.7	183.5
Boragin.																					
CORDIA ALLIODORA	Pardillo	433	1	2	1	1	6	22.9	34.2	0.13	6449	3.67	61	9876	ε						
— EXALTATA	Alatrique blanco	8	1	1	3	1	4	77.9	26.0	−0.73	7589	7.71	411	9909							
— NODOSA	— peludo	9	1	2	3	1	5	105.6	27.3	−0.24	7041	91.84	824	9427							
— VIRIDIS	Caujaro	131	1	2	1	1	4	32.9	34.2	0.05	6582	1.85	127	9927							

Appendix 2 (continued)

		1	2	3	4	5	6	7	8	a	b·10⁴	s²S/Ll	s²	r·10⁴	9	10	11	12	13	14	15
LEPIDOCORDIA PUNCTATA	Guatacare negro	273	1	2	3	1	6	27.6	24.9	−0.22	6667	3.08	47	9680	ε						
Burser.																					
DACRYODES aff. RORAIMENSIS	Aracho blanco	23	3	2	3	1	?	21.5	30.7	−0.18	7130	0.79	44	9912	x	9.53	14.2	204.9			
PROTIUM DECANDRUM	Azucarito	35	3	2	3	1	5	24.8	40.1	0.35	7163	3.01	99	9878	x	3	0	74.4			
-id- ?	Caraño blanco	99	3	2	3	1	4	40.2	27.9	−0.35	7019	2.93	126	9837	x	6.53	17.9	262.5			
— insigne? neglectum?	Caraño	98	3	2	4	1	5	16.2	33.4	−0.42	7495	0.63	29.2	9895	xx	11.41	14.2	184.8			
— neglectum	Azucarito blanco	37	3	2	3	1	5	60.5	37.6	3.29	6717	7.96	517	9923	xx	7.35	22.4	444.7			
— sp.	Sipuede	529	3	2	1	2	5	18.1	24.9	0.87	6427	0.58	20.3	9862		7.93	10.5	143.5			
TETRAGASTRIS PANAMENSIS	Aracho	22	3	2	1	1	?	28.8	25.7	−1.87	7541	3.20	54	9713	x	9.97	12.5	287.1			
-id-	Caraño negro	100	3	2	3	1	5	42.2	42.2	−1.59	7663	5.84	317	9911	xx	10.07	17.7	425.0			
?	Maro	379	3	2	3	1	1	41.9	35.6	−2.45	8202	5.96	223	9865	x	8.67	10.2	363.3			
Capparid.																					
CAPPARIS AMPLISSIMA	Burro muerto	53	1	2	1	1	5	34.1	13.0	4.83	5702	1.52	19.5	9616							
— GUAGUAENSIS	Toco negro	551	1	2	3	1	6	130.0	39.6	−1.92	6799	36.05	2657	9934							
CRATAEVA TAPIA	Toco blanco	550*	6	2	3	1	6	68.5	34.4	1.52	6213	36.01	539	9671		3	0	205.5			
Celastr.																					
GOUPIA GLABRA	Congrio blanco	162	1	1	1	1	5	24.3	32.4	0.08	6399	1.74	62	9859	ε−						
Chrysobalan.																					
COUEPIA GLANDULOSA	Merecure Terán	390	1	2	3	1	5	21.2	27.6	0.79	7063	0.98	34.2	9861							
HIRTELLA AMERICANA	Terciopelo blanco	546	1	1	3	1	4	154.8	29.5	2.90	7327	22.09	2084	9947							
— DAVISII	Ceniza negra	138	1	2	3	1	6	38.4	32.2	−0.53	7246	2.37	153	9925	x						
LICANIA ALBA	Hierro	298	1	1	3	1	4	150.1	34.5	18.20	6094	424.55	2683	9204							
LICANIA DENSIFLORA	Hierrito	296	1	1	3	1	1	58.7	40.2	0.85	6758	6.22	556	9946							
— HYPOLEUCA	Hierrito blanco	297	1	1	3	1	4	85.1	37.9	0.47	7457	24.73	1042	9885	x						
— PARVIFRUCTA	Hierrito blanco	294	1	2	3	1	5	17.7	28.5	−1.01	7400	0.54	25.5	9896	x						
— RUFESCENS	Hierrillo negro	295	1	1	1	1	1	36.8	41.2	−0.73	6606	3.47	230	9927	ε						
— sp.	Pilón nazareno	464	1	1	3	1	5	11.5	48.1	0.15	6897	0.45	30.6	9929							
PARINARI EXCELSA	Merecurillo	391	1	1	3	1	5	16.2	29.6	−0.08	7223	1.27	23.0	9729	x						
Clusiac.																					
CARAIPA RICHARDIANA	Hicaco	289	1	1	3	1	3	59.6	33.0	−1.47	7845	6.28	386	9921	ε−						
CLUSIA	Copey	168	1	3	5	1	4	81.4	44.7	−1.78	6976	32.92	1324	9879							
RHEEDIA aff. SPRUCEANA	Cozoiba	174	1	3	3	1	?	54.5	33.3	−5.99	7389	3.85	329	9944							
RHEEDIA sp.	Cozoiba picuda	177	1	3	3	1	3	250.6	36.1	−16.98	8013	191.90	8193	9890	ε						
SYMPHONIA GLOBULIFERA	Mangle amarillo	362	1	3	3	1	4	10.3	20.1	0.33	6586	0.46	4.3	9485	x						
TOVOMITA BREVISTAMINA	Coloradito	154	1	3	3	1	3	61.6	48.4	0.06	6597	3.86	888	9979	x						
VISMIA GUIANENSIS	Lacre	321	1	3	3	1	6	20.5	30.2	0.26	6899	1.38	38.3	9826	x						
VISMIA sp.	— amarillo	322	1	3	1	1	5	12.3	23.6	−0.23	5711	0.23	8.4	9861							
Combret.																					
BUCHENAVIA CAPITATA	Pata de danto redondo	436	1	2	5	1	4	8.1	23.1	0.95	5841	0.22	3.5	9690							
Dichapetal.																					
TAPURA GUIANENSIS	Jabón	305	1	1	3	1	4	44.7	31.8	0.47	7265	3.71	202	9911	ε						
Eben.																					
DIOSPYROS LISSOCARPIOIDES	Carboncito	108	1	1	4	1	4	170.8	24.7	−8.43	7850	53.05	1785	9855							
— MELINONII	Moradito	400	1	2	3	1	5	19.5	43.9	0.06	7231	1.34	73	9912	ε						
Elaeocarp.																					
SLOANEA GUIANENSIS?	Espina de erizo	213	1	2	3	1	6	54.2	27.2	−0.04	6773	16.62	217	9623	ε−						
Euphorb.																					
CHAETOCARPUS SCHOMBURGKIANUS	Cacho	64	1	1	1	1	1	47.6	11.7	2.17	6842	8.44	30.9	9867							
CONCEVEIBA GUIANENSIS	Nicolás	406	1	2	3	2	4	59.5	54.7	1.39	6582	8.55	1059	9961	x						
CROTON MATOURENSIS	Canelo	84	1	2	3	1	4	139.4	27.4	−1.84	6988	24.56	1462	9916	x						
DRYPETES VARIABILIS	Kerosen	318	1	2	3	1	4	58.6	41.5	−4.44	7907	5.21	591	9958							
HIERONYMA LAXIFOLIA	Aguacatillo	4	1	2	3	1	4	75.7	29.8	0.91	6918	7.22	509	9933	x						

		1	2	3	4	5	6	7	8	a	b · 10⁴	s² S/LI	s²	r · 10⁴	9	10	11	12	13	14	15
PERA SCHOMBURGKIANA	Pilón rosado	467	1	1	3	1	5	20.6	33.5	0.87	6710	0.64	47	9932							
PIRANHEA LONGEPEDUNCULATA	Caramate	95	6	2	3	1	6	37.9	24.3	1.55	5781	12.16	211	9718		3	0	113.7			
POGONOPHORA SAGOTIA	Flor de Mayo	215	1	1	3	1	6	89.1	33.8	−5.23	7631	6.21	909	9967	ε−						
PAUSANDRA FLAGELLORACHIS	Manglillo	364	1	2	5	2	5	181.2	41.6	−0.16	6401	21.03	5683	9981	x						
Fab. Caesalpin.																					
BROWNEA LATIFOLIA	Rosa de montaña	500	2	2	3	1	5	42.1	22.6	−0.44	7342	2.75	91	9853	xx	12.76	22.0	537.2			
CRUDIA GLABERRIMA	Arepito rebalsero	33	3	2	3	1	5	21.8	40.0	0.73	7428	2.04	76	9866	x	7.27	13.2	158.5			
— OBLONGA	Algarrobo rebalsero	14	3	2	3	1	5	18.9	68.7	−0.27	7225	0.53	169	9985		7.00	18.0	132.3			
HYMENAEA COURBARIL	Algarrobo	13	2	2	3	1	4	11.1	17.7	1.19	6416	0.57	3.9	9260		2	0	22.2			
MORA EXCELSA	Mora rebalsera	399	2	2	3	1	4	52.3	27.8	−0.07	7757	3.62	1568	9914	x	2	0	104.6			
PELTOGYNE PORPHYROCARDIA	Zapatero	582	2	2	3	1	3	20.7	21.5	−0.80	7323	1.33	19.8	9670		2	0	41.4			
— sp.	— negro	583	2	2	3	1	4	65.9	35.5	3.20	6515	8.60	546	9924		2	0	131.8			
SCLEROLOBIUM PANICULATUM	Guamillo	237	2	2	3	1	4	36.3	39.4	1.47	7588	1.87	205	9956	ε-	6.27	18.2	227.6			
— sp.	— rojo	239	2	2	3	1	4	40.9	28.9	−0.61	7112	3.99	138	9855		12.53	18.3	512.5			
INGA ALBA	Guamo colorado	249	2	2	3	1	5	21.1	62.0	−0.17	6686	1.67	172	9950		6.13	14.7	129.3			
— CAPITATA	— negro	255	2	1	1	1	6	20.9	59.6	−0.08	6499	0.50	155	9984		4	0	83.6			
— RUBIGINOSA	— terciopelo	261	2	2	3	1	4	75.8	40.2	7.91	6301	28.05	930	9854	x	3.67	18.0	278.2			
— sp.	— caraota	246	2	2	1	1	5	39.0	56.3	−0.17	6501	2.26	482	9976		4	0	156.0			
PARKIA PENDULA	Cascarón	126	4	2	11	1	10	7.83**													823.7
PENTACLETHRA MACROLOBA	Clavellino	144	4	2	11	1	10	14.35**													381.2
PIPTADENIA PSILOSTACHYA	Yigüire	578	4	2	8	1	10	5.56**	15.5							46.73	12.9		16.27	9.5	42.3
PITHECELLOBIUM BASIJUGUM	Curarina chiquita	187	4	2	8	1	6	72.8**	42.4	−3.91	8103	68.60	952	9646		23.87	18.7		13.80	12.2	239.8
— cf. CLAVIFLORUM	Hueso de pescado blanco	303	4	2	8	1	6	151.6	41.9	−20.83	8962	116.37	4041	9860		20.07	19.3		7.60	14.5	231.2
— jupunba	Samán montañero	506	4	2	8	1	4	4.76	47.0	25.20	6715	11.2	500	9896		6.33	36.2		4.73	23.5	142.5
STRYPHNODENDRON PURPUREUM	Jose fina	317	4	2	8	1	5	0.874	41.2	1.29	7817	27.04	1293	9898		20.83	19.9		21.10	11.7	384.1
Fab. Papilion.																					
ALEXA IMPERATRICIS	Leche de cochino	339	3	2	3	1	4	159.1	38.6	4.11	7639	82.35	3785	9894	xx	8.40	12.7	1336.4			
ANDIRA RETUSA	Pilón rebalsero	466	3	2	3	1	5	19.1	26.8	−0.20	7700	1.72	26.2	9677		8.90	4.5	170.0			
DIPLOTROPIS PURPUREA	Cóngrio	160	3	2	3	1	4	25.8	36.4	−0.42	7389	0.76	88	9957		5.48	28.2	141.4			
DIPTERYX ODORATA	Sarrapia	522	3	1	3	1	4	65.3	32.5	−1.60	8041	4.05	450	9955		6.77	12.1	442.1			
HYMENOLOBIUM sp.	Alcornoque montañero	11	3	2	4	1	4	4.8	33.5	0.20	7729	0.16	2.6	9698		16.00	11.2	76.8			
LONCHOCARPUS SERICEUS	Jebe	309	3	2	3	1	5	7.5	46.8	−0.11	6745	0.12	12.2	9951		5	0	37.5			
PTEROCARPUS ROHRII	Sangrito alado	512	3	2	3	1	5	30.1	41.9	−0.83	7544	2.97	160	9910	xx	5.73	21.4	172.5			
— sp.	Sangrito	511	3	2	3	1	5	20.5	34.5	−1.34	7689	1.03	50	9897		6.67	13.0	136.7			
Flacourt.																					
BANARA NITIDA	Cayenito	132	1	1	3	2	5	45.2	34.7	0.53	7122	3.70	286	9937	x						
CASEARIA GUIANENSIS	Rastrojero	497	1	2	1	2	6	27.2	31.4	1.03	6399	1.32	73	9910							
— RUSBYANA?)	Palmito	420	2	3	2	2	5	14.5	51.8	0.83	6394	2.04	56	9821		8.57	9.5	124.3			
LAETIA PROCERA	Jobo macho	315	1	1	4	1	6	64.1	34.0	−2.38	8785	5.46	475	9942							
?	Máspara	381	1	2	3	1	4	82.0	34.3	−0.50	7394	27.51	810	9835	xx						
Guttiferae see Clus																					
Humir.																					
SACOGLOTTIS CYDONIOIDES	Ponsigue montañero	473	1	1	3	2	5	48.6	26.4	1.63	7075	11.67	164	9651	xx						
Lacistem.																					
LACISTEMA AGGREGATUM	Don Juan	206	1	1	3	1	4	46.4	23.1	−2.51	7709	3.49	115	9854	x						
Laur.																					
ANIBA EXCELSA	Laurel Rollet	335	1	1	1	1	3	154.2	34.3	1.31	6434	51.63	2806	9908							
— RIPARIA	— amarillo	325	1	2	1	1	5	71.2	34.0	−0.69	7116	12.85	585	9893	ε−						
BEILSCHMIEDIA CURVIRAMEA	Aguacatillo Moises	5	1	2	3	1	4	39.4	36.3	−1.46	6839	3.73	205	9912							
NECTANDRA GRANDIS	Laurel	324	1	2	1	1	1	36.6	32.7	−0.27	7125	0.64	143	9955							
OCOTEA DUOTINCTA	Laurel verde	337	1	2	1	1	4	52.3	25.5	3.00	6167	7.19	178	9804							
— MARTIANA	— baboso	326	1	2	3	1	6	73.7	35.3	0.98	6836	19.11	676	9863	x						

		1	2	3	4	5	6	7	8	a	$b \cdot 10^4$	s^2 S/Ll	s^2	$r \cdot 10^4$	9	10	11	12	13	14	15
— NICARAGUENSIS	— blanco	328	1	2	5	1	4	80.7	32.2	0.82	6335	18.12	676	9870	ε—						
— cf. SUBALVEOLATA	— paragüito	332	1	2	3	1	5	31.8	28.7	−0.45	7088	2.46	83	9858	x						
Lecythid.																					
COURATARI MULTIFLORA	Tampipio	540	1	2	3	2	5	26.8	30.2	0.73	6305	2.42	66	9814							
ESCHWEILERA CHARTACEA	Guacharaco amarillo	233	1	1	3	1	5	20.5	36.0	−0.35	6818	0.66	54	9939							
— CORRUGATA	— rosado	234	1	2	3	1	4	32.2	24.4	1.77	6922	2.68	62	9788							
— GRATA	Cacaito	57	1	2	3	1	5	21.4	36.8	0.52	7060	0.99	62	9922	x						
ESCHWEILERA ODORA	Cacao	58	1	1	3	2	5	35.8	37.8	0.06	7707	2.08	181	9944	ε						
— SUBGLANDULOSA	Majagüillo	349	1	2	3	1	4	81.2	24.6	1.99	7248	16.55	400	9798							
— cf. TRINITENSIS	Majagüillo negro	351	1	1	4	1	4	38.6	41.1	−2.86	7985	1.36	252	9974	ε						
— sp.	— erizado	350	1	1	3	2	1	119.0	28.2	−2.41	7842	21.39	1128	9908	ε—						
LECYTHIS DAVISII	Tinajito	549	1	1	3	2	4	22.9	27.5	1.44	6906	3.22	40	9590	xx						
Leguminosae see Fabaceae																					
Malpigh.																					
BYRSONIMA AERUGO	Manteco de agua	368	1	3	3	1	4	58.8	28.4	−4.19	7218	7.74	278	9865							
Melastom.																					
CARAPA GUIANENSIS	Carapa	102	2	2	4	1	?	227.0	35.3	7.92	7170	179.3	6436	9865		6.17	20.2	1400.8			
GUAREA SCHOMBURGKII	Carapillo	104	2	2	3	1	5	57.4	33.2	0.73	6852	17.70	363	9753	x	4.73	35.6	271.5			
TRICHILIA PROPINGUA	Bizcochuelo amarillo	46	3	2	3	1	5	23.5	48.7	−0.33	7510	2.27	131	9915	x	10.30	14.1	242.0			
— SCHOMBURGKII	Suipo	530	3	2	3	1	5	81.6	35.9	1.65	7243	17.62	859	9900	x	5.43	20.3	443.1			
— SMITHII	Mijarro	396	3	2	3	1	5	71.6	48.2	2.80	6810	24.32	1192	9892	ε	5.60	18.8	401.0			
Mor.																					
CLARISIA RACEMOSA	Cajimán	78	1	1	3	1	5	25.9	24.3	1.45	7751	1.45	6.3	9821	x						
FICUS sp.	Matapalo	382	1	2	3	1	5	93.5	31.8	0.38	7645	4.70	888	9973							
HELICOSTYLIS TOMENTOSA	Charo macho	197	1	1	3	1	1	9.1	37.9	−0.42	7754	0.17	11.8	9929	x						
?	— negro	198	1	1	3	1	6	86.5	29.4	−4.18	7957	9.42	645	9930	xx						
Myristic.																					
VIROLA SEBIFERA	Cuajo negro	182	1	1	4	1	4	178.4	29.3	1.97	8449	74.20	2724	9869							
IRYANTHERA sp.	Cuajo grande	181	1	1	1	1	4	138.7	35.4	−3.83	7372	4.35	2415	9991							
Myrt.																					
CALYCORECTES sp.	Terciopelo	545	1	3	1	1	3	211.1	37.0	−0.02	7326	77.17	6107	9939							
EUGENIA ANASTOMOSANS	Guayabo montañero	285	1	3	3	1	5	77.2	32.6	2.17	6598	7.46	632	9941							
— COMPTA	Guayabito blanco	280	1	3	3	1	5	38.6	23.1	0.51	6986	3.01	80	9816	ε—						
MYRCIA AMAZONICA	Curtidor	188	1	3	1	1	4	19.7	33.8	−1.38	6541	1.94	44	9786							
?	Guayabillo negro	277	1	3	1	1	5	14.0	24.3	1.63	4869	0.82	11.6	9656	xx						
?	Guayabito piedrero	283	1	3	3	1	4	17.8	26.6	−0.81	7320	0.64	22.5	9860	x						
?	— — chiquito	0	1	3	1	1	4	6.1	31.0	0.05	6157	0.20	3.6	9730							
?	Guayabito zaba	284	1	3	1	1	4	13.8	43.0	−0.86	6237	0.42	34.4	9939							
Nyctagin.																					
TORRUBIA CUSPIDATA	Casabe	120	1	4	1	1	5	24.9	38.0	1.59	6175	2.29	90	9876							
— sp.	Casabe blanco	123	1	4	3	1	4	107.8	27.4	−0.85	6477	34.78	872	9806							
Ochn.																					
OURATEA SAGOTII	Pilón amarillo	462	1	1	3	1	1	77.9	30.3	−3.88	7649	6.99	559	9928							
Opil.																					
AGONANDRA	Aceituno	3	1	2	2	1	4	11.8	36.5	3.50	4824	15.25	4.3	8110							
Polygon.																					
COCCOLOBA sp.	Arahuequillo	28	1	2	3	1	4	123.5	35.8	−3.57	7388	15.46	1956	9960	ε						
-id-	Uvero negro	562	1	2	10	1	1	1622.0	41.1	−73.00	8678	7522	445089	9915							
Quiin.																					
QUIINA GUIANENSIS	Cola de pava chiquita	151	1	4	3	1	4	49.6	49.6	−2.03	7245	2.55	607	9979	x						
Rhizophor.																					

Appendix 2 (continued)

		1	2	3	4	5	6	7	8	a	$b \cdot 10^4$	s^2 S/LI	s^2	$r \cdot 10^4$	9	10	11	12	13	14	15
CASSIPOUREA GUIANENSIS	Mamoncillo blanco	357	1	3	1	1	4	76.3	30.5	2.49	7649	9.54	543	9815	ε						
Ros. see also Chrysobalan.																					
PRUNUS SPHAEROCARPA	Menta berti	388	1	1	3	1	5	26.8	24.1	−1.78	7024	1.60	42	9813							
Rub.																					
AMAIOUA GUIANENSIS	Cacho de venado	66	1	4	3	1	4	59.4	31.6	−3.98	7271	5.19	353	9926	xx						
CHIMARRHIS MICROCARPA	Carutillo	117	1	4	3	1	7	150.5	34.9	−1.78	6434	30.92	2764	9946	ε						
DUROIA	Conserva	165	1	4	3	1	4	223.8	49.5	−6.32	7306	22.97	12286	9990	x						
FARAMEA TORQUATA	Pata de grulla	437	1	4	3	1	2	95.1	36.3	−6.50	7862	5.74	1195	9976							
GENIPA AMERICANA	Caruto montañero	118	1	4	6	1	6	186.0	39.2	−8.07	6722	63.99	5326	9942							
Rut.																					
?	Erizo	208	2	2	3	1	4	36.2	31.0	0.46	7141	1.17	127	9955	ε	9.80	15.5	354.8			
?	Mapurite	372	3	2	3	1	4	23.5	41.2	−0.74	7341	1.92	94	9901		8.73	19.5	205.2			
Sabiae.																					
MELIOSMA HERBERTII	Carruache	119	1	2	1	1	5	47.1	39.0	0.77	6282	8.42	337	9879	ε−						
Sapot.																					
CHRYSOPHYLLUM AURATUM	Caimito morado	76	1	2	3	1	4	78.7	39.9	−1.93	7652	28.33	984	9860	ε						
ECCLINUSA GUIANENSIS	Chicle	199	1	1	1	1	4	16.7	29.6	−0.03	6257	0.68	24.5	9865	ε						
MANILKARA BIDENTATA	Purguo	485	1	2	3	1	3	70.3	21.3	3.23	6818	11.88	224	9742							
-id- ?	Purguo morado	487	1	2	3	1	3	57.5	34.9	−1.97	7429	4.21	402	9949							
sp.	Pendare	451	1	2	3	1	3	47.9	25.3	−0.64	7106	3.50	146	9883							
POUTERIA cf. ANIBAEFOLIA	Caimito blanco	74	1	2	3	1	4	22.0	42.3	−1.53	7467	0.76	86	9912	x						
— EGREGIA	Purgüillo amarillo	481	1	2	3	1	5	34.2	30.0	−0.43	7192	2.23	105	9890	ε						
— cf. TRILOCULARIS	Rosado	502	1	2	3	1	5	71.0	43.9	0.50	6875	6.23	970	9969	x						
— VENOSA	Bámpara	38	1	1	6	1	5	52.7	34.5	1.12	6741	3.74	331	9945							
— sp.	Caimito negro	77	1	2	3	1	4	21.2	28.1	0.23	6654	8.79	35.5	9880	ε−						
?	Capurillo negro	589	1	2	6	1	5	28.8	35.1	0.19	6682	1.95	102	9908							
?	Pendarito	452	1	2	3	1	3	28.2	22.8	−1.45	7645	1.26	41.4	9852	ε						
?	Purgüillo	480	1	2	3	1	4	20.4	34.9	0.56	7049	1.43	51	9864	ε−						
Sapind.																					
TALISIA RETICULATA	Cotoperiz montañero	172	2	2	3	1	5	46.4	35.6	0.32	6956	3.40	273	9940		12.20	24.9	566.1			
TOULICIA GUIANENSIS	Carapo blanco	105	3	2	3	1	5	106.6	52.8	−0.96	7219	70.7	3163	9891	ε	15.57	18.4	1659.8			
Simaroub.																					
SIMABA MULTIFLORA	Congrillo	159	3	2	3	1	3	18.3	42.5	0.39	6734	0.61	60	9949		4.50	35.4	82.3			
SIMAROUBA AMARA	Cedro blanco	134	3	2	3	1	4	18.7	26.4	−0.73	7878	0.57	24	9885		14.93	29.4	279.2			
PICRAMNIA MACROSTACHYA	Tortolito	555	3	2	1	1	6	21.8	34.1	−1.46	6950	2.74	55	9758		6.5	?	141.7			
Stercul.																					
STERCULIA PRURIENS	Majagua	348	1	2	3	1	5	83.8	54.6	−2.97	7963	7.58	2091	9983	ε						
Tiliac.																					
APEIBA ECHINATA	Onotillo	415	1	2	3	2	5	61.7	35.8	1.61	6715	11.03	488	9886							
— sp.	Cabeza de negro	56	1	1	3	2	6	125.3	32.7	−4.81	7805	20.17	1667	9879	ε						
Verben.																					
VITEX STAHELII	Totumillo	556*	6	4	1	1	6	32.3	42.4	1.41	5756	3.08	187	9920	ε−	4.60	13.3	148.6			
Viol.																					
PAYPAYROLA LONGIFOLIA	Gaspadillo negro	226	1	2	1	1	4	49.6	33.4	0.67	5947	2.00	274	9963							
RINOREA RIANA	Gaspadillo marrón	225	1	1	3	1	6	47.1	33.8	−2.47	7006	4.62	253	9912	xx						
?	Gaspadillo blanco	224	1	2	1	1	5	30.1	42.4	−0.82	5854	0.88	163	9973							
Vochysiac.																					
ERISMA UNCINATUM	Mureillo	404	1	4	6	1	4	39.0	40.2	0.08	6751	2.21	246	9955							
QUALEA DINIZII	Guarapo	266	1	2	4	1	4	19.9	26.1	1.40	7151	2.73	27.1	9540	x						
VOCHYSIA LEHMANNII	Canelito	82	1	4	6	1	5	24.9	28.7	0.32	6834	2.19	51	9790							

Appendix 3. Comparison of characteristics and measurements on leaves at various stages of development

1 — Total height of the trees in m; A higher than 20 m; a between 10 and 20 m; aa below 10 m * DBH in cm.
2 — Mean ratio of length to width of the leaves.
3 — Range of 2.
4 — Mean form factor K.
5 — Range of 4.
6 — Length of drip-tip in mm. For ε xx, see codes in Text.
7 — Range of 6.
8 — Mean product length × width in cm^2 (in bold when young stage is smaller than adult).
9 — Coefficient of variation of 8 in %.
10 — Difference of the means of column 8: O non-significant; + significant $P = 0.05$; ++ significant $P = 0.01$.
11 — Mean leaf area cm^2.
12 — Coefficient of variation of 11 in %.
13 — Mean number of leaflets.
14 — Density in cg/cm^2.
15 — Density d_i in g/cm^3.
16 — Leaf thickness in microns (data from I. Roth).
? — No observation.

Appendix 3

VERNACULAR NAMES Scientific names in Appendix 2	1	2	3	4	5	6	7	8	9	10	11	12	13	14	15	16
AGUACATILLO (Euphorb.)	A	1.86	1.66–2.18	0.664	0.62–0.72	7.9	0–13	108	30		75.7	30		1.880	0.696	270
	A							110	35	O						
	0.5	1.61	1.54–1.69			10		192	51	+						
ALATRIQUE BLANCO (Boragin.)	A	2.05	1.36–2.73	0.729	0.66–0.79	4.3	0–7	103	25					1.471	0.287	513
	1	2.77	2.22–3.57			5		**77**	23	++						
ALETÓN *SLOANEA GUIANENSIS*	A	2.21	1.8–2.6			5		36	35							185
ALGARROBO	35	2.33	2.06–2.86	0.719	0.64–0.81	0		15.4	16		11.1	18	2	1.201	0.600	200
	A	2.26	1.85–2.61			ε		17.4	18	+						
ALGARROBO REBALSERO	27	2.42	1.20–3.33	0.711	0.63–0.77	0		26.6	68		18.9	69	7.0	0.942	0.645	146
	1.5	4.39	3.5–5.8					43.7	47	+						
ANAMÚ *CASEARIA* (*JAVITENSIS*?)	6	2.23	1.76–2.83			15		141	37		109	37		0.799		
ANISILLO *PIPER RETICULATUM*	1.5	2.83	2.46–3.50					80	30							
ANISILLO REDONDO *PIPER* sp.	3	1.20	1.07–1.49			13	0–35	158	17		128	19				
ANONCILLO (Annon.)	26	2.69	2.07–3.61	0.725	0.64–0.80	0		70.3	38		51.7	42		0.674	0.358	188
	4	2.79	2.47–3.10			0		102	36	++						169
	0.3	6.77	5.47–8.43			0		**21.3**	25	++						
ARACHO (Burser.)	A	2.99	2.38–3.63	0.664	0.57–0.77	6.7	2–10	40.6	23		28.7	26	10.0	1.360	0.602	226
	A	2.67	2.13–3.64					30.4	34	++			9.8			

Appendix 3 (continued)

VERNACULAR NAMES Scientific names in Appendix 2	1	2	3	4	5	6	7	8	9	10	11	12	13	14	15	16
AZUCARITO (Burser.)	A	2.71	2.24—3.10	0.666	0.62—0.73	7.5	5—13	34.2	40		24.8	40	3	1.676	0.741	226
	A	2.69	2.06—3.33					32.8	37	o			3			
	4	3.16	2.55—3.79				12—20	118	53	++			5			
	1	2.79	2.10—3.50				5—10	129	36	++			5			
AZUCARITO BLANCO (Burser.)	A	2.37	1.73—3.05	0.673	0.61—0.74	8.5	5—12	85	39		60.5	38	7.35	1.535	0.767	200
	A	2.15	1.78—2.57			7		73	32	o			7.17			
BAMPARA (Sapot.)	31	1.87	1.35—2.13	0.690	0.64—0.75	0		76.6	35		52.7	34		1.308	0.707	185
	3	3.31	2.94—3.71			10		250	37	++						
BARAMÁN (Bombac.)	76*	2.11	1.64—2.64					41	26				1			
CATOSTEMMA COMMUNE	10*	2.41	2.00—2.90					137	55	++			3			
BIZCOCHUEIO AMARILLO	20	2.94	2.00—3.88	0.710	0.65—0.82	4.8	0—12	31.8	47		23.5	49	10.3	1.131		
(Meliac.)	2	3.93	3.00—4.71			15		53.2	30	++			12.0			
CABEZA DE ARAGUATO																
SLOANEA GRANDIFLORA	0.6	2.34	2.02—2.62			10		175	41							
CACAITO (Lecythid.)	A	2.34	2.00—2.57	0.666	0.61—0.73	7.0	2—12	29.5	37		21.4	37		1.211	0.631	192
	A	2.73	2.13—3.16					32.4	34	o						
	A							20.7	39	+						
	2	3.65	3.11—4.00			20		25.1	48	o						
CACAO (Lecythic.)	A	2.68	2.22—3.30	0.733	0.67—0.79	5.6	0—10	46.3	37		35.8	38		1.034	0.458	226
	A	2.00	1.50—2.47					67.7	33	++						
CACHICAMO (Clus.)	A	2.27	1.56—2.68			0		92.7	23		71.6	25				
CALOPHYLLUM BRASILIENSE	A	1.63	1.37—2.15			0		29.5	42	++						
CACHO (Euphorb.)	A	2.88	2.40—3.35	0.719	0.61—0.80	0		66.4	38		47.6	37		2.293	0.507	452
	A	3.21	2.85—3.55					99.1	34	++						
	A	2.84	2.39—3.10					39.8	33	++						
	shoot	2.86	2.35—3.25					162.2	28	++						
CAIMITO MORADO (Sapot.)	14	2.12	1.81—2.38	0.710	0.65—0.91	7.6	0—14	105	39		78.7	40		1.816	0.685	285
	7	2.58	2.08—3.03			13	6—20	187	30	++						
CAIMITO NEGRO (Sapot.)	20	2.61	2.14—3.00	0.646	0.57—0.70	4.2	0—10	31.6	27		21.2	28		0.987	0.580	170
	2	2.46	2.22—2.88			10		46.0	37	+						
CAJIMÁN (Mor.)	A	2.67	2.17—3.09	0.731	0.66—0.80	5.3	0—8	33.3	23		25.9	24		1.515		
	A	2.39	2.00—3.39			5	4—6	42.9	35	+						
	3	3.28	2.90—3.83			15		49.5	34	++						
	2	3.37	2.61—4.33			20		**16.2**	41	++						
CANJILÓN NEGRO (Apocyn.)	A	2.32	1.90—3.08	0.702	0.67—0.76	0		106	27		74.4	27		1.574	0.425	370
	A	2.39	2.06—2.84			0		85.2	28	+						
(Fab. Pap.)	1.5	3.00	2.62—3.55			0		158.8	48	+						
CANELITO NEGRO *ANDIRA*	A	1.46	1.31—2.03			0		19.0	28				6.4			256

Appendix 3 (Continued)

VERNACULAR NAMES Scientific names in Appendix 2	1	2	3	4	5	6	7	8	9	10	11	12	13	14	15	16
CAPA DE TABACO (Lecythid.) *COURATARI PULCHRA*	5	2.60	2.20—3.02			ε		187	36							113
CARAMACATE (Euphorb.)	A	2.52	2.10—2.91	0.600	0.55—0.72	ε		62.7	39		37.9	24	3	0.657	0.291	226
	A	2.45	2.04—3.06					38.0	37	++						
	0.6	2.70	2.45—3.06			ε		38.4	28							
CARAÑO (Burser.)	A	3.21	2.59—4.35	0.598	0.53—0.67	16.5	12—22	22.1	32		16.2	33	11.4	1.125	0.694	162
	A	3.43	2.83—4.30					36.9	27	++			8.6			
CARAÑO NEGRO (Burser.)	23	3.31	2.33—4.23	0.611	0.50—0.71	22.9	16—32	57.4	40		42.2	42	10.0	0.887	0.974	90
	2	3.35	2.57—4.21					172	39	++			8.0			
CARAPA (Meliac.)	A	242	1.88—3.63	0.747	0.66—0.89	0		306	36		227	35	6.2	1.088		
	A	2.51	1.85—3.24					119	27	++			14.8			
	0.7	2.79	2.37—3.14					**91.7**	32	+			4			
CARAPO BLANCO (Sapind.)	A	2.69	2.32—3.01	0.717	0.69—0.77	4.8	2—7	146	53		106.6	53	15.6			
	A	2.32	1.64—3.20			ε		93.6	41	++			19.0			
	8	2.70	1.93—3.27			ε		175	40	○			?	1.375	0.941	146
	4	2.75	1.66—3.28			0		157	49	○			14.1			
CARUTILLO (Rubiac.)	40*	2.46	2.15—2.86	0.606	0.56—0.67	11.4	0—18	237	34		150	35		0.942	0.413	228
	A	2.29	2.01—2.65					98	29	++						
CARUTO MONTAÑERO (Rub.)	27	2.60	2.09—3.10	0.637	0.58—0.68	0		289	38		186	39		0.935	0.799	117
	6	2.97	2.23—3.63			0		595	48	++						
CASABE BLANCO (Nyctagin.)	5	2.47	1.81—2.81	0.637	0.55—0.69	0		168	27		108	27		0.875	0.332	263
	5							100	30	++						
CAUCHO *SAPIUM* sp. (Euphorb.)	0.5	3.81	3.20—4.14			10		56.4	29							
CEDRO BLANCO (Simaroub.)	A	2.75	1.95—3.03	0.753	0.66—0.81	ε		24.6	25		18.7	26	14.9	2.190	0.548	399
	A	2.66	2.05—3.21			0		25.9	29	○			16.0			
	1.5	2.66	1.85—3.42			ε		36.7	23	++			28.2			
CEDRO DULCE *ERIOTHECA* (Bombac.)	1	2.53	2.00—3.33			10		119	35				5			
CENIZA NEGRA (Chrysobalan.)	24	3.12	2.70—3.68	0.656	0.59—0.71	10.0	7—15	53.8	31		38.4	32		0.929	0.411	226
	A	2.90	2.32—3.28			0		30.5	29	++						
COLA DE PAVA (Rut.)	2	5.30	4.56—6.00			xx		594	32							
ERYTHROCHITON BRASILIENSIS	2	5.13	4.28—6.00			xx		588	32	○						
CONGRILLO (Simaroub.)	A	2.26	1.94—2.90	0.690	0.64—0.75	0		26.5	43		18.3	42	4.5	1.312	0.279	470
	A	2.58	2.13—2.93			0		11.7	43	++			3.6			
	0.4	3.04	2.48—3.63			ε		20.7	29	++			9.3			
CÓNGRIO (Fab. Papilion.)	A	2.18	1.76—2.94	0.723	0.66—0.76	0		35.5	36		25.8	36	5.5	1.614	0.476	339
	A	2.13	1.71—2.57			0		35.5	28	○			5.9			
COTOPERIZ MONTAÑERO (Sapind)	A	2.78	1.80—3.72	0.646	0.55—0.68	10.7	3—19	66.2	36		46.4	36	12.2	0.995	0.829	120
	6	2.92	2.48—3.43			25.8	13—30	130.8	26	++			?			68

Appendix 3 (Continued)

VERNACULAR NAMES Scientific names in Appendix 2	1	2	3	4	5	6	7	8	9	10	11	12	13	14	15	16
CUAJO (Myristic.)	A	4.20	2.80—6.66			0		44.2	22							200
VIROLA SURINAMENSIS	5	3.29	2.92—3.84			10		214	24	++						
CHAPARRILLO PELUDO	4	2.55	2.00—3.00			0		85.7	54							
(Malpigh. ?)	4	2.62	2.00—3.11			ε		43.6	30	++						
CHARO MACHO (Mor.)	A	2.22	1.59—2.71	0.686	0.60—0.76	3.7	0—7	12.3	40		9.1	38		1.721		
	4	2.66	2.39—3.09			13	10—16	51.8	32	++						
	5	2.26	2.09—2.50			10		52.8	23	○						
CHARO NEGRO (Mor.)	26	2.80	2.14—3.35	0.703	0.66—0.75	13.3	0—21	114	28		86.5	29				
	A	3.10	2.65—3.63					102	33	○						
CHICLE (Sapot.)	A	3.24	2.69—4.18	0.624	0.57—0.72	ε		26.7	29		16.7	30		1.990	0.802	248
	A	2.98	2.57—3.33			ε		46.1	27	++						
	a	3.44	2.75—4.00			ε		81.6	32	++						
CHUPÓN *CHRYSOPHYLLUM* sp.																
(Sapot.)	9	3.26	2.60—3.80			0		129	44							
FRUTA DE BURRO MORADO																
(Annon.)	2	3.72	3.25—4.50			10		15.9	21							
GASPADILLO MARRÓN (Viol.)	3	2.73	2.40—3.13	0.576	0.51—0.62	18.6	0—27	70.7	32		47.1	34		0.810	0.378	214
	1.5	2.32	2.15—2.64			12	10—15	77.1	30	○						
GASPADILLO NEGRO	a	4.33	3.30—5.28	0.603	0.57—0.64	0		82.3	34		49.6	33		0.991	0.360	275
	3	3.37	3.00—4.00			0		115	32	+						
GUACHARACO AMARILLO																
(Lecythid.)	A	2.51	1.89—3.13	0.668	0.60—0.83	0		30.6	35		20.5	36		1.373	0.822	167
	A	2.25	1.76—2.64				7—10	57.1	35	++						
	1	2.93	2.38—3.30			15		77.5	36	+						
GUACHIMACÁN (Apocyn.) *BONAFOUSIA UNDULATA*	1	2.78	2.11—3.40			10		114.3	38							
GUACIMILLO (Ulm.) *TREMA MICRANTHA*	1.5	2.94	2.44—3.66			20		43.3	30							
GUÁCIMO (Stercul.) *GUAZUMA ULMIFOLIA*	1.5	2.34	2.09—2.62			0		51.0	19							
GUAMILLO (Fab. Caesalpin.)	A	2.23	1.54—2.58	0.779	0.68—0.85	20	0—12	45.9	41		36.3	39	6.3	1.716	0.524	327
	A	1.94	1.46—2.31					39.7	38	○			11.6			
	A	2.43	2.00—2.81					33.2	37	++			12.5			
GUAMILLO ROJO (Fab. Caesalpin.)	A	2.48	1.89—2.91	0.697	0.61—0.74	0		58.4	28				12.5	1.504	0.580	259
	A	2.47	2.14—2.90			0		62.5	40	○			9.6			
GUAMO CINTA (Fab. Mimos.) *INGA LATERIFLORA*	2	7.83	6.25—9.50					8.4	16				20.0			
GUAMO RABO DE MONO																
(Fab. Mimos.)	22	2.81	1.93—3.73	0.627	0.53—0.70	0		58.3	67		36.9	67	9.3	1.059	0.784	135
	A	2.74	2.17—3.29					63.4	46	○			7.2			

Appendix 3 (Continued)

VERNACULAR NAMES Scientific names in Appendix 2	1	2	3	4	5	6	7	8	9	10	11	12	13	14	15	16
GUAMO VERDE. *INGA* sp. (Fab. Mimos.)	A	2.07	1.83—2.48					66.9	43				4			213
	1	3.26	2.82—3.85					226	27	++			2			
GUARAPO (Vochys.)	A	2.64	1.85—3.36	0.704	0.57—0.92	7.2	0—10	25.9	27					1.708	0.495	345
	A	2.90	2.00—3.50			7		17.9	19	++						
GUAYABILLO BLANCO (Myrt.) *MYRCIARIA FLORIBUNDA*	1	2.64	2.22—3.18			5		6.7	33							
GUAYABITO BLANCO (Myrt.)	A	2.55	2.19—3.00	0.669	0.58—0.73	6.5	0—10	54.6	23					1.312		
	A	2.12	1.68—2.37					23.4	26	++						
	4	2.53	2.16—3.09			10		80.6	35	++						
GUAYABITO NEGRO (Myrt.) *MYRCIA SPLENDENS*	2	2.30	1.91—2.83			15		40.7	32							
HICACO (Clus.)	23	2.27	1.47—3.01	0.743	0.67—0.84	3.8	0—14	77.8	32		59.6	33		1.722		
	6	2.73	2.28—3.33				0—10	118	23	++						
HIELILLO BLANCO (Apocyn.)	A	2.55	2.07—3.01	0.701	0.66—0.75	0		105	27		74.6	26		1.490		
	A					0		28.7	20	++						
	9	3.29	2.88—3.80				5—10	110	25	o						190
	6	2.46	2.00—2.77				?	78.7	42	+						
HIERRILLO BLANCO (Chrysobalan.)	24	2.07	1.67—2.76	0.615	0.55—0.67	8.4	2—12	25.3	27		17.7	28		1.088		
	A	2.71	2.13—3.31					11.2	32	++						
	0.5	2.91	2.50—3.23					16.1	45	++						
HIERRITO (Chrysobalan.) *LICANIA DENSIFLORA*	A	2.28	1.88—2.81	0.689	0.66—0.72	0		85.6	40		58.7	40		1.598	0.523	305
	A	2.14	1.82—2.50			ε		69.4	43	o						
	10	3.56	3.04—4.06			15		77.3	42							
	7	2.78	2.13—3.60			ε		113	25	++						
HIERRO (Chrysobalan.)	A	2.74	2.02—3.29	0.707	0.66—0.77	0		216	36		150	34				
	A	2.37	1.85—3.08					106	29	++						
	10	2.84	2.31—3.29			20		314	20					1.589	1.025	155
	3	2.53	2.00—2.87			20		**241**	43							
HIERRITO BLANCO (Chrysobalan.)	A	2.13	1.41—2.54	0.721	0.65—0.81	5.8	0—10	113	38		85.1	38		1.207	1.326	91
	4	2.28	1.86—2.75			15	10—20	185	53	++						
JABÓN (Dichapetal.)	A	2.70	2.23—3.00	0.694	0.63—0.74	6.3	0—10	60.9	32		44.7	32		1.179	0.351	335
	A	3.07	2.66—3.74					61.2	31	o						
	3	3.22	2.92—3.64			12	10—15	71.7	25	+						270
JEBE (Fab. Papilion.)	25	2.00	1.50—2.60	0.665	0.60—0.87	ε		11.2	46		7.5	47	5.0	1.005	0.561	179
	0.5	2.23	1.90—2.83			25	20—30	39.6	40	++			4.0			
JOBO MACHO (Flacourt.)	A	3.22	2.53—3.62	0.802	0.73—0.85	7.4	5—11	75.6	33		64.1	34		1.099		
	A	3.07	2.28—3.58					75.9	19	o						
	0.5	2.13	1.80—2.48			12	10—15	66.9	35	o						

Appendix 3 (continued)

VERNACULAR NAMES Scientific names in Appendix 2	1	2	3	4	5	6	7	8	9	10	11	12	13	14	15	16
KEROSEN (Euphorb.)	24	1.92	1.50—2.71	0.727	0.65—0.77	ε		79.8	38		58.6	41		1.689	0.461	366
	A	1.86	1.49—2.42					40.6	29	++						
LACRE AMARILLO (Clus.)	A	3.04	2.50—3.50	0.565	0.52—0.61	0		21.9	23		12.3	24		1.383		
	4	2.77	2.40—3.21			0		47.8	35	++						
LACRE NEGRO (Clus.)	{ 6	2.88	2.30—3.29					211	36							
VISMIA MACROPHYLLA	{ 2	2.34	2.00—3.00					335	33	++						
LAUREL PARAGÜITO (Laur.)	A	2.11	1.59—2.56	0.645	0.59—0.76	7.8	0—17	45.5	28		31.8	29		1.758		
	2	3.12	2.33—3.71			10		**31.4**	29	++						
LAUREL VERDE (Laur.)	33	3.47	2.76—4.00	0.657	0.62—0.78	0		79.9	27		52.3	25				
	5	3.68	2.94—4.36			0		115.7	37	++						
LECHE DE BURRA? (Apocyn.)	5	2.60	2.28—3.10			18		25.9	28							
LECHE DE COCHINO (Fab. Papilion)	A	2.48	1.93—3.02	0.723	0.66—0.77	15.4	10—20	203	39		159	39	8.4	1.021	0.451	226
	2	2.45	2.00—2.95			15	10—20	**130**	39	++			6			
MACHO *SWARTZIA* sp. (Fab. Caesalpin)	A	2.91	2.42—3.33			ε		154	19							
MAJAGUA (Stercul.)	A	1.97	1.70—2.39	0.742	0.66—0.78	3.6	0—6	109	53		83.8	55		1.567	0.631	248
	A	2.04	1.75—2.26					92.6	43	○						
	1						10—20 (35—190)									
MAJAGÜILLO	A	2.43	1.95—3.19	0.698	0.60—0.75	13.6	5—15	109	25		81.2	25		2.202	0.673	327
	10	3.30	2.64—4.04			15		268	32	++						
	7	2.96	2.72—3.18			13	12—15	259	27							196
	3	2.31	1.83—2.64			10		199	32	+						
MAMONCILLO BLANCO (Rhizophor.)	18	2.00	1.63—2.35	0.757	0.67—0.85	6.6	0—12	96.5	30		76.3	30		1.649		
	5	2.41	2.15—2.86			15		155	21	++						
MANGLE AMARILLO (Clus.)	24	2.71	2.21—3.09	0.629	0.56—0.72	5.2	0—10	15.1	20		10.3	20		1.402		
	A	2.66	1.84—3.30					21.4	48	+						
MANGLILLO (Euphorb.)	a	3.00	2.56—3.45	0.606	0.57—0.64	14.6	0—30	283	41		181	42		1.016	0.793	128
	4	3.52	3.20—3.85			20		**211**	44	+						
MANIRA *SLOANEA ROBUSTA* (Elaeocarp.)	1	2.94	2.72—3.13				20—30	293	22							
MANTECO DE AGUA (Malpigh.)	A	2.48	2.05—2.80	0.671	0.61—0.74	0		87.1	26		58.8	28		1.498	0.768	195
	1.5	2.88	2.51—3.33			10		414	28	++						
MAPOLO (Apocyn.)	10	3.25	2.59—4.50	0.623	057—0.69	12.5	6—21	222	26		145	24		0.582	0.786	74
	A	2.16	1.78—2.59			0		155	49	++						175
MARO ? (Burser)	A	2.02	1.61—2.65	0.687	0.58—0.80	10.9	0—22	54.1	33		41.9	36		2.024	0.854	237
	1	1.99	1.58—2.28			7	5—10	105	57	+						
MASAGUARO (Fab. Mimos.)																

Appendix 3 (Continued)

VERNACULAR NAMES Scientific names in Appendix 2	1	2	3	4	5	6	7	8	9	10	11	12	13	14	15	16
STRYPHNODENDRON POLYSTACHYUM	3	2.19	1.72—2.44			15		48.2	24							
MERECURE MONTAÑERO (Chrysobalan)	1	2.88	1.57—3.54				10—15	157	36							260
PARINARI RODOLPHII	0.6	4.35	3.55—5.10				10—15	51.2	50	++						
MERECURE TERÁN (Chrysobalan.)	25	2.44	2.02—3.20	0.733	0.62—0.78	0		28.9	28		21.2	28		1.239	0.548	226
	1	2.83	2.30—3.33			10		75.4	23	++						
MERECURILLO (Chrysobalan.)	33	1.97	1.60—2.29	0.659	0.57—0.80	5.6	0—9	22.5	29		16.2	30.		1.304	1.003	130
	A	1.82	1.62—2.04			4		24.2	33	o						
	A	2.01	1.71—2.36			4		16.0	31	++						
MERGUO *GUSTAVIA AUGUSTA* (Lecythid.)	1	3.07	2.58—3.64			ε		267	45							156
MORA (Fab. Caesalpin)	A	2.15	1.38—2.73					44.4	45				4			
DIMORPHANDRA GONGGRIJPII	1	2.54	2.33—2.87			10		72.4	38	++			4			
MORADITO (Eben.)	20	2.12	1.54—3.25	0.674	0.49—0.74	58	1—14	26.9	44		19.5	44		1.868	0.732	255
	9	2.92	2.53—3.40			17		45.4	40	++						140
MUREILLO (Vochys.)	A	2.22	1.98—2.71	0.677	0.63—0.73	ε		57.6	40		39.0	40		2.003	0.707	283
	A	2.52	1.96—3.10					36.7	28	++						
NICOLAS (Euphorb.)	18	2.46	2.20—2.80	0.634	0.55—0.70	9.0	0—18	88.2	56		59.5	55		1.469	1.300	113
	7	2.54	2.03—3.02					202	37	++						100
NICUA (Flacourt.)	A	2.66	2.30—3.03	0.683	0.62—0.73			41.9	23		28.5	22		0.881		
HOMALIUM RACEMOSUM	1	3.06	2.76—3.30					74.7	27	++						
NISPERO (Fab. Caesalpin.)	A	2.17	1.83—2.80	0.690	0.54—0.74	1.5	0—4	21.7	28		15.8	25		0.961		203
LECOINTEA AMAZONICA	A	2.32	2.00—3.03					35.3	19	++						
OJO DE GRULLA. (Euphorb.) *MARGARITARIA NOBILIS*	2	2.68	2.37—2.88			15		60.9	30							
ONOTILLO (Tiliac.)	A	2.71	2.14—3.63	0.685	0.60—0.75	0		89.4	36		61.7	36		1.256	0.794	226
	3	2.67	2.34—3.27				20—35	163	32	++						
	1	2.23	1.19—2.54				15—25	108	40	++						
PALO DE MARIA (Fab. Caesalpin.) *TACHIGALIA PANICULATA*	3	2.88	2.15—3.57				20	155	25				12			113
PATA DE DANTO (Combret.)	6	2.10	1.65—2.52			0		36.6	31							152
TERMINALIA AMAZONIA	1	2.36	1.50—2.66			0		12.7	31	++						
PATA DE PAUJI (Euphorb.)	A	2.81	2.37—3.58	0.711	0.58—0.84	9.8	6—14	31.5	21		24.7	24				
MABEA PIRIRI	6	2.40	1.94—2.75			8	6—10	38.7	31	+						122
	3	2.89	2.60—3.33			20		68.2	33	++						
PENDANGA BLANCA ? (Myrt.)	2	2.43	2.00—3.00			20		75.1	41							
PENDARE (Sapot.)	A	2.00	1.74—2.34	0.699	0.63—0.74	0		68.3	25		47.9	25		2.665	0.720	370
	A	2.49	2.16—2.78					100.9	30	++						

Appendix 3 (Continued)

VERNACULAR NAMES Scientific names in Appendix 2	1	2	3	4	5	6	7	8	9	10	11	12	13	14	15	16
	A	2.57	2.23—3.01					137	28	++						
PENDARITO (Sapot.)	A	2.46	2.14—2.91	0.684	0.62—0.73	5.8	3—9	38.8	21		28.2	23		1.886	0.639	295
	shoot	2.82	2.30—3.14			10		160	25	++						
PICA PICA MORADA (Chrysobalan.)	A	2.40	2.02—3.00	0.721	0.60—0.84	4.2	3—5	33.8	30		25.7	50		1.630		
HIRTELLA RACEMOSA	4	2.72	2.50—3.00			10		47.1	27	++						
PILÓN AMARILLO (Ochn.)	A	2.35	1.86—2.77	0.725	0.67—0.78	0		107	29		77.9	30		1.634	0.592	276
	4	3.90	3.25—4.75					101	32	o						
PILÓN NAZARENO (Chrysobalan.)	A	2.00	1.48—2.47	0.695	0.65—0.78	0		16.5	48		11.5	48		1.639	0.329	497
	A	2.05	1.72—2.46			0		16.3	17	o						
PILÓN REBALSERO (Fab. Papilion)	A	2.14	1.31—3.00	0.763	0.63—0.93	0		25.0	26		19.1	27	8.9	1.164	0.831	140
	1	2.92	2.20—3.60			0		**16.2**	32	++			?			
PILÓN ROSADO (Euphorb.)	A	2.17	1.66—2.54	0.706	0.63—0.80	0		29.4	35		20.6	33				192
	A	2.02	1.68—2.35					18.2	27	+						
	A	2.09	1.66—2.55					21.8	34	++						
PONSIGUE MONTAÑERO (Humir.)	A	2.54	2.20—3.02	0.667	0.61—0.76	11.8	7—19	66.4	27		48.6	26		1.500	0.545	275
	A	2.37	2.10—2.68			?		49.8	21	++						
PURGÜILLO ? (Sapot.)	A	2.26	1.80—2.87	0.666	0.58—0.75	2.6	0—7	29.7	33		20.4	35		1.678	0.633	265
	A	1.93	1.60—2.56			1		19.0	26	++						
PURGÜILLO BLANCO (Bignon.) *TABEBUIA STENO CALYX*	1	3.95	3.33—4.76			10		190	47							
PURGUO (Sapot.)	26	2.80	2.45—3.46	0.718	0.64—0.84	0		98.4	22		70.3	21		2.642	0.800	330
	A	2.87	2.12—3.45			0		85.0	35	o				1.892	0.620	305
	3	3.50	3.16—3.72			15		203	20	++						
PURGUO MORADO (Sapot.)	A	2.09	1.67—2.49	0.716	0.68—0.75	0		80.1	33		57.5	35				
	A	2.20	1.94—2.46			0		79.5	25	0						
RABO DE CANDELA ? (Acanth.)	3	2.41	2.22—2.83			0		145	28							
RASTROJERO (Flacourt.)	9	2.99	2.63—3.37	0.669	0.59—0.74	ε		41.0	32		27.2	31		0.700		
	1	2.62	2.37—3.00			5		39.5	34	o						
ROSADO (Sapot.)	25	2.97	2.48—3.53	0.660	0.60—0.74	8.2	0—16	102	44		71.0	44		1.453		
	0.3	2.72	2.50—2.95			10		147	52	o						
ROSA DE MONTAÑA (Fab Caesalp.)	a	2.69	2.08—3.20	0.609	0.55—0.67	21.7	15—31	57.9	22		42.1	23	12.8	1.046	0.590	178
	5	2.80	2.20—4.00			20		55.8	44	o			10.0			
SAMAN MONTAÑERO (Fab. Mimos.)	23	1.47	1.07—1.80	0.710	0.59—0.84	0		6.71	49		4.76	47	29.9	1.476	0.569	259
	A	1.53	1.23—1.76			0		7.28	64	o			26.4			
SAN JOSÉ (Bignon.)	A	2.90	1.92—4.00	0.615				0.758	74		0.466	57	393			146
JACARANDA OBTUSIFOLIA		2.90	2.20—4.33		same tree			0.716	44	o	0.447	34				
SANTA MARIA (Polygon.)	A	2.80	2.42—3.07	0.720	0.68—0.78	0		204	21					1.156	0.466	248
TRIPLARIS SURINAMENSIS	A	2.13	1.88—2.54					201	28	o						

Appendix 3 (Continued)

VERNACULAR NAMES Scientific names in Appendix 2	1	2	3	4	5	6	7	8	9	10	11	12	13	14	15	16
SAQUIYAK VERDE? (Melastom.)	2	1.95	1.68—2.22				10—15	208	17							
SIPUEDE (Burser.)	A	3.04	2.50—3.73	0.680	0.61—0.74	0		26.7	26				7.9	1.119	0.643	174
	A	2.80	2.30—3.58			0		38.7	34	++			5.5			
	0.5	3.11	2.70—4.00					22.9	39	o			6.4			
SUIPO (Meliac.)	A	2.14	1.46—2.54	0.702	0.59—0.78	7.5	0—11	110	36		81.6	36	5.4	1.115		
	aa	2.99	2.13—3.57			30		323	33	++			7			
SUNSUN (Araliac.)	A	2.89	2.50—3.27	0.696	0.66—0.72	13.0	6—25	119	48		88.5	47	10.1	1.886	0.628	300
	A	3.43	2.70—4.14			?		43.7	48	++			9.9			
	10	2.84	2.48—3.24			22		345	30	++			9.6			
	0.5			FIRST LEAF SIMPLE				58	32				4.0			
TERCIO PELO (Myrt.)	20	3.65	2.65—4.35	0.734	0.68—0.79	0		288	37		211	37		1.617		
	2	3.04	2.73—3.40				3—4	313	27	o						
TINAJITO (Lecythid.)	A	2.74	2.30—3.43	0.634	0.52—0.70	13.9	10—19	31.0	28		22.9	28		1.526	0.794	192
	A	2.60	2.23—3.05			10		20.3	30	++						
	A	2.71	1.78—3.66			10		22.0	32	++						
TORTOLITO (Simaroub.)	10	2.79	2.13—3.28	0.648	0.55—0.76	ε		33.5	31					0.991	0.586	169
	6	2.57	2.00—2.95			0		28.9	32	o						180
TOTUMILLO (Verben.)	32	2.87	1.86—3.75	0.594	0.53—0.67	ε		53.7	44		32.3	42	4.8	0.965	0.393	245
	1	2.69	2.25—3.16			10		94.8	45	++			5(or4)			
UVERO BLANCO *COCCOLOBA* sp. (Polygon.)	1	2.75	2.23—3.22			0		454	36							
YAGRUMO MORADO (Mor.) *CECROPIA* sp.	0.5	2.01	1.77—2.33			ε		176	35							
YAGRUMO MONTAÑERO (Mor.) *CECROPIA SCIADOPHYLLA*	1	2.24	1.90—2.44			10		135	33							
YARAYARA *DUGUETIA* sp. (Annon.)	5	3.16	2.83—3.42			15		193	32							
YARAYARA AMARILLA CHIQUITA (Annon.) *DUGUETIA PYCNASTERA*	4	2.62	2.40—2.92			20		146	31							
YARAYARA NEGRA (Annon.)	A	3.05	2.30—3.60	0.732	0.66—0.78	8	0—21	145	30		110	29		1.385	0.695	199
	3	3.12	2.52—3.69			10		210	39	+						
ZAPATERO (Fab. Caesalpin.)	A	1.84	1.53—2.08	0.706	0.65—0.77	0		29.4	20		20.7	21	2	1.760	0.519	339
	7	2.47	2.01—2.87			ε		120	41	++			2			270
	aa	2.99	2.80—3.33			ε		72	25	++			2			

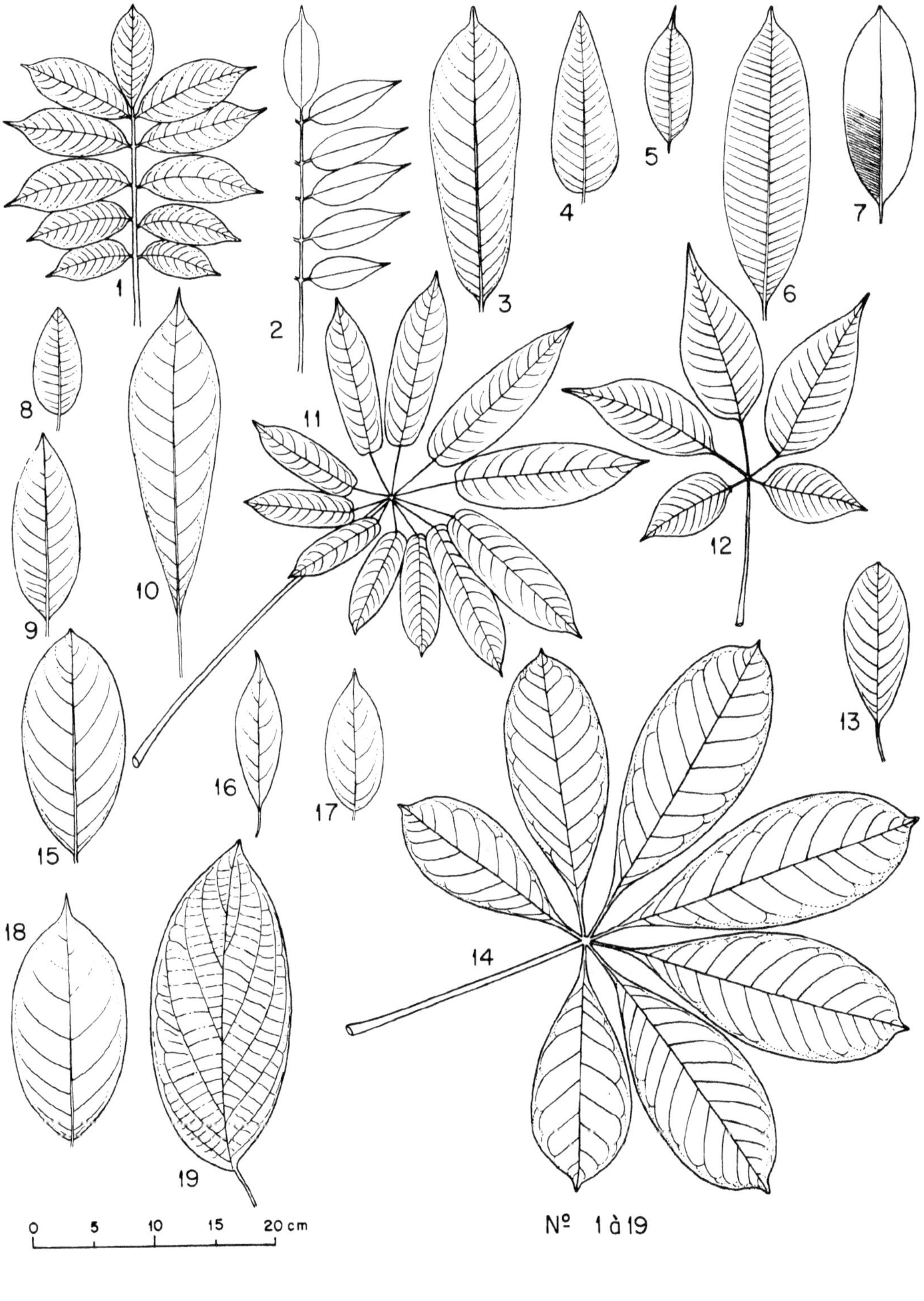

N° 1 à 19

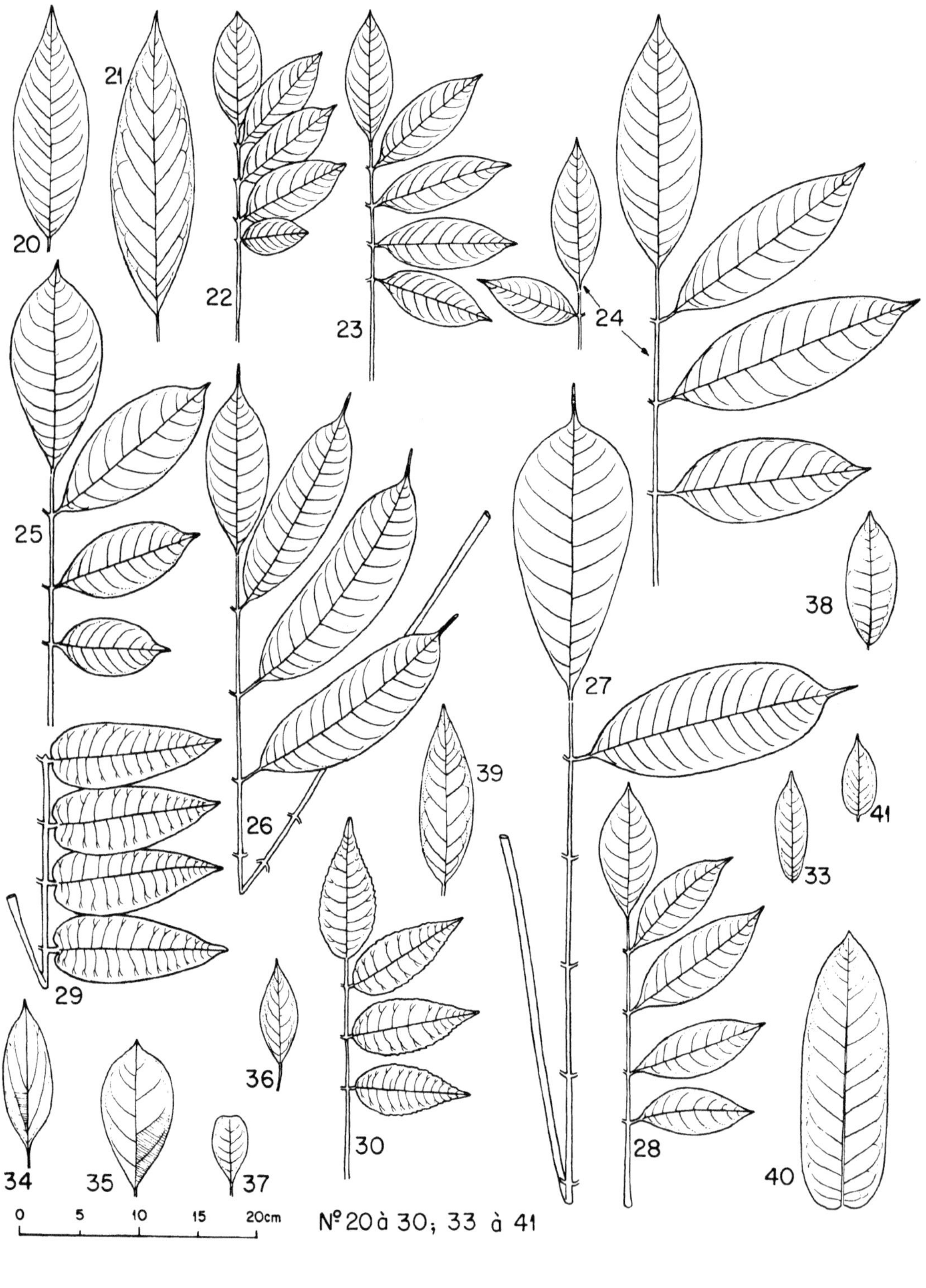
20
21
22
23
24
25
26
27
28
29
30
33
34
35
36
37
38
39
40
41
0 5 10 15 20cm
Nº 20 à 30; 33 à 41

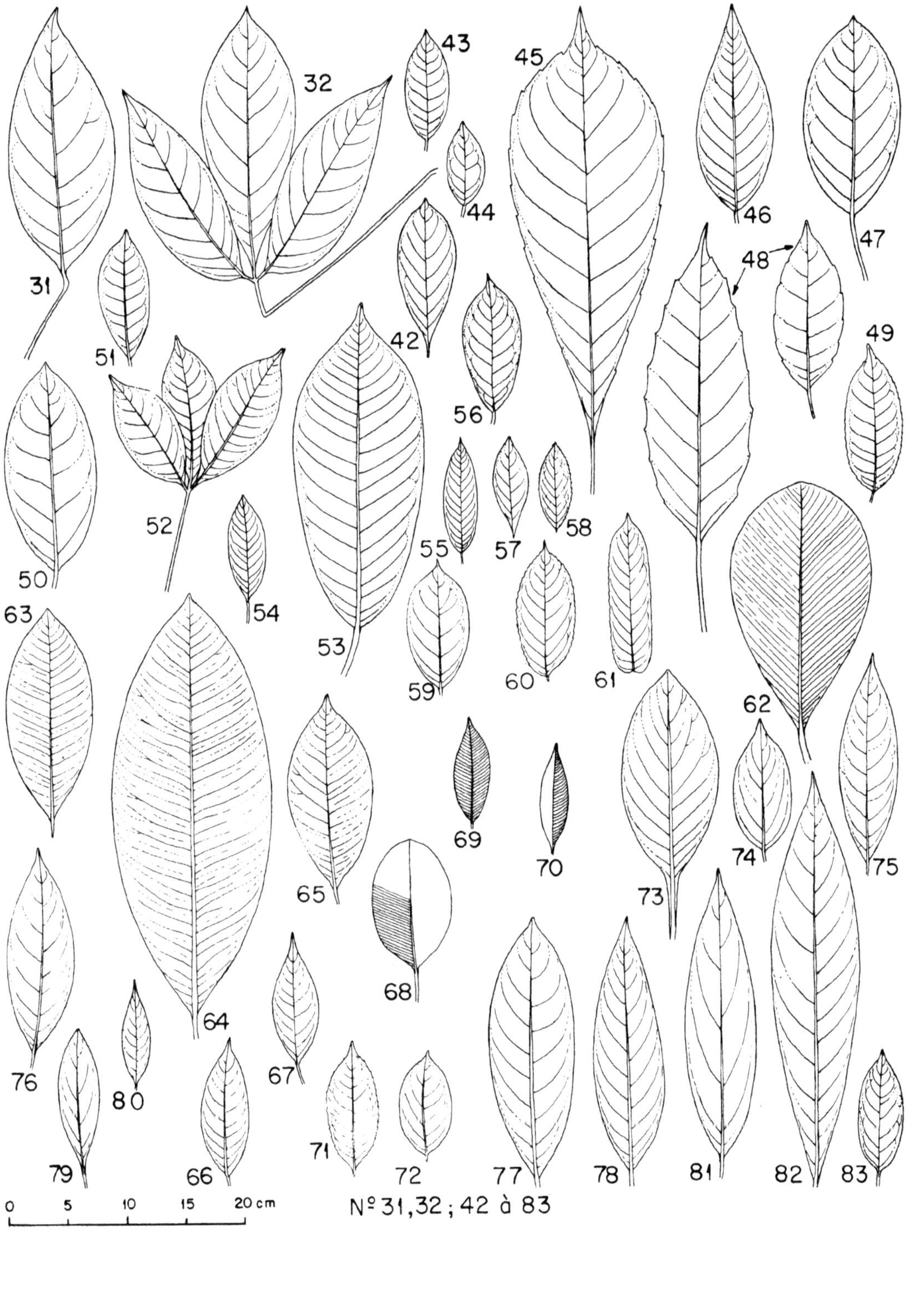
31
32
43
44
45
46
47
48
49
50
51
52
53
42
54
55
56
57
58
59
60
61
62
63
64
65
66
67
68
69
70
71
72
73
74
75
76
77
78
79
80
81
82
83
0 5 10 15 20 cm

N° 31, 32; 42 à 83

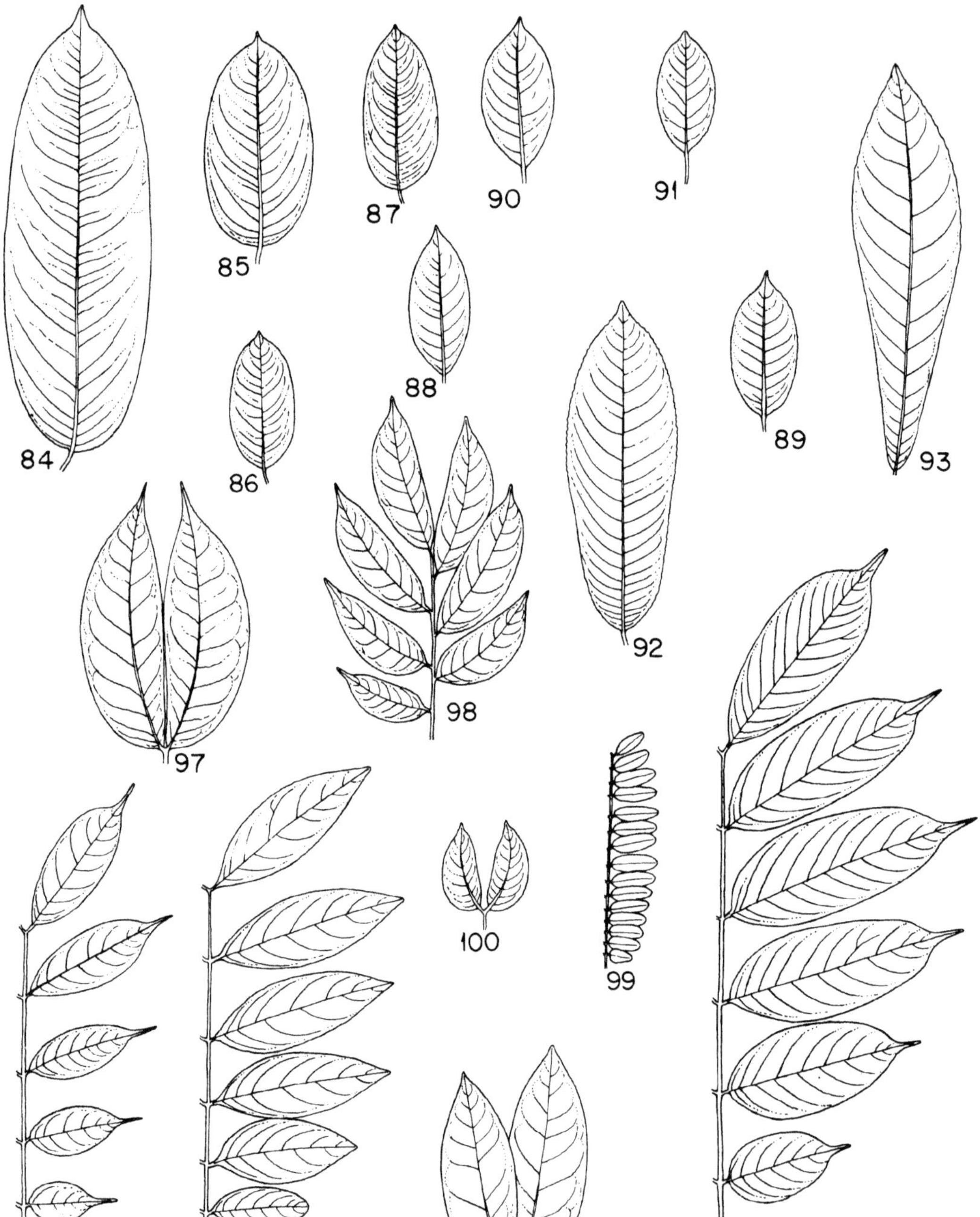
84
85
86
87
88
89
90
91
92
93
94
95
96
97
98
99
100
101
0 5 10 15 20cm
Nº 84 à 101

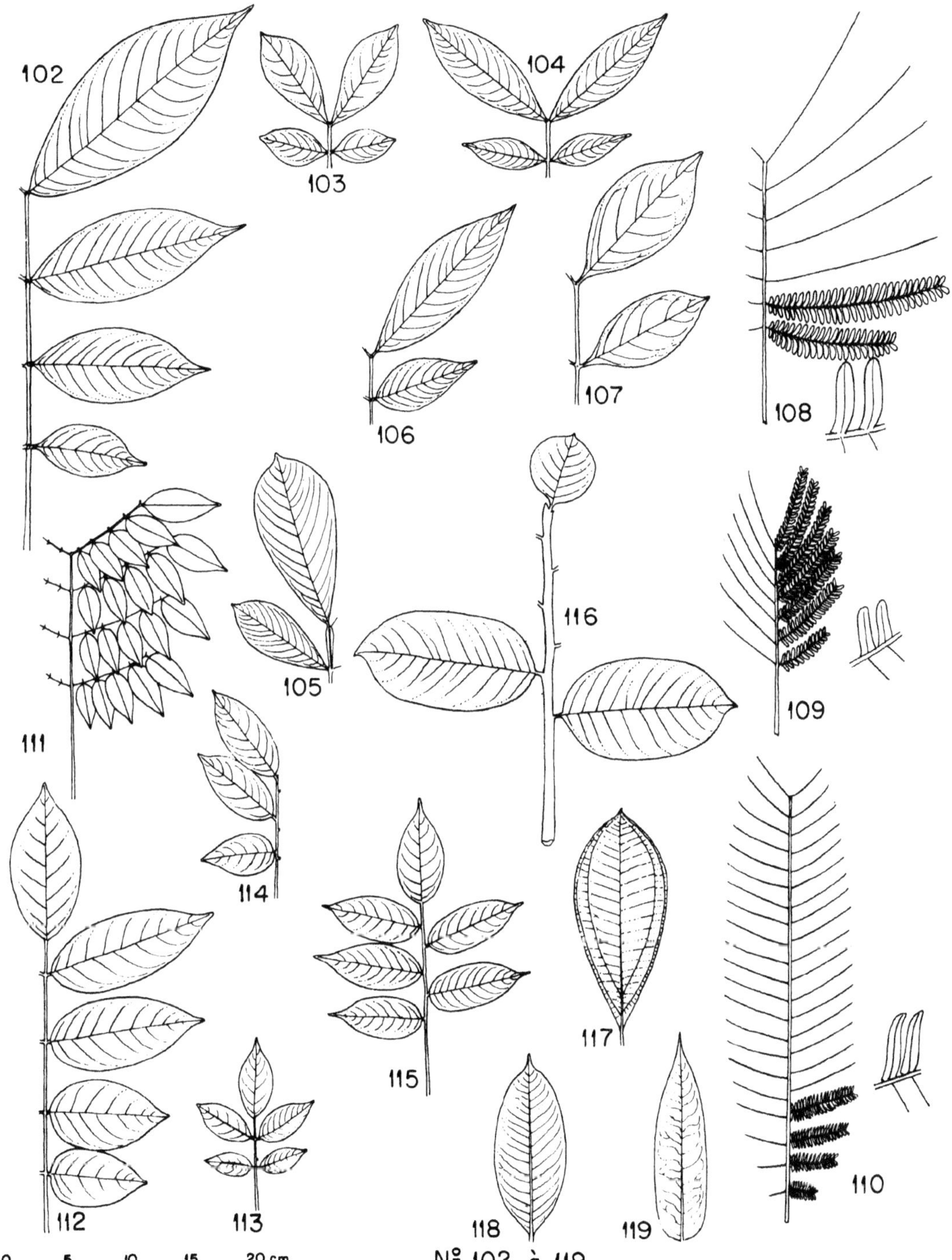
102
103
104
105
106
107
108
109
110
111
112
113
114
115
116
117
118
119
0 5 10 15 20 cm

Nº 102 à 119

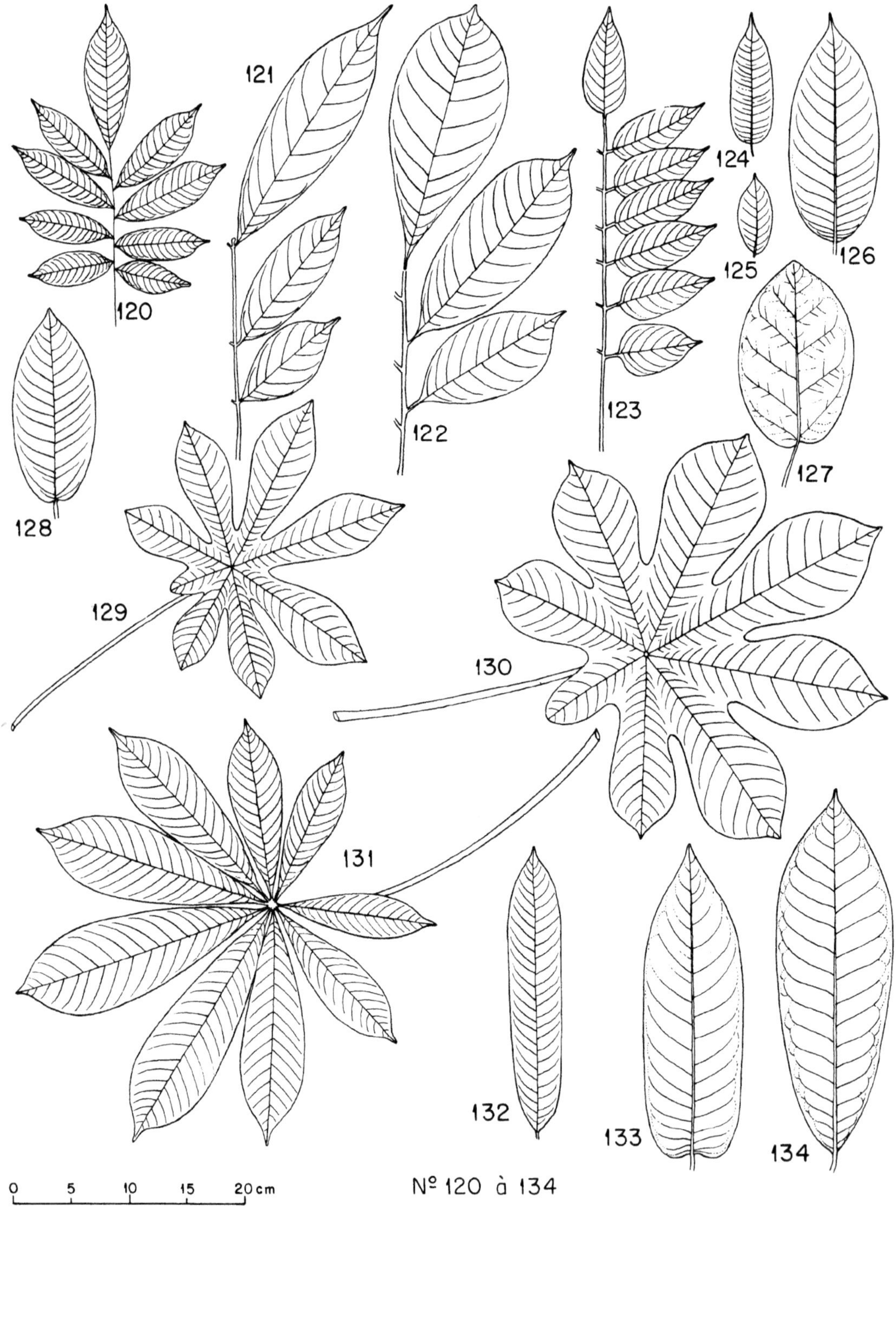
121
120
128
122
123
124
125
126
127
129
130
131
132
133
134
0 5 10 15 20 cm

Nº 120 à 134

135
136
137
138
139
140
141
142
143
144
145
146
147
148
149
150
151
152
153
154
155
156
157
158
159
160
161
162
163
164
165
166
167
168
169
0 5 10 15 20 cm

Nº 135 à 169

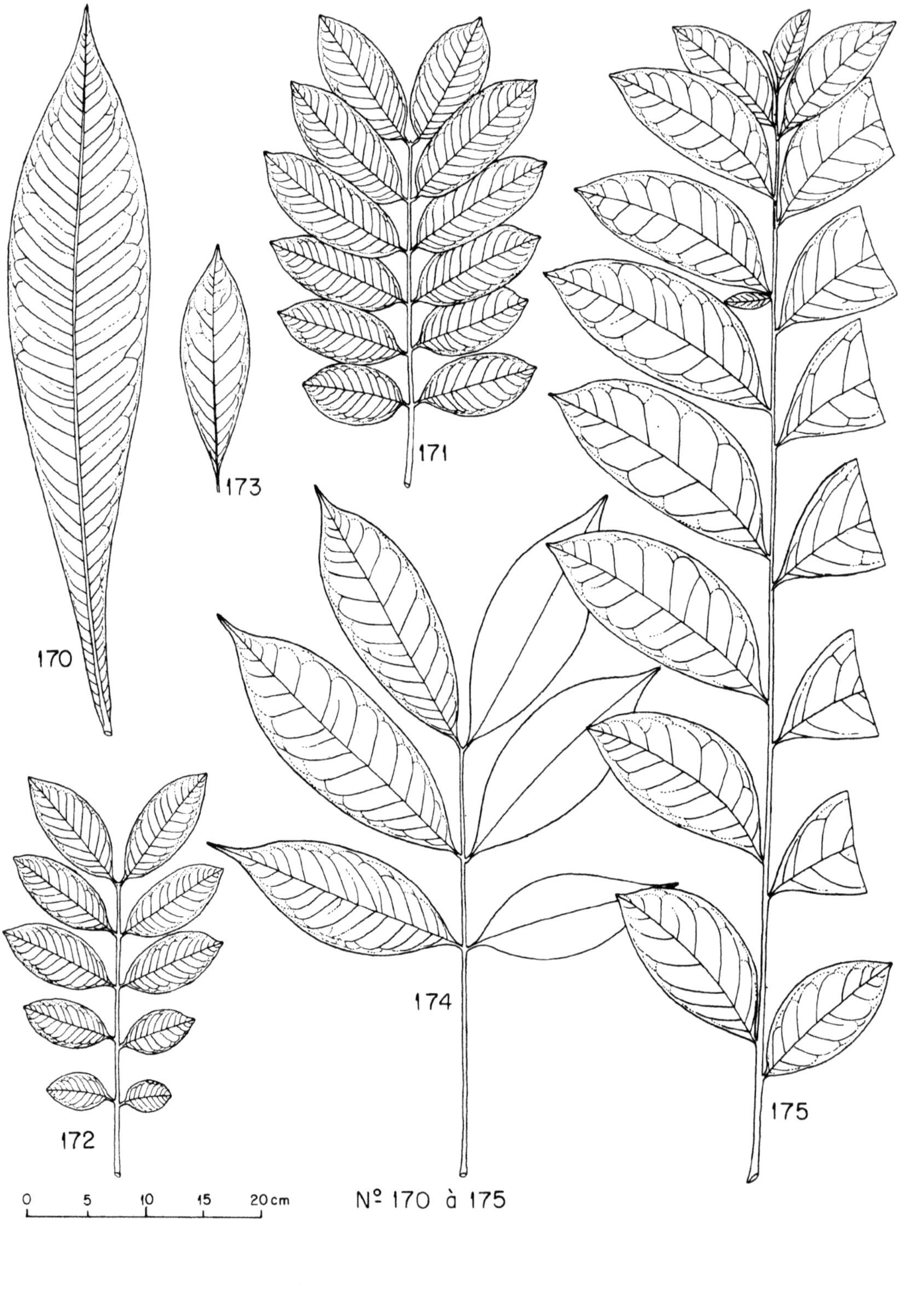

0 5 10 15 20 cm

Nº 170 à 175

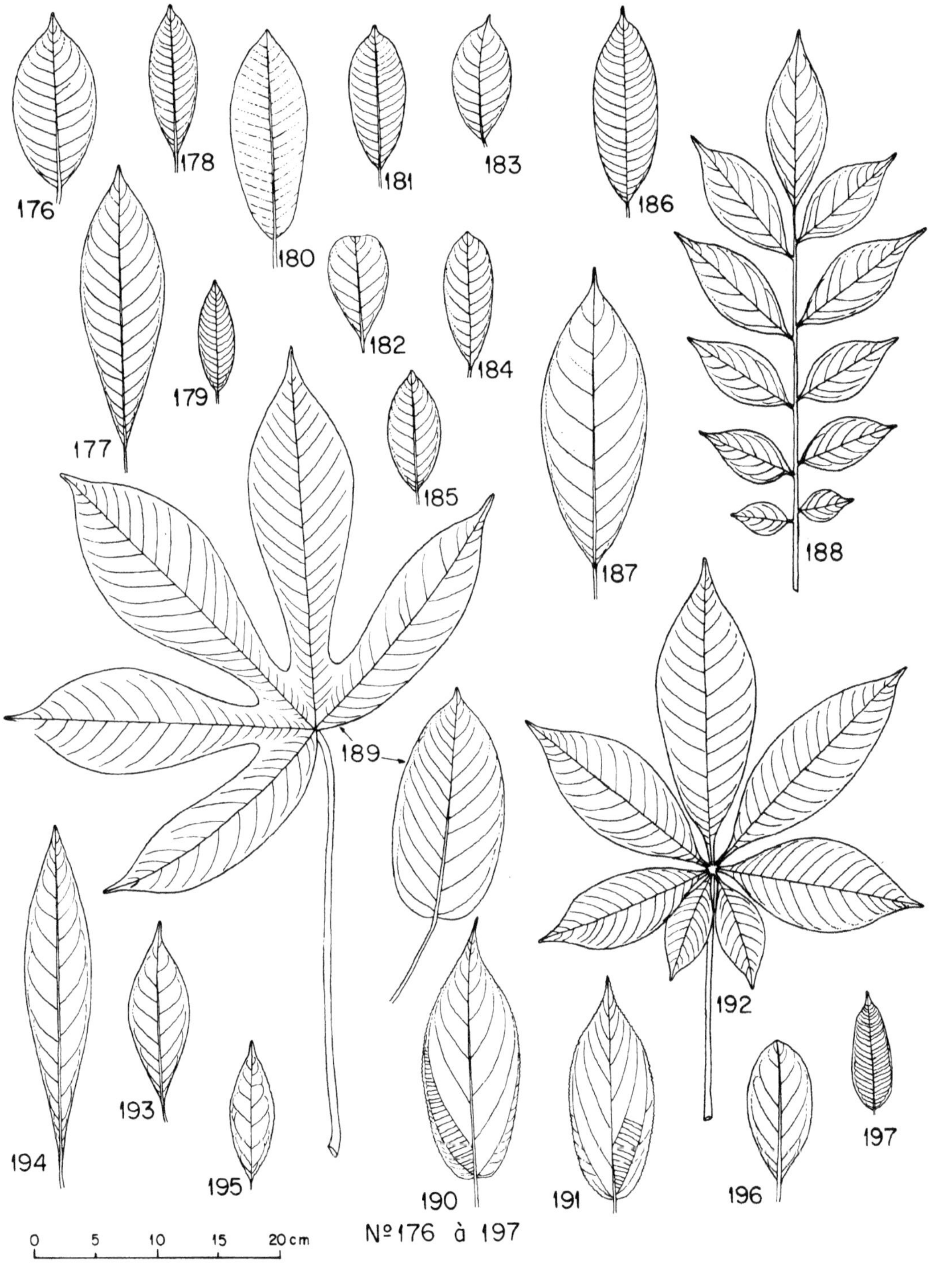
176
178
180
181
183
186
177
179
182
184
185
187
188
189
192
194
193
195
190
191
196
197
0 5 10 15 20 cm

N° 176 à 197

REFERENCES

Ashby, E. (1949): De la forme de feuilles et de leur rapport avec l'âge physiologique des plantes. Endeavour 8 (29): 18—25. 10 figs.

Ashton, P. S. (1964): Ecological studies in the mixed dipterocarp forests of Brunei State. Oxford forestry memoirs, 25, 75 pp., 70 figs. (17 tabl. + Annex).

Asprey, G. F. and Loveless, A. R. (1958): The dry evergreen formations of Jamaica. 2. The raised beaches of the north coast. J. Ecol. 46: 547—570.

Baker, J. R. (1938): Rain forest in Ceylon. Kew Bull: 9—16.

Bouchon, J. (1973): Précision des mesures de superficie par comptage de points. Ann. Sci. Forest. 32 (2): 131—134.

Cain, S. A., De Oliveira Castro, G. M., Pires, J. M. and Da Silva, N. T. (1956): Application of some phytosociological techniques to Brazilian rain forest. Amer. J. Bot. 43 (10): 911—941.

Cain, S. A. and De Oliveira Castro, G. M. (1959): Manual of vegetation analysis. 325 pp., 45 figs., 74 tabl. Harper N.Y.

Cooper, A. W. (1960): A further application of length-width values to the determination of leaf-size classes. Ecology 41 (4): 810—811, 3 tabl.

Delf, E. M. (1912): Transpiration in succulent plant. Ann. Bot. 26: 409—442.

Duvigneaud, P., Smet, S., Kiwak, A. and Mesotten, G. (1951): Ecomorphologie de la feuille chez quelques espèces de la "laurisilve" du Congo méridional. Bull. Sc. Roy. Bot. Belgique 84: 91—95.

Gessner, F. (1956): Die Wasseraufnahme durch Blätter und Samen 215—246 in W. Ruhland (ed.) Handbuch der Pflanzenphysiologie. Bd III. Springer Verlag.

Gessner, F. (1956): Träufelspitzen. Bd III. ditto.

Givnish, T. J. (1978): On the adaptive significance of compound leaves, with particular reference to tropical trees 351—380 in P. B. Tomlinson and M. H. Zimmermann. Tropical trees as living-systems, 675 pp. Cambridge University Press.

Huber, B. (1925): Die Beurteilung des Wasserhaushaltes der Pflanze. Jb. Wiss. Bot. 64: 1—120.

Hickey, L. J. (1973): Classification of the architecture of dicotyledonous leaves. Amer. J. Bot. 60 (1): 17—33, 107 figs.

Lawrence, G. H. (1951): Taxonomy of vascular plants. Macmillan Co. N.Y. 823 pp.

Loveless, A. R. and Asprey, G. F. (1957): The dry evergreen formations of Jamaica. 1. The limestone hills of the south coast. J. Ecol. 45: 799—822.

Mouton, J. A. (1963): De la possibilité d'identifier les feuilles des espèces des phanérogames ligneuses de la Côte d'Ivoire. Mem. inédit, 337 pp. dactyl., 112 figs.

Mouton, J. A. (1966): Sur la systématique foliaire en paléobotanique. Bull. Soc. Bot. Fr. 113 (9): 492—502, 4 figs., 3 tabl.

Mouton, J. A. (1967): Architecture de la nervation foliaire. C.R. 92è Congr. Nat. Soc. Sav. Sect. Sciences T 3: 165—176, 30 figs.

Mouton, J. A. (1967): Les types biologiques foliaires de Raunkiaer. Etat actuel de la question. Mém. Soc. Bot. Fr. Colloque de Montpellier 1965.

Mouton, J. A. (1972): Une nouvelle méthode d'isolement de la nervation des feuilles d'arbres. Bull. Soc. Bot. Fr. 119: 581—590, 2 pl.

Mouton, J. A. (1979): Inventaire des arbres tropicaux de Côte d'Ivoire dont la nervation foliaire a été isolée. Bull. Soc. Bot. Fr. 126. Lettres bot. (3): 361—372, 2 pl.

Müller-Stoll, W. (1947): Der Einfluss der Ernährung auf die Xeromorphie der Hochmoorpflanzen. Planta (Berlin), 35: 225—251.

Parkhurst, D. F. and Loucks, O. L. (1972): Optimal leaf size. J. Ecol. 60 (2): 505—537.

Payne, W. W. (1978): A glossary of plant hair terminology. Brittonia, 30 (2): 239—255, 3 pl., 20 ref.

Petit Betancourt, P. (1964): Relaciones fisionómicas de las hojas de 2 asociaciones vegetales. Tesis Turrialba, 60 pp. illustr.

Raunkiaer, C. (1916): Om Bladstørrelsens Anvendelse: den biologiske Plantegeografi. Bot. Tidsskr., 33: 225—240.

Raunkiaer, C. (1918): Recherches statistiques sur les formations végétales. Biologiske Meddelelser, 1 (3): 1—80.

Raunkiaer, C. (1934): The life forms of plants and statistical plant geography. (Being the collected papers of Raunkiaer), XIV + 632 pp. Oxford.

Richards, P. W. (1952): The tropical rainforest. An ecological study. 450 pp. Cambridge University Press.

Rollet, B. (1974): L'architecture des forêts denses humides sempervirentes de plaine. 298 pp., 155 tabl. + Appendices, 35 figs., 8 pl. CTFT. Nogent-sur-Marne.

Stahl, E. (1893): Regenfall und Blattgestalt. Ann. Jard. Bot. Buitenz. 11: 98—182.

Stehlé, H. (1945—46): Les types forestiers des Iles Caraïbes. Caribb. Forester, 6, Suppl: 273—468; 7, Suppl.: 337—709.

Tasaico, R. (1959): La fisionomía de las hojas de árboles en algunas formaciones tropicales. Tesis de grado. Inter-american Institute of Agricultural Sciences Turrialba. Costa Rica, 86 pp.

Thoday, D. (1933): The terminology of "xerophytism". J. Ecology, 21 (1): 1—6.

Vareschi, V. (1980): Vegetationsökologie der Tropen. E. Ulmer. Stuttgart.

Watson, D. J. (1937): The estimation of leaf area in field crops. J. Agric. Sci. 27: 474—483.

Webb, D. A. (1959): A physiognomic classification of Australian rain forest. J. Ecol. 47 (3): 551—570.

CHRISTIANE HÖGERMANN

LEAF VENATION

CONTENTS

B. Rollet et al.: Stratification of Tropical Forests as seen in Leaf Structure, Part 2, pp. 77–183.

1. INTRODUCTION

This study of the leaf-venation of some tropical plant-families was initiated by Prof. Dr Ingrid Roth. In her book "Stratification of Tropical Forests as seen in Leaf Structure", Roth investigated nearly 50 anatomical leaf characteristics in 230 species; the venation patterns, however, were not included. Roth gave approximate estimates of vein density as well as measurements of vein spacing taken from leaf-transsections. These observations have paved the way for a further examination of the same material with regard to the venation patterns and to the exact vein density, the more so since Pykköö (1979) in her investigations on the same material had already given new impetus to this field of study.

The present work, therefore, is divided into a taxonomical part, in which the venation patterns of leaves are studied, whilst attempting to use taxonomical characteristics of pattern development for their determination. At the same time, an attempt has been made to develop a description and classification key. In the second part of the investigations, the vein density of leaves belonging to the different strata of the forest and of leaves of young plants was to be determined and compared in order to establish the ecological signification of the venation network.

As far back as the early 19th century, the taxonomical value of leaf venation was used scientifically as a criterion for the determination and classification of fossil plants (cf. the investigations of von Ettinghausen). At the beginning of the present century, studies concerning the eco-physiological function of veins, which were frequently very costly in terms of the apparatus required, were well to the fore (cf. the papers of Stocker, Foster, Pray, and Wylie).

Nearly all these investigations refer to single species (partly from one family) or to a sample collection of dicotyledons. As a result, there is an absence of other findings which would allow the examination of the venation of different families of plants from the same location. Such opportunities of comparison only arise, if more extensive plant material is available.

The object of my studies of the venation of some Sapotaceae, Lauraceae and Euphorbiaceae from the tropical rain forest of Venezuelan Guyana was to find valid taxonomical and ecological criteria for their examination, thereby producing a comprehensive synthesis.

I was able to carry out my own examinations on Dr Rollet's leaf collection, which Prof. Dr Roth had used previously for her book "Stratifications of Tropical Forests as seen in Leaf Structure" (Roth, 1984).

Dr Rollet, who had established this collection in 1964 (cf. Section 2.1), published extensive studies of this plant material ("Etudes quantitatives d'une forêt dense humide sempervirente de plaine en Guayane Vénézuélienne", 1969a, b) as well as M. Pykköö ("Morphology and Anatomy of Leaves from some Woody Plants in a Humid Tropical Forest of Venezuelan Guyana", 1979).

Fundamental findings on the order of different venation patterns or their phylogenetic and histogenetic development were established by the researches of von Ettinghausen (1861), Glück (1919), Troll (1939), Wylie (1939, 1943a, 1946b, 1950), Foster (1952) and Mouton (1967, 1976).

In his work "Blatt-Skelete der Dikotyledonen"

printed in 1861 in Vienna, Constantin von Ettinghausen published extensive morphological studies of the dicotyledon venation. The techniques of "Naturselbstdruck" allowed the exact illustration of skeletonized samples of almost all kinds of venation by means of a stamp technique in lead plates, and their minute description. Von Ettinghausen's definition of "veins" and "nervation", which is nowadays normally called "venation", indicates that anatomical, morphological and taxonomical aspects were already well to the fore, in spite of the fact that at that time the ecological significance of veins had not yet been investigated.

Die in den Blattgebilden der Pflanzen sich ausbreitenden Gefäßbüdel nennt man *Rippen, Adern* oder gewöhnlich *Nerven*, und die Vertheilungsweise derselben bezeichnet man mit dem Ausdrucke *Nervation.* (von Ettinghausen, 1861, p. XIII)

However, my own work is based on the more detailed definitions by K. Esau in her "Plant-Anatomy":

Die Nervatur (Venation), also das Leitbündelmuster, ist ein charakteristisches Merkmal des Blattes. In der Botanik bezieht sich der Begriff Nervatur manchmal auf ein Leitbündel oder eine Gruppe dicht zusammenliegender Bündel, manchmal auch auf das mit den Leitbündeln assoziierte nichtleitende Gewebe. (Esau, 1969, p. 316)

Here, the term "vein" means a vascular bundle, i.e. a group of closely associated strands, which indicates that K. Esau includes fundamental results from anatomical-morphological studies.

Troll's definition of venation (1939) finally comprises the function of the venation network:

Die *Nerven* oder *Adern* des Blattes, in ihrer Gesamtheit als *Nervatur* (Nervation, Venation) bezeichnet, haben die doppelte Aufgabe, der Stoffleitung zu dienen und die Aussteifung der Blattfläche zu besorgen. (Troll, 1939, p. 1044)

Von Ettinghausen attaches great importance to the taxonomical value of leaf venation:

Die Vertheilungsweise der Nerven in den Blättern und blattartigen Organen der Pflanzen ist durchaus nicht unregelmäßig, zufällig oder verworren. Die Gesetzmäßigkeit derselben ersieht man bald aus der vergleichenden Untersuchung und Bestimmung ihrer Verhältnisse, [. . .]. (von Ettinghausen, 1861, p. XIII)

In comparison Troll's "Vergleichende Morphologie der Höheren Pflanzen" (1939) emphasizes mainly the morphological aspects.

To date, the taxonomical significance of leaf venation has undergone few changes. Apart from defining classification, evolutionary aspects are given prominence, Foster (1952) regards von Ettinghausen's investigations (1854—1872) as fundamental for the definition and classification of vascular bundle patterns:

These classical studies, although devoid of morphological and phylogenetic conclusions, nevertheless represent the first attempt to devise a descriptive classification of venation systems based on the course of the main veins in the lamina. (Foster, 1952, p. 752)

In his paper "Foliar venation in angiosperms from an ontogenetic standpoint" (1952), Foster calls for a critical reflection on the seven venation types by von Ettinghausen (1861, p. XVI f.):

. . . from an ontogenetic, as well as from a broad comparative standpoint. (Foster, 1952, p. 752)

In this paper, Foster himself dealt with the histogenetical development of the venation of *Quiina pteridophylla* (Radlk.) Pires — Quiinaceae and *Liriodendron tulipifera* L. — Magnoliaceae.

In earlier papers, Foster had also studied the comparative morphology and histogenesis of the venation of different Quiinaceae. *Quiina acutangula* Ducke (1950a), *Touroulia guianensis* Aubl. and *Froesia tricarpa* Pires (1950b), *Lacunaris* Ducke (1951).

As the leaf venation is a characteristic which has been conserved over millions of years by means of petrifaction, it constitutes the only possibility for paleobotanists to classify fossil leaf remains. In Sturm's taxonomical classification of fossil leaves of Lauraceae from the Messel mine near Darmstadt, the leaf venation and epidermal structures proved to be a suitable criterion for determination and comparison with recent forms (Sturm, 1971). According to Mouton (1966a), the leaf with its characteristic vein course holds a key position with regard to the determination of fossil leaf remains. Hickey (1979), Dilcher (1974), Hill (1980) and Mouton (1966a, 1967, 1976) have worked out determination keys, by means of which not only the other anatomical-morphological characteristics of the leaf architecture, but also the venation types can be precisely described and classified. Hill supports the standardized criteria of investigation for the description of leaf types, which allows the employment of computer programs.

Because of the great difficulty in classifying the family of the Sapotaceae, it is necessary to find characteristics which ensure a clear demarcation of Sapotaceae in general and of the species within the families. In his thesis "Über den anatomischen Bau des Blattes in der Familie der Sapotaceae und dessen Bedeutung für die Systematik", published 1892, Holle had already made a contribution to the systematical classification of Sapotaceae by means of leaf anatomy with special regard to the vascular bundles. The following quotation by H. J. Lam (1939), also underlines the important position that this plant family has occupied for many years in taxonomy:

> The Sapotaceae have of old a bad reputation among classificators because of the extreme complexity in which the single features are distributed over its genera, species and individuals. It is, in particular, extremely difficult to find satisfactory generic delimitations and, consequently, the family yields a rich field both for lumpers and for splitters. (Lam, 1939, p. 509)

The extent to which ecological questions gave rise to critical reflections about ecological-morphological experimental techniques in the middle of the 1920s, is shown in H. Fitting's paper "Die ökologische Morphologie der Pflanzen im Lichte neuerer physiologischer und pflanzengeographischer Forschungen" (1926). Fitting criticizes the one-sided attitude of ecologists, who only included the living conditions of plants in the dominant problem of that time of the coordination of morphological symptoms of adaptation with the inducing ecological factors. He requires:

> Zum Verständnis der Pflanzenformen muß es ja doch wohl in allererster Linie darauf ankommen, die Unterschiede zwischen ihren Lebens*äußerungen* in Abhängigkeit von den besonderen Standortsbedingungen zu erforschen. (Fitting, 1926, p. 9)

R. B. Wylie restricted his investigations to the venation of different specimens of dicotyledons solely to ecological-physiological criteria. In a series of papers, Wylie concentrated on the functional significance of the major and minor veins with regard to conduction. It is the minor venation which is mainly responsible for the quick healing of wounds and for the supply of material to wounded parts of the lamina. The experiments were carried out on artificially wounded leaves of *Syringa vulgaris* L. — Oleaceae (Wylie, 1938).

Using leaves from herbaceous and woody dicotyledons from Iowa and woody plants from the Pacific coast, Wylie demonstrated a connection between the organization of mesophyll and vein spacing ("vein separation"), Wylie (1939). He used the same material for clarifying the function of leaf tissue with a reduced chlorophyll content (epidermis, bundle-sheath extensions). These tissues contribute to the supply of the lamina and, to a certain extent, provide an additional system of transport in mesomorphic leaves (Wylie, 1943b).

Although the major veins in mesomorphic leaves constitute only 5% of the whole vein length per cm^2 of the leaf area, they carry out the transport of materials in the lamina. Just like the minor veins, they are capable of conductive overload (Plymale and Wylie, 1944).

Observations on foliar leaves of five species of dicotyledons confirm this capacity of the major and minor vascular bundles for conduction. At the same time, the living tissue between them exercises a decisive influence on the conducting capacity and thus on the spatial stretching of the vascular system ("vein spacing") (Wylie, 1946b).

The results of the abovementioned papers definitely disagree with overvascularization of the lamina (Wylie, 1946a).

In a further study, Wylie (1952) dealt with the bundle-sheath extensions devoid of chlorophyll which can appear on both sides of minor vascular bundles in dicotyledons. The functional significance of these tissues lies in their supportive role in the transport of materials between the bundles.

On 5—18 foliar leaves of five species of Solanaceae, Gupta (1961a) was able to establish an inversely proportional dependence between the number of vein islets in unit area[1)] and veinlet terminations[1)] to the area of the blade.

Hall and Melville (1951/54), on the other hand, worked out the diagnostic value of the veinlet-termination numbers (= number of veinlet terminations[1]) on the leaves of a total of 11 medicinal plants and called for the application of this criterion within a broader botanical field.

Manze (1967) determined the density of veins (in terms of a number of veins which cross a measuring section of 1 cm) on recent leaf material of a single species, i.e. ecologically closely related species of one genus. The values for the leaves from different climates were compared with those of fossil leaves. It was the aim of the investigation to prove whether the vein density is valid as a paleontological criterion. Manze finally concluded from the results that the comparison of vein density under the conditions mentioned, does allow conclusions to be drawn concerning the climate of past earth eras.

Geiger and Cataldo (1969) used the vein length/cm^2 [1] in order to explain the supply of mesophyll with organic compounds. The different intensity of the transport of major and minor vascular bundles was investigated in *Beta vulgaris* L. var. *Klein Wanzleben*. The measurements show that the access of bundles of higher orders to the mesophyll is 13 times higher in comparison to the major veins. Geiger and Cataldo also proved that cells of the minor veins which are rich in organelles store up assimilates.

It could therefore be presumed that the vascular bundle system is of some ecological significance. Since the work of Zalenski (1902) and Shields (1950), it is generally known that xeromorphic leaves possess a denser vascular bundle network.

The aim of the present study is to prove the taxonomical and ecological significance of leaf venation by means of special descriptive and planimetric criteria. At the same time the different forms of venation will be considered from two aspects: firstly the age of the trees in question (adult leaves of young plants and adult trees), secondly their origin from different strata of the rain forest.

[1] The same criteria have also been applied in this investigation.

2. MATERIAL AND METHODS

2.1. Origin of the investigated leaf material

The present material (species of the families Sapotaceae, Lauraceae, Euphorbiaceae) was collected by Dr. Bernard Rollet.

The taxonomical investigations of the species by means of their venation were carried out with the help of total preparations from the margin and the middle of the leaf, while the data for the ecological significance of the vascular bundles were taken from a prepared cutting of the margin and the middle.

The material collected by Rollet allowed the examination of leaves of young trees as well as leaves of adult trees.

"Leaves of adult trees" means adult leaves of adult trees, whilst "leaves of young trees" are adult leaves of young trees (cf. Roth, 1984).

Investigated material

A total of 39 species from 19 genera are distributed among the families of Sapotaceae, Lauraceae, and Euphorbiaceae as follows:

1. Sapotaceae: 15 species (including three without scientific names and three species, which probably can all be classified as *Manilkara bidentata*) from four genera,
2. Lauraceae: 11 species (including the two not yet identified species) from five genera,
3. Euphorbiaceae: 13 species (with one unidentified) from 11 genera.

Regarding the number of different species, the Sapotaceae take first place, while the leaf material of the Euphorbiaceae originates from comparatively numerous genera, and nearly every species represents a different genus.

A list of the investigation material including family, genus, species, vernacular name, herbarium number and height of tree, precedes the explanations of each family.

Age of trees and stratification

As already mentioned in Section 2.1, the tropical species of trees are not only classified systematically, but also with regard to their age and height.

The stage of age is related to the two age groups of the adult foliar leaves of young plants (leaves of young trees) and adult foliar leaves of adult trees (leaves of adult trees). The stratification is divided into three layers: "A" (trees of at least 30m in height), "a" (trees between 10 and 29m) and "aa" (trees and shrubs under 10m).

As the young plants belong almost exclusively to the lowest stratum "aa" (exceptions: leaves of young trees of *Ocotea nicaraguensis, Beilschmiedia curviramea* — Lauraceae) it follows that classification of age and height do, to some degree, run a parallel course.

While the foliar leaves of trees and shrubs from "aa" belong to the shade-leaf group, more mesomorphic leaves are found in "a". In the upper stratum there is a preponderance of sun leaves. Consequently, the increasing height of trees is connected with a transition from the hygromorphic shade leaf to the xeromorphic sun leaf (cf. Roth, 1984).

2.2. Sample preparations for the taxonomical and ecological investigations

Production of preparations from the margin and the middle

The leaf fragments which are to be investigated come from the margin and middle of the leaf. They are taken from the middle of each leaf and are submerged for the night in 70% alcohol in order to remove any resinous substances. After having been rinsed in distilled water, they are divided according to their surface quality and put into Petri dishes containing "Domestos" chlorine bleaching solution. The bleaching process terminates when the brown colour has completely disappeared from the fragments, and they have turned white. After having rinsed them again in aqua dest., they are stained with "Methylene blue" in order to make the vascular bundles visible. The time required for staining also depends on the composition of the leaf surface; it should soak for at least 45—60 minutes. The fragments must be stained in such a way that the vascular bundles stand out clearly in a dark blue colour against the light-blue leaf tissue. If the stain is too strong, a treatment with absolute alcohol or a weak acid can improve the quality for differentiation purposes. After rinsing, the preparations are dehydrated for 3—4 days in isopropanol vapour, then mounted in isopropanol (the medium must be renewed once) for 30 minutes and using xylor (for at least 1 hour) embedded in "Eukitt". (For the techniques for the production of permanent preparations, see Gerlach, 1977.)

Production of total preparations

Numerous techniques are to be found for isolating the vascular bundle network in the technical literature. These techniques have not yet been described as widely as those for the preparation of cuttings (eco-physiological techniques of analysis). Total preparations of leaves can only provide a general view of special morphological characteristics, they are suitable for histological studies. For the classification of the course of the veins as a taxonomical criterion, they are, however, indispensable.

The frequently applied method of maceration which, as is well known, artificially imitates the natural processes of decomposition, is sometimes used in a variety of ways in order to bring out the venation pattern (cf. Roth, 1964).

A very precise, but also complicated method for the isolation of venation is described by J.-A. Mouton (1972) in his paper "Une nouvelle méthode d'isolement de la nervation des feuilles d'arbres". This "méthode originale pour isoler les nervations des feuilles des plantes ligneuses tropicales" (Mouton, 1972, p. 581) is based on the following principle: after a treatment with a KOH solution, the strength of which varies according to the composition of the leaf surface, a pressure process follows in the autoclave. The leaf tissue, which has now become loose, can be removed carefully with a fine brush. Finally, the isolated leaf skeleton must be bleached and stained.

My own method for the production of total preparations, described below, has the advantage that the leaves are conserved as a whole — the tissue is not removed — so that the course of the vascular bundles in the lamina can be analysed without any displacements between the secondaries due to missing supporting tissues. These total preparations are easily handled, without any risk of breakage.

Just as with the preparations of the leaf cuttings, the dried leaves are treated with 70% alcohol and rinsed afterwards. Depending on the leaf size and thickness, the bleaching takes 7 days to 4 weeks. As soon as the leaves, submerged in the chloride solution, have lost their chlorophyll colour or, as in this case, the dried specimens have lost their brown colour and brown traces, they are carefully taken out of the liquid, washed several times with aqua dest. or sprayed if the leaves are very fragile. They are then placed on a glass plate and smeared with "Methylene blue" with the help of a paint brush. The more compact leaves can be immediately stained with the "Methylene blue" solution. As soon as the venation stands out clearly in a dark blue colour against the leaf tissue the staining has finished. A second, careful treatment with aqua dest. follows. The preparations must be dried between several layers of absorbent paper in a dry press for at least 1 week.

Since the dried, parchment-like leaves tend to curl towards the leaf surface, it is advisable to stick them onto self-adhesive tape and then to cut them out, leaving an edging of several millimetres from the leaf margin. By using illuminated milk glass or a fluorescent screen, you can analyse the leaf venation in transmitted light.

For the photographic presentation of the leaf venation, you need a camera with a reproducing stand and a skylight filter as well as a fluorescent screen. A black and white film with a low sensitivity (slow speed, fine grain) is most suitable (e.g. Ilford Pan F).

With the help of the negatives, which are put into an enlarger or into slide frames and then projected onto a white screen, it is now possible to analyse details in the course of the veins.

Since the foliar leaves are dry preparations, they are best kept in fold-up file wallets.

The total preparations of the leaves of young and of adult trees (cf. the pictures) reveal information about the behaviour of the primaries and secondaries as well as about the formation of the margin. The adhesive labels at the side of each serve as a scale comparison. The vernacular name and the size of the lamina (in cm^2) are noted on these labels. All photographic illustrations are approximately 2.2-fold reductions of the original leaf size.

2.3. Discussion of alternative methods of producing preparations

In order to clarify the relation between the tissue structure of the leaf and the vein spacing, R. B. Wylie uses transverse and longitudinal sections of leaf cuttings. For the production of the preparations, the fresh leaf cuttings are conserved in 3% formalin and then embedded in paraffin (Wylie, 1939) or fixed in FAA (fluoroacetic acid) before being embedded; they are then treated with fluoroacetic acid (Wylie, 1946b). Longitudinal sections through the lamina are particularly suitable for the investigation of the tissue structure and for the measurement of the spatial distribution of the veins (Wylie, 1946a). For the determination of the leaf thickness — or rather that of single tissue layers — Wylie employs transverse sections (1946b).

In his investigation into the role of the epidermis and the spongy parenchyma in the conduction of organic materials, Wylie uses a solution of hexacyanoferrate(II) together with iron(III) chloride as the indicator 'Prussian blue'. Leaves, treated in such a manner can be examined with the help of sections (Wylie, 1943).

A. S. Foster recommends a method that uses four steps in order to obtain cuttings: clarifying the leaf cuttings in 5% NaOH; dehydration in an increasing amount of alcohol, staining with alcoholic safranin solution, then embedding in "Clarite" (Foster, 1950). A very similar method served Foster for the production of sections of leaf cuttings which he had previously soaked in paraffin (Foster, 1950, 1952).

B. Gupta has developed a more extensive procedure in order to determine the "absolute vein islet numbers" and "absolute veinlet-termination numbers" (Gupta, 1961a, p. 65). First the trichomes are removed from the leaf cuttings of the middle between the midrib and margin of the dried, pressed leaves in order to eliminate disruptive factors, before clarification by boiling in a concentrated solution of chloral hydrate is carried out. Hydrochloric acid is required for the removal of calcium oxalate crystals. Before embedding in 40% glycerin, the cuttings can be stained with Gentian violet.

T. R. Pray and K. Lems also studied the vascular bundles with the help of tissue sections. Pray (1963) produced longitudinal sections in order to illustrate the histogenesis of veinlet terminations in angiosperm leaves. In Lems' (1964) investigations into the scope of phylogenetic studies in Ericaceae leaves, the production of preparations follows Foster's (1952) methods.

In their paper "Leaf structure and translocation in Sugar Beet" D. R. Geiger and D. A. Cataldo (1968) present several methods: making the vascular bundles visible with the help of fuchsin—sulphurous acid, ultrastructural investigations in 2μ thick sections and freeze-dried autoradiography with $^{14}CO_2$.

The methods used by Wylie (1943) and Geiger and Cataldo (1968) were ruled out for my own investigations as they need living leaf tissues as the original material. I produced my cutting preparations with a technique based on that described by Foster (1950). It was modified and tested by Prof. Dr Roth (1964) and her team, before I finally used it, after having modified it once again. One advantage of my method is that it facilitates a very efficient method of working, and that it is only slightly susceptible to technical errors.

3. RESULTS

The following sections first present those criteria of research which I applied in order to evaluate the leaf venation of some tropical Sapotaceae, Lauraceae, and Euphorbiaceae from a taxonomical and ecological point of view.

The taxonomical descriptions of the venation as well as the recorded measurements within the investigation material of one family are divided into taxonomical and ecological headings. The Sapotaceae is a family whose species can only be discriminated from each other with difficulty, so here I performed an additional sample taxonomical investigation. The results of this work, therefore, commence with the taxonomical and ecological characteristics of the Sapotaceae, followed by those of the Lauraceae and Euphorbiaceae.

3.1. Investigation criteria

Taxonomical investigations

For the taxonomical classification of the species with the help of their venation pattern, I mainly used total preparations for the study of the total vein course. The preparations from the margin reveal information about the veinlet terminations at the leaf margin. The total preparations and those from the middle show the hierarchy of the lateral veins and the form of the vein islets.

I consider the following investigational criteria to be of taxonomical significance:

1. general survey of the "*venation form*":
   - "specification" of the venation type,
   - "form of the margin".
2. "*Patterns of the primary veins*":
   - "relative thickness" in comparison to the secondary veins,
   - division of the leaf area ("leaf symmetry"),
   - course and thickness of the primary vein from the base to tip ("course"),
   - termination at the leaf tip ("termination").

As the venation form includes a number of

primary veins — in the leaves of the sample collection there is without exception a single primary vein functioning as medianus — I considered it unnecessary to allow an extra section for this criterion. (For the characteristics of the primaries, cf. von Ettinghausen, 1861, pp. XVIIf.)

3. "*Patterns of the secondary veins*":
   - "angle of emergence" from the medianus, measured in the middle of the leaf in all samples, carried out using the Sapotaceae as an example,
   - "position" of the secondaries in relation to each other according to the nomenclature for the description of leaf positions,
   - comparison of the thickness of the secondaries with that of the primary vein of the leaf ("relative thickness to the primary vein"),
   - comparison of the thickness of the secondary veins in each leaf ("relative thickness to each other"),
   - "direction and course" from the emersion of the medianus to the margin as a decisive factor for the venation type,
   - ending of the veinlet termination at the margin of the leaf ("termination").

(For the characteristics of the secondary veins, cf. von Ettinghausen, 1861, pp. XVIf. and Mouton, 1967, pp. 167ff.)

4. "*Patterns of the veins of higher orders*":
   - data of the degree of ramification which can be clearly identified on preparations from the middle and on total preparations,
   - testing of the preparations from the middle with regard to freely ending veinlet terminations; if present, details of the degree of ramification of "blind" terminations; symbols "+" = present, and "–" = not present,
   - description of the vein islets ("pattern of the vein islets"): I have defined as vein islet the smallest area of the lamina which is enclosed by the vascular bundles or their ramifications (cf. Hill, 1980 and pictures 4–19). As the size of the meshes is to be seen from the number of "vein islets/cm^2" (cf. Section 2.2), I waived the criterion at this point and confined myself to the exterior form and the discrimination of the vein islet pattern. In the description of the pattern, I attached great importance to the peculiarity of the vascular bundle pattern of those species which possess a very remarkable and typical network of meshes.

According to Pray (1954), the midrib, the secondary and tertiary veins form the "major venation", while the "minor venation" includes the quaternary veins and their further ramifications.

Ecological investigations

The data which I used for the ecological evaluation are taken from the preparations of the middle — and, in doubtful cases, also from the total preparations — and from the dried original leaf material. The preparations of the middle come from two or three examples of the species. The measurements on the cuttings are based on 10 values. I employed the two or three preparations according to the law of random selection. The data of the dried leaves (total preparations) represent at least 10 and at most 15 leaves of a species, depending on the sample collection. In some cases, which I will particularize in the various sections, I had only one leaf at my disposal.

The following measurements were carried out:

1. "*Leaf area*": Calculation of the average surface area of all leaves belonging to the species.
2. "*Number of secondaries/leaf*": My whole sample collection served as a basis for this count; the secondary veins of each leaf were counted, although I ignored very fine secondaries which end some way from the margin.
3. "*Area-index*": This term means the part of the lamina supplied by one secondary vein. The area of the lamina — if not yet indicated on the leaf by Dr. Rollet — was measured with the "Morphomat 10" (Zeiss production) and then related to the number of secondaries, giving a

quotient of area of the leaf and the number of secondaries.

4. "*Mean distance of secondaries*": All samples were used for this measurement of the distance between the secondary veins adjacent to the medianus. Then the average was calculated, excluding the very small secondaries.
5. "*Vein islet number/cm²*": I obtained these data from the two or three preparations of the middle. With the help of a transparent foil with a millimetre scale, I counted out the islets on 1 mm^2 or, in the case of very large meshes, on 4 mm^2 with the lowest magnification of the microscope. This result was projected to 1 cm^2. Ten counts were carried out on each preparation. (For the method of counting vein islets, cf. Gupta and Kundu, 1965.)
6. "*Ramifications/islet*" (cf. Hill, 1980): Ramifications are those vascular bundles which branch off directly into the vein islet. Further branches of the ramifications are excluded from this research criterion (cf. picture 13).
7. "*Veinlet terminations/cm²*": The veinlet terminations, counted on 1 mm^2 on the preparations of the middle by means of the transparent foil, are calculated for 1 cm^2 (cf. pictures 15, 16, 18, 19).
8. *Vein length/cm²*": As a basis for the determination of the vein length I employed black and white photographs (13 × 18 cm), showing parts of the preparations from the middle. The photographs were taken with a "Leitz SM-Lux" camera + Wild camera body "MPS 51" and steering gear type MPS 55. The "Agfaortho 25" proved to be the best negative material. I magnified the venation of the middle with an achromatic lens 4/0.12 connected with an eyepiece, magnifying 6.3- or 8-fold. The factor of the camera was 0.32. In order to compensate for the staining in "Methylene blue" (cf. Section 2.5.) I used a yellow filter. As a stencil for a measured area of 1 mm^2 I took a photograph of a transparent foil with millimetre scale according to the magnification of the preparation. The magnification of 1 mm^2 was cut out from the photograph of the graph paper. With the help of this 'window' I could mark 1 mm^2 on the photograph and then measure it. According to the law of random selection I selected fields on the photographs and worked with three stencils for each species in order to obtain more precise data. The "Morphomat 10", an apparatus which is mainly applied in cytological studies, guarantees an exact measurement of distances (as well as areas and other functions). With each stencil I went over the vascular bundles three times, corresponding to nine measurements on each picture. Finally, I calculated the average of all data for 1 cm^2 in compliance with the ca. 50-fold (in some cases ca. 60-fold) linear magnification of the photograph.

From Tables 7, 11, 20, 24, 33 and 37 it can be concluded which component of the venation, especially of the minor venation, has had a major influence on the "vein islets/cm²", "ramifications of the islets" or the "veinlet terminations/cm²".

9. "*Conductive volume/cm²*": On the same photographs which had served for the determination of the vein length/cm^2 five cross-section measurements were taken of the thickest vascular bundles (without midrib), of those veins which make up the meshes, and of the freely ending terminations. The multiplication πr^2 × vein length/cm^2 yields the "gross conductive volume" for 1 cm^2 of the lamina. In accordance with the formula πr^2 the cross-section should be circular. But since various types of vascular bundles exist among the vascular plants, I chose this overriding formula. Pictures 20 to 37 show that, within the sample collection, the diameters also differ considerably, so that it may be assumed that within one family the deviation from the cross-section is almost equal.

Although this method offers a representative average of calibers of vascular bundles belonging to different orders and is also very rational — since the photographs may be used repeatedly — the data include a considerable error in measurement which, however, is probably on the same scale in all three families: as random tests in 6 Sapotaceae, 7 Lauraceae and 5 Euphorbiaceae reveal, the amount of sclerenchyma and/or phloem makes up roughly between ca. 67% and ca. 93% of the total diameter. These factors occasioned by the anatomy of the leaf, as well as the different conductive capacities and line resistances play a significant role. However, their influence cannot be determined at this point.

Any further detailed data from the diameters of the vascular bundles are placed under "net conductive volume" next to the approximate values (cf. Tables 9, 13, 22, 26, 35).

In order to obtain more reliable information about the conductive volume/cm^2, it is necessary to study cross-sections of at least three regions of

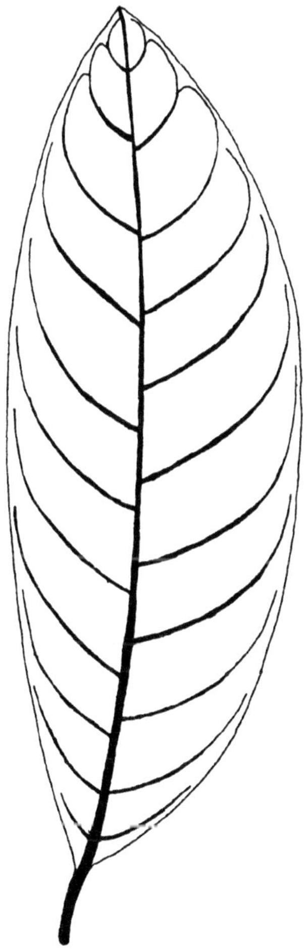

Fig. 1. *Nervatio camptodroma* (camptodromous venation).

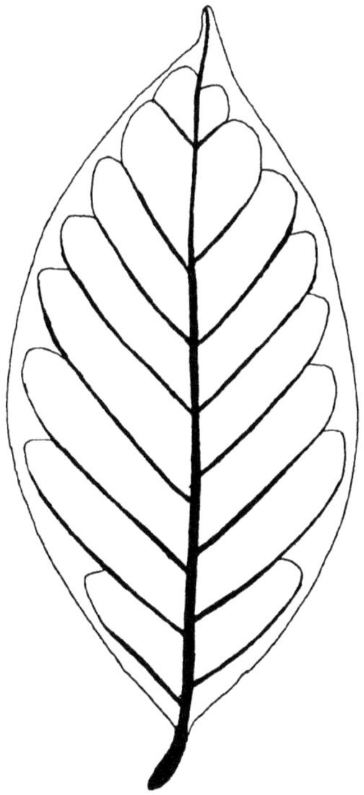

Fig. 2. *nervatio brochidodroma* (brochidodromous venation).

the lamina (base, middle, tip) or, even better, five regions (base, between base and middle, middle, between middle and tip, and tip). On the preparations the xyleme is only measured in five characteristic vascular bundles of different calibres.

In my research I regard the "gross conductive volume" or "net conductive volume" as a characteristic of the venation which can be standardized, without taking the conductive capacity into account. For the evaluation of the water balance of a plant, living, transpiring foliar leaves as well as the root system and the bundles in the stem or trunk must be investigated with regard to their water balance (cf. Huber, 1924). In view of the probable varying conductive capacity of leaves of young and adult trees, such a detailed study would be of great value.

Criteria for the description of leaf venation
Characteristics of venation

The description of the course and the ending of

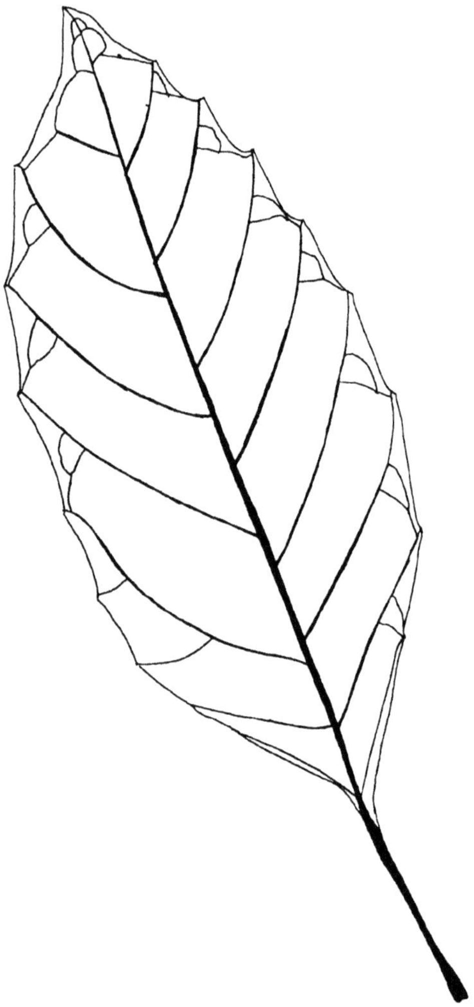

Fig. 3. nervatio craspedodroma (craspedodromous venation).

the primary and secondary veins has been found to be the most suitable approach to the taxonomical classification of the species by means of the course of the veins. In order to guarantee a universal nomenclature, by which the types of lateral veins can hierarchically be classified, I take as a basis Mouton's definition (1967, p. 166):

La nervation primaire est constituée par l'ensemble des nervures issues directement du pétiole; la nervation secondaire est l'ensemble des nervures les plus fortes se raccordant à la nervation primaire; la nervation tertiaire se raccorde à la secondaire.

Instead of the terms "secondary", "tertiary" and "quaternary veins" the synonyms "lateral veins of first", "second" and "third order" are used. As the nerves branching off the petiolus are generally called "primary veins", I have opted for the nomenclature "secondary", "tertiary" etc. with regard to the different degrees of ramification. In my view this nomenclature is preferable, as there is no ambiguity.

Moreover the term "lateral vein" is not always clearly defined. Whilst von Ettinghausen describes the more weakly formed primary veins as "lateral veins", as opposed to the medianus developed from the primary vein (e.g. *Ficus benghalensis* L. — Moraceae, which possesses several primaries, — cf. von Ettinghausen (1861) pp. 1077ff.), W. Troll and K. Napp-Zinn, on the other hand, consider primary lateral veins as branches of the midrib (= sole primary vein): the reduction of the number of vascular bundles in the midrib brings about the formation of the lateral veins of the first order (Napp-Zinn (1973) p. 664 — vol. VIII, 2 A).

In his classification of the venation-types von Ettinghausen (1861) summarizes the behaviour of primaries, secondaries and tertiaries. He introduces seven subdivisions:

1. nervatio craspedodroma (craspedodromous venation) — picture 3
   1.1. single secondaries ending in the margin of the leaf
   1.2. combined secondaries ending in the margin
2. nervatio camptodroma (camptodromous venation) — picture 1
   2.1. nervatio brochidodroma (brochidodromous venation) — picture 2
   2.2. nervatio dictyodroma (dictyodromous venation)
   2.3. real comptodromous venation
3. nervatio hyphodroma (hyphodromous venation)
4. nervatio parallelodroma (parallelodromous venation)
5. nervatio campylodroma (campylodromous venation)
6. nervatio acrodroma (acrodromous venation)
   6.1. complete acrodromous venation

Table 1. Survey of the venation types (main types), appearing in Sapotaceae, Lauraceae and Euphorbiaceae.

Venation type	Number of primary veins	Behaviour of the secondary veins
nervatio camptodroma (camptodromous venation) (fig. 1)	1	The secondaries do not reach the margin in their course, they run towards it in a curve. Before reaching the margin, they form loops, disperse reticularly or end along the margin of the leaf.
nervatio brochidodroma (brochidodromous venation) (fig. 2)	1	The secondary veins form anastomoses, which lead to a distinct formation of loops.
nervatio craspedodroma (craspedodromous venation) (fig. 3)	1	All secondaries, at least the stronger ones or their branches, end in the margin of the leaf.

The composition of this table follows von Ettinghausen (1861) p. XVI.

6.2. incomplete acrodromous venation
7. nervatio actinodroma (actinodromous venation)
   7.1. nervatio actinodroma marginalis (actinodromous ramifications in the margin of the leaf)
   7.2. nervatio actinodroma imperfecta (incomplete actinodromous venation)
   7.3. nervatio actinodroma flabelliformis (fan-shaped actinodromous venation)
   (cf. von Ettinghausen (1861) pp. XVIf.)

J.-A. Mouton also undertakes a classification of venation types, as found in his paper "Architecture de la nervation foliaire" (1967). His method of classification is based on von Ettinghausen's work. In contrast to von Ettinghausen he defines separate types for the primary, secondary and tertiary veins. Mouton differentiates according to the degree of hierarchy of the lateral vein:

1. the *primary veins*: 6 possibilities for their course with possible further subdivisions:
   1.1. *single primary* vein/medianus
   1.2. longitudinal *parallelodromous*
   1.3. *compylodromous*
   1.4. *dichotomous*
   1.5. complete and incomplete *acrodromous*
   1.6. complete, composed, incomplete, deformed foot-like *actinodromous*
2. the *secondary veins*: 3 main groups with numerous subordinated groups:
   2.1. *craspedodromous*, composed craspedodromous
   2.2. *camptodromous*, dentate, camptodromous with grawns
   2.3. dentate, curved, at the margin veined *brochidodromous*
3. the *tertiary veins* and the areas of the tissue enclosed by them: 6 partly subdivided forms:
   3.1. transverse *parallelodromous*
   3.2. vertical, bended, sinuate *plagiodromous*
   3.3. polygonale, irregular *reticulate*
   3.4. *fan-shaped*
   3.5. *crest-like*
   3.6. centripetal, opposed splayed

(The main types are underlined. The nomenclature is translated from French according to Mouton (1967) pp. 167f.)
A second difference between Mouton and von Ettinghausen becomes evident in the classification of the venation types. Von Ettinghausen defines the "nervatio brochidodroma" (brochidodromous venation) as being subordinate to the "nervatio camptodroma" (camptodromous venation), whereas Mouton describes these types separately with subdivisions to show they are characteristics of the secondaries.

The venation patterns concerning the investigated species of the Sapotaceae, Lauraceae and Euphorbiaceae are listed in Table 1. I classified the venations of the varying species on the basis of these main types.

In those representatives of a genus whose vein islet pattern is not sufficient for the unequivocal separation of the species, the cross-sections of the midrib can be taken into consideration for the demarcation: although the vein islet patterns of the three *Manilkara* species (pictures 72, 74, 75) differ clearly from the patterns of the other species of the sample collection, a positive identification of "Purguo", "Purguo blanco" and "Purguo morado" is made possible by the comparison of the cross-sections of the midrib, which differ considerably from the cross-sections of the medianus of the remaining Sapotaceae (pictures 63—65). The discrimination of *Pouteria egregia* and *P. venosa* (pictures 60, 61) and of *P. aff. anibaefolia* and *P. sp.* (pictures 59, 62) is also rather doubtful when using the venation patterns, because the patterns are very similar among these species. They can be distinguished by means of the cross-sections of the midrib.

Characteristics of the primary and secondary veins. The categories for the characterization of the behaviour of primary and secondary veins were chosen following Constantin von Ettinghausen's work "Blatt-Skelete der Dikotyledonen", (1861) and I applied them, as described by von Ettinghausen, to the primaries and secondaries (cf. ibid. pp. XIV and XVII).

Von Ettinghausen proposes a differentiation between the characteristics of the veins according to their absolute and relative thickness, the direction, the course and the termination at the tip or the margin of the leaf.

The criterion of the absolute thickness of primaries, secondaries and veins of higher orders seems to me to be of no purpose in my investigation, as there is no absolute thickness of vascular bundles. Moreover one should define exactly in which part of the unequally thick vein, which is mostly thinner towards the tip or the margin, the measurement should be carried out, or whether an average should be calculated. The relative thickness of the primary vein is a characteristic of taxonomic relevance if the form of the vascular bundles developed as medianus is compared with the different species within the families. Whereas the course of the primary vein through the lamina may differ in the leaves of one species, the direction of the medianus is characteristic for each species. The direction means the origin and the termination of a vein. Also, the manner of ending at the tip, the margin or any other region of the lamina may serve as a valid taxonomical feature.

The direction, course and termination of the secondaries are particularly important for taxonomical identification, because these criteria justify their classification under the types of venation, described above.

With the exception of the category "leaf symmetry" the characteristics defining the primary vein can be applied to the secondaries.

The information in the tables of the taxonomical part of this work refer comparatively to the species of the three families.

3.2. Key to the description and classification of vein islet patterns and types of ramifications of the veinlet terminations

(Compiled on the basis of the types occurring in the investigated Sapotaceae, Lauraceae and Euphorbiaceae.)

After the comparison of all venation patterns in the investigated material, seven subdivided criteria of separation were drawn up for patterns of vein islets and types of ramification.

I. *Characterization of the vein islets*

A. *Regularity in form and size*:
This criterion describes whether the islet pattern shows uniform form and size or whether it is irregular.

B. *Form of the vein islets*:
Inasmuch as islet forms cannot be classified as "longish", "roundish", "polygonal" or "squares" (cf. Roth (1984) p. 392), they are called "typical of the species".

C. *Size of the vein islets*:
The nomenclature of the differently-sized groups comply with the sizes of meshes which occur in the whole sample collection.

D. *Arrangement of the vein islets*:
"Single", "two-ranked" or "with several ranks" means the position of the vascular bundle islets between the straight running vascular bundles of subordinated degrees of ramification. The characteristic "without principle of order" differs considerably from "typical of the species". While the former is a pattern which can neither be included under 1—4, nor considered invariable to the species, the term "typical of the species" indicates that the characteristic in question is unmistakable and determinative for the species.

II. *Characterization of the ramifications of the veinlet terminations*

E. *Form of the ramifications*:
The terms "dendroid" and "cocktread-like" were chosen for the description and nomenclature of extremely typical kinds of ramification. The "cocktread" differs from the "tree-type" in the sharp bending of the freely ending terminations and the more angular ramifying vascular bundles in contrast to the more curved "treelike" ramification.

F. *Thickness of the ramifications*:
The specification of the thickness of the ramifications refers to the relation between the thickness of the bundles forming the islets, and the ramifications branching off into the meshes.

G. *Frequency of ramifications*:
This characteristic serves as a description of the "overall impression", which the ramifications of the meshes produce (branches of 1st, 2nd, 3rd order etc. plus free terminations) with regard to the density of ramification. The venation systems of the investigated species are once again referred to as standard.

Key to the description and classification:

A. *Regularity in form and size*:
1. regular
   examples: *Chrysophyllum sp., Manilkara* — Fig. 4.
2. irregular
   examples: *Chrysophyllum auratum, Ocotea martiana* — Fig. 5
3. alternating of regular and irregular parts
   examples: *Pouteria cf. trilocularis, P. sp.*

B. *Form of the vein-islets*:
1. elongated
   examples: leaf of a young tree of *Manilkara bidentata*/"Purguo", "Pendarito" — Fig. 6
2. roundish
   no examples within the sample collection
3. irregular polygonal
   examples: *Pouteria aff anibaefolia, P. egregia* — Fig. 7
4. "squares" (compare Roth, 1984, p. 392)
   examples: *Ocotea duotincta, Piranhea longepedunculata* — Fig. 8
5. typical of the species
   examples: "Capurillo negro", *Aniba riparia*

C. *Size of the vein islets*:
1. comparatively large
   examples: leaf of a young tree of *Mabea piriri, Pausandra flagellorhachis* — Figs. 6, 9
2. comparatively medium-large
   examples: *Manilkara bidentata, Ocotea nicaraguensis* — Fig. 10
3. comparatively small
   examples: *Pouteria venosa, Hieronyma laxifolia* — Fig. 11
4. mixture of the characteristics 1 to 3
   examples: *Ecclinusa guianensis*, "Purguillo"

D. *Arrangement of the vein islets*:
1. single
   examples: no examples in the investigated material
2. two-ranked
   examples: *Manilkara, Pogonophora sagotii* — Figs. 12, 19
3. with several ranks
   no examples within the sample collection
4. reticulate
   examples: *Chrysophyllum sp., Pouteria venosa* — Fig. 4
5. without principle of order
   examples: *Beilschmiedia curviramea, Mabeae taquiri* — Fig. 18
6. typical of the species
   examples: *Sapium sp.*, leaf of a young tree of *Drypetes variabilis*

E. *Form of ramification*:

1. simple
   examples: *Chrysophyllum sp., Piranhea longepedunculata* — Fig. 13
2. dichotomous
   no examples in the sample collection
3. mixture of simple and dichotomous ramification
   examples: *Manilkara*, "Laurel canelo" — Figs. 14, 19
4. dendroid
   examples: "Purguillo", *Drypetes variabilis* (leaf of the adult tree) — Fig. 15
5. cocktread-like
   examples: *Ocotea martiana*, leaf of a young tree of *Drypetes variabilis*
6. mixture of the characteristics 1 to 4
   examples: *Ocotea aff. subalveolata, Conceveiba guyanensis*
7. typical of the species
   examples: leaf of a young tree of *Manilkara bidentata*/"Purguo", *Chaetocarpus schomburgkianus*

F. *Thickness of the ramifications*:

1. as thick as the vascular bundles forming the islets
   examples: *Manilkara bidentata*, "Laurel canelo" — Figs. 17, 19
2. thinner than the bundles forming the islets
   examples: *Manilkara sp., Mabea taquiri* — Fig. 18
3. mixture of the characters 1 and 2
   examples: leaves of young trees of *Ocotea duotincta* and *O. nicaraguensis*

G. *Frequency of ramifications*:

1. comparatively well ramified
   example: *Mabea taquiri, Pera schomburgkiana* — fig. 18
2. comparatively medium ramification
   examples: *Manilkara bidentata*/"Purguo blanco", *Croton matourensis* (leaf of a young tree) — Fig. 19
3. comparatively little ramified
   examples: *Chrysophyllum sp., Piranhea longepedunculata* — Fig. 13
4. no ramifications
   no examples in the investigated material

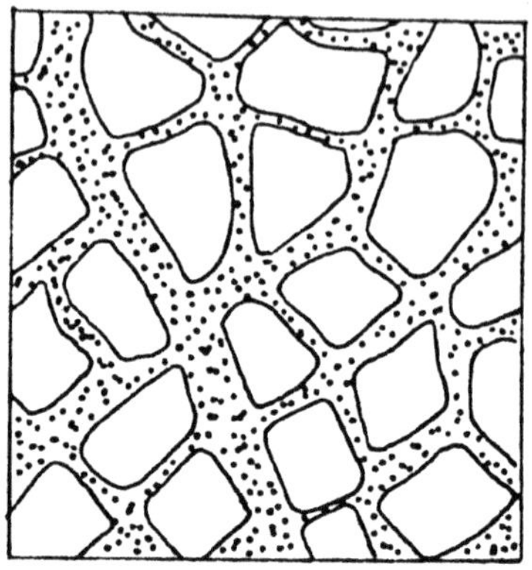

Fig. 4. A 1, D 4
Chrysophyllum spec.
- Sapotaceae -

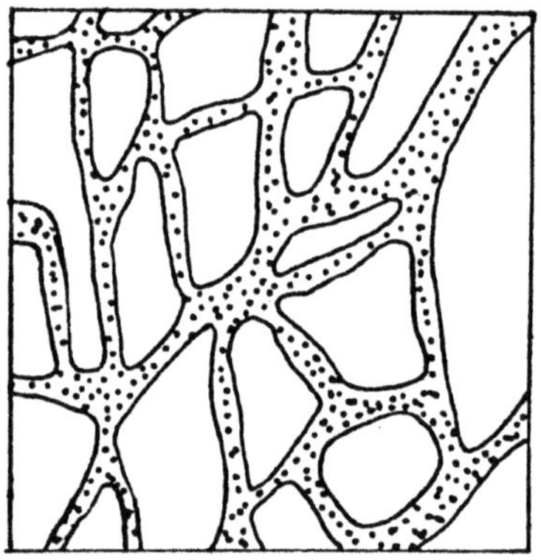

Fig. 5. A 2
Chrysophyllum auratum
- Sapotaceae -

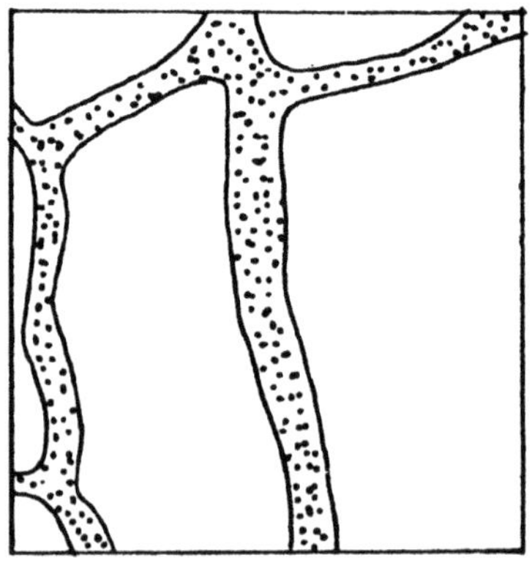

Fig. 6. B 1, C 1
Manilkara bidentata/
"Purguo" (leaf of a young tree)
- Sapotaceae -

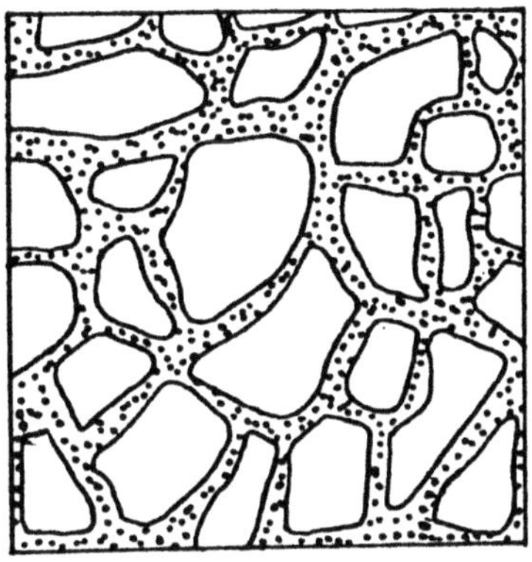

Fig. 7. B 3
Pouteria aff. anibaefolia
- Sapotaceae -

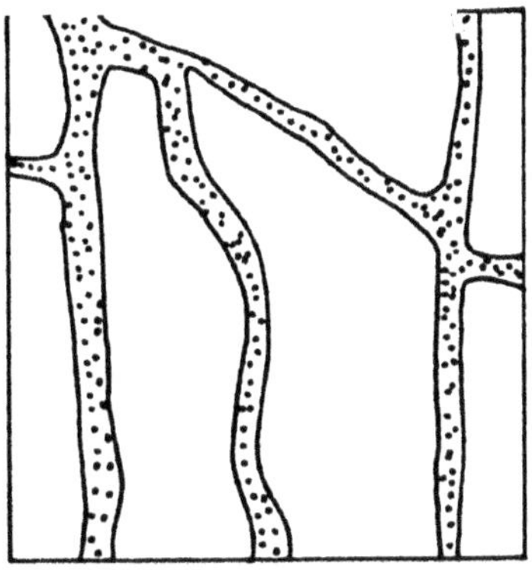

Fig. 9. C 1
Mabea piriri
- Euphorbiaceae -

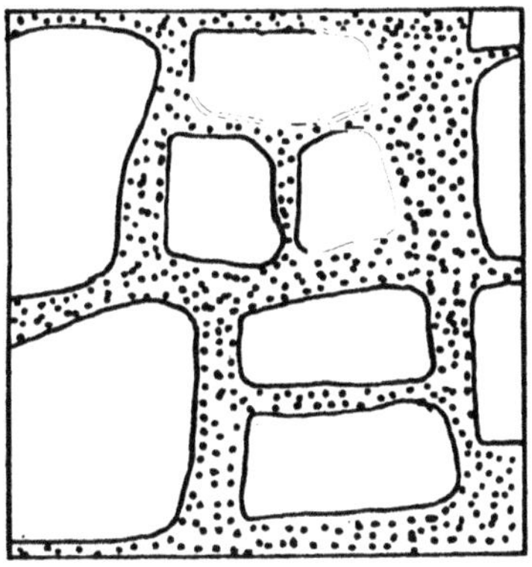

Fig. 8. B 4
Ocotea duotincta
- Lauraceae -

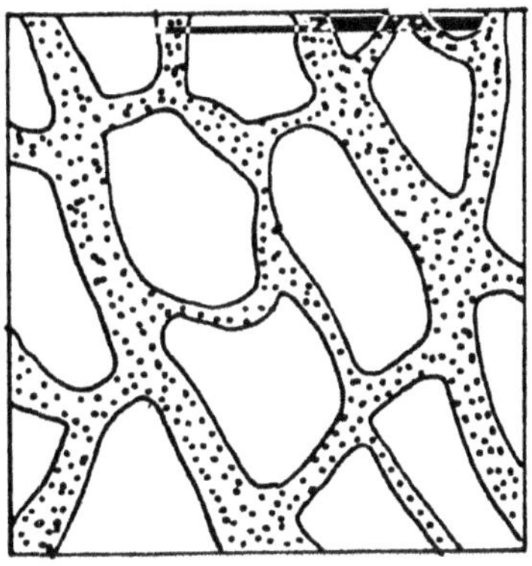

Fig. 10. C 2
Manilkara bidentata/
"Purguo morado"
- Sapotaceae -

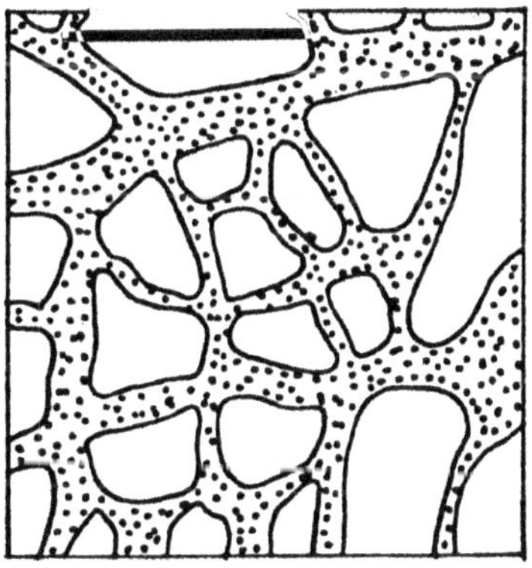

Fig. 11. C 3
Pouteria venosa
- Sapotaceae -

1 mm

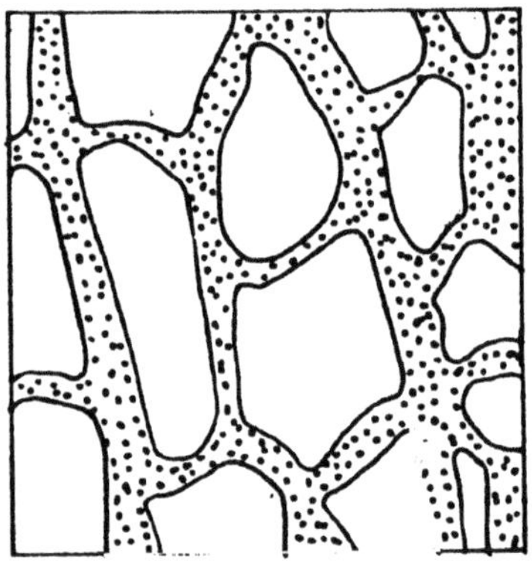

Fig. 12. D 2
Manilkara bidentata/
"Purguo blanco"

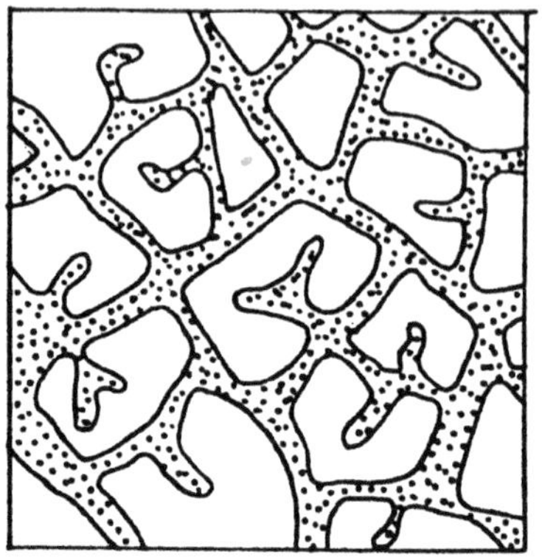

Fig. 14. E 2, E 3
"Laurel canelo"
- Lauraceae -

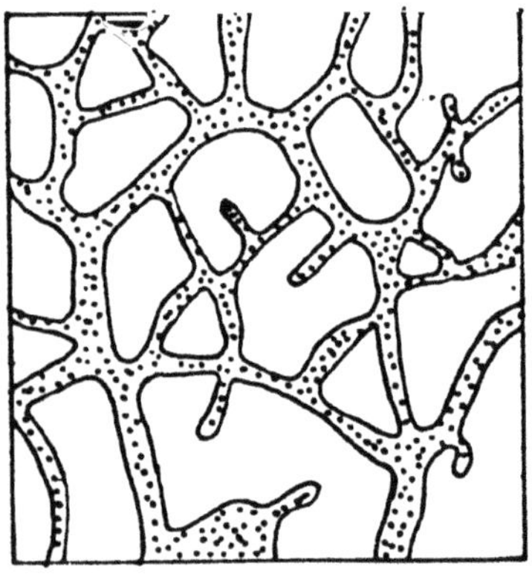

Fig. 13. E 1, G 3
Piranhea longepedunculata
- Euphorbiaceae -

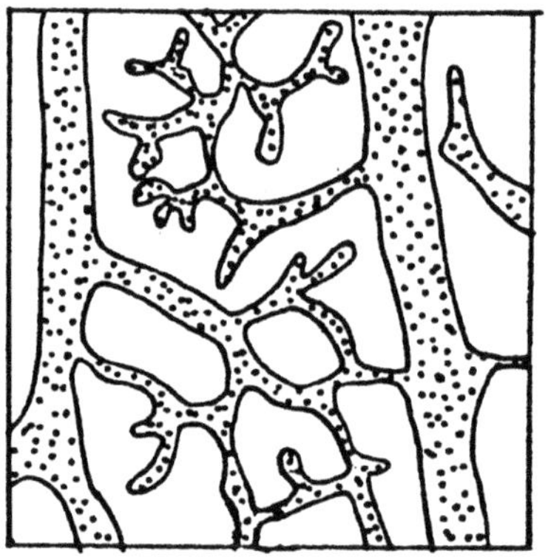

Fig. 15. E 4
"Purguillo"
- Sapotaceae -

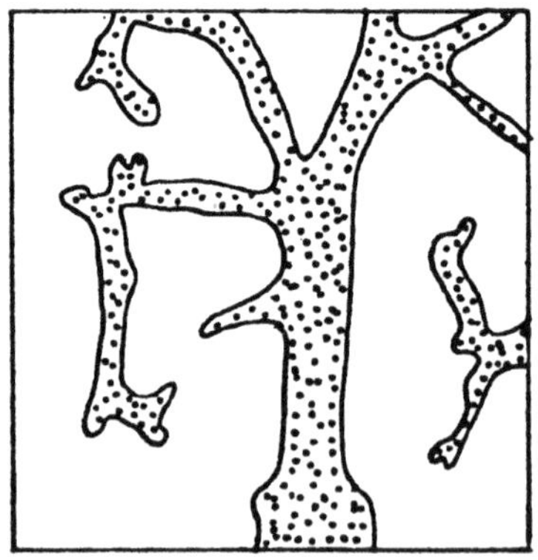

Fig. 16. E 5
Drypetes variabilis
(leaf of a young tree)
- Euphorbiaceae -

1 mm

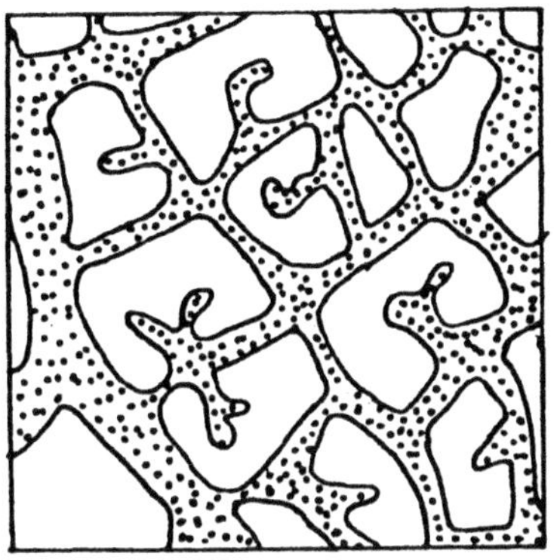

Fig. 17. F 1
"Laurel canelo"
- Lauraceae -

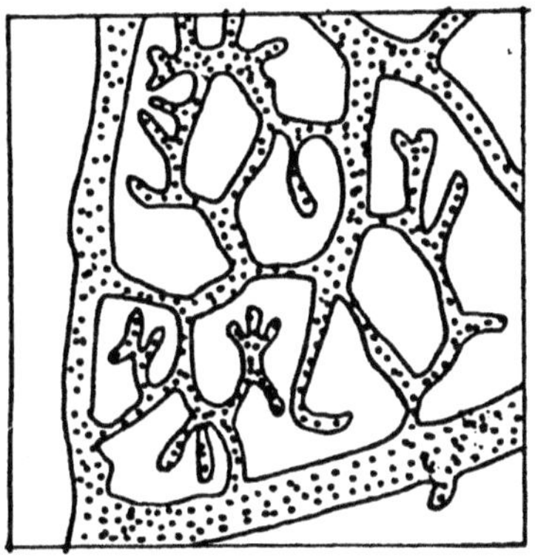

Fig. 18. F 2, G 1
Mabea taquiri
- Euphorbiaceae -

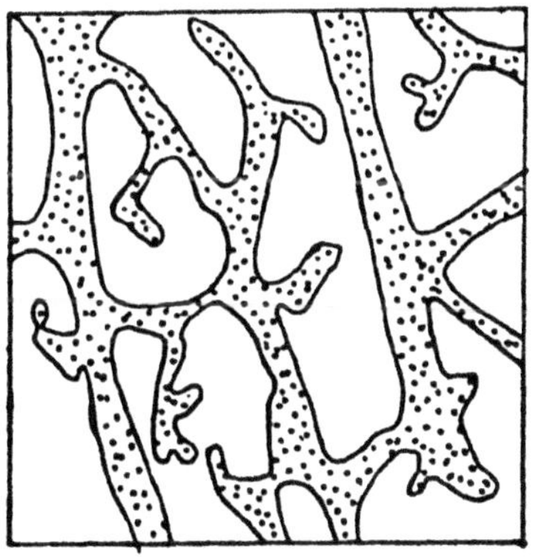

Fig. 19. D 2, E 3, F 1, G 2
Manilkara bidentata /
"Purguo blanco"
- Sapotaceae -

1 mm

3.3. Sapotaceae

Taxonomical utilization of the characteristics compared with leaf anatomy

Taxonomical characteristics of the Sapotaceous venation

All Sapotaceous species investigated are distinguishable by the presence of one primary vein, acting as medianus. Therefore, further types of veins such as "Basalnerven" ("basal veins"), "Hauptnerven" ("main veins" = stronger primaries), "Nebennerven" ("joint veins") or "Zwischennerven" ("intermediate veins" = weaker primaries) as well as "Quernerven" ("transversal veins" = connecting veins between two contiguous primaries) are missing (German nomenclature and definition according to von Ettinghausen (1861) p. XIV). Connecting veins and "Außennerven" ("outer veins") appear in different types.

A second common characteristic of all species is the closed venation. This appears either as the camptodromous form (pure or dentate camptodromous) or brochidodromous type (pure or dentate brochidodromous).

Pouteria cf. trilocularis and "Purguillo" differ from the other Sapotaceous representatives as to their fine vascular ramifications of higher orders, which end freely at the leaf margin.

The medianus has a pronounced development in comparison with the secondary veins. In all sample leaves it passes through the lamina from the base to the tip. Apart from a few exceptions (*Pouteria cf. trilocularis, P. aff. anibaefolia*, "Pendarito") it divides the leaf blade into two equal parts. Along its course the medianus of several species reveals a slight curve. This may be limited to the base or tip or to the region from the base to the middle. Apart from *Manilkara sp.*, where the thickness is constant, the medianus narrows regularly from the leaf base to the tip. In *Pouteria venosa* and *Manilkara* it ends directly at the tip. The medianus of "Capurillo negro" has a ramified subtilization at the tip. (Nomenclature of the terminations of the primary according to von Ettinghausen (1861) pp. XVII ff.)

Table 2. Investigated leaf material.

Family	Genus	Species	Vernacular name	Tree height	Individuals/ number of the leaves studied	No.[1]
Sapotaceae	Pouteria	aff. anibaefolia (A. C. Smith) Baehni	Caimito blanco	a	94/12	D– 45 (74)
	Pouteria	egregia Sandw.	Purguillo amarillo	A	274/10	X– 153 (481)
	Pouteria	cf. trilocularis Cronquist	Rosado	a	722/14	D– 85 (502)
	Pouteria	venosa (Mart.) Baehni	Bámpara	A	103/11	D– 148 (38)
	Pouteria	spec. Aubl.	Caimito negro	a	614/13	X– 93 (77)
	Manilkara	spec. Adans.	Pendare	A	129/10	(451)
	Manilkara	bidentata (A. DC.) A. Chev.	Purguo	A	387/11	(485)
			Purguo	6 m	1	(485)
			Purguo morado?	A	387/9	(487)
			Purguo blanco	A	387/11	D– 99
	Ecclinusa	guianensis Eyma	Chicle	A	1658/15	X– 73 (199)
			Chicle	1.5 m	1	X– 73 (199)
	Chrysophyllum	auratum Mig.	Caimito morado	A	12	(76)
			Caimito morado	7 m	7	(76)
	Chrysophyllum	spec. L.	Chupón	9 m	28/10	23 (201)
			Pendarito	A	28/10	D– 377 (452)
			Capurillo negro	A?	7	(489)
			Purguillo	A	795/11	D– 165 (480)

This survey only includes those sample leaves in which all investigation criteria could be carried out. The number of secondary veins and their mean distances could have been measured on further examples, whose area of the lamina was, however, mostly indeterminable.

[1] D, X and L mean the exact region of origin of the leaf material: D = Rio Grande (80–90 km east of Upata); X = El Paraiso (40–50 km north-east of Upata); L = El Dorado (20–50 km south- to southeast of El Dorado).

The first number indicates both the number of the collection and the number of the felled trees. The number in brackets refers to the list of the collected species, where the species are classified alphabetically according to their vernacular name.

The angles of emergence of the secondaries from the medianus – measured in the middle of the leaf – range from ca. 75° (*Manilkara bidentata*/"Purguo", *Pouteria aff. anibaefolia*) to ca. 53° ("Capurillo negro"). The highest values and thus the most obtuse-angled emergences are to be found (in descending order) in *Manilkara bidentata*/"Purguo" (75.25°) and *Pouteria aff. anibaefolia* (75.08°) for *Manilkara bidentata*/"Purguo blanco" (71.67°), "Purguillo" (71.58°), *Manilkara bidentata*/"Purguo morado" and *Pouteria sp.* (both 70.17°). A mean angle of emergence between 60° and 70° is present in *Ecclinusa guianensis* (63.83°), *Pouteria venosa* (68.17°), *P. egregia*

Table 3. Venation type, patterns of the primary veins.

No.	Species	Venation pattern: Name	Venation pattern: Formation of the margin	Patterns of the primaries: Relative thickness	Patterns of the primaries: Leaf symmetry	Course	Termination
1	Pouteria aff. anibaefolia "Caimito blanco"	camptodromous	closed	fine	almost equally large areas	towards the base slightly curved; regular narrowing	ending at the tip
2	Pouteria egregia "Purguillo"	〃	〃	〃	〃	towards the tip slightly curved; regular narrowing	〃
3	Pouteria cf. trilocularis "Rosado"	〃	〃	thick	conspicuously unequal areas	from the base towards the middle slightly curved; regular narrowing	〃
4	Pouteria venosa "Bampára"	〃	〃	〃	equally large areas	straight; regular narrowing	〃
5	Pouteria sp. "Caimito negro"	〃	〃	fine	〃	〃	〃
6	Manilkara sp. "Pendare"	brochidodromous	〃	thick	〃	straight, towards the tip slightly curved; uniform thickness	〃
7	Manilkara bidentata "Purguo"	〃	〃	〃	〃	straight; regular narrowing	〃
8	Manilkara bidentata "Purguo morado"	〃	〃	〃	〃	〃	〃
9	"Purguo blanco"	〃	〃	〃	〃	〃	〃
10	Ecclinusa guianensis "Chicle"	camptodromous	〃	fine	〃	at the tip slightly curved	〃
11	Chrysophyllum auratum "Caimito morado"	〃	〃	thick	〃	straight; regular narrowing	
12	Chrysophyllum sp.	〃	〃	〃	〃	〃	〃

Table 3 (Continued)

No.	Species	Venation pattern: Name	Venation pattern: Formation of the margin	Patterns of the primaries: Relative thickness	Patterns of the primaries: Leaf-symmetry	Course	Termination
	"Chupón" (9 m high)						
13	"Pendarito"	brochidodromous	″	thick	almost equally large areas	towards the tip slightly curved; regular narrowing	″
14	"Capurillo negro"	camptodromous	″	fine	″	straight; regular narrowing	ramified subtilization
15	"Purguillo"	brochidromous	″	″	equally large areas	″	ending at the tip

Table 4. Patterns of the secondary veins.

No.	Species	Angle of emergence	Position	Rel. thickness to the primary	Rel. thickness among each other	Direction and course	Termination
1	Pouteria aff. anibaefolia "Caimito blanco"	70.99—75.08 −79.17 (±5.45%)	slightly alternate opposite, towards the base more alternate	finer	the thickest ones in the middle, finer towards the base and tip	straight to slightly curved course until short of the margin: nervatio camptodroma (pure to dentated camptodromous); along the course slight narrowing	loops among the secondaries short of the margin
2	Pouteria egregia "Purguillo amarillo"	60.56—65.83 −71.10 (±8.01%)	opposite, partly alternate	″	″	slightly curved course until short of the margin: nervatio camptodroma (pure to dentated camptodromous); along the course slight narrowing	″
3	Pouteria cf. trilocularis "Rosado"	55.14—60.0 −64.86 (±8.1%)	slightly alternate opposite to alternate	″	″	curved course until short of the margin: nervatio camptodroma (pure camptodromous); along the course slight narrowing	loops among the secondaries, delicate loops, which can hardly be distinguished from the margin
4	Pouteria venosa	65.2—68.17 −71.14	opposite	″	finer towards the tip	″	Only at the tip distinctive loops

Table 4 (Continued)

No.	Species	Angle of emergence	Position	Rel. thickness to the primary	Rel. thickness among each other	Direction and course	Termination
	"Bámpara"	(±4.36%)					among the secondaries; in the middle and at the base the loops continue to the margin
5	Pouteria sp. "Caimito" negro"	64.3—70.17 —76.04 (±8.37%)	slightly alternate opposite, partly alternate	finer	the thickest ones in the middle, finer towards the tip and the base	straight to slightly curved course until short of the margin: nervatio camptodroma (pure to dentate camptodromous); along the course slight narrowing	loops among the secondaries short of the margin
6	Manilkara sp. "Pendare"	59.25—62.25 —65.25 (±4.82%)	opposite	remarkably finer	several finer secondaries between 2 finer ones	straight course until short of the margin: nervatio brochidodroma (curved brochidodromous); along the course no narrowing	distinctive anastomoses and loops among the thicker and finer secondaries
7	Manilkara bidentata "Purguo"	73.42—75.25 —77.08 (±2.43%)	"	"	regular change between thicker and finer ones	straight, towards the tip slightly curved course until short of the margin: nervatio brochidodroma (curved brochidodromous); along the course no narrowing	"
8	Manilkara bidentata "Purguo morado"	66.15—70.17 —74.19 (±5.73%)	"	"	"	"	"
9	"Purguo blanco"	64.35—71.67 —78.99 (±10.21%)	"	"	"	"	"
10	Ecclinusa guianensis "Chicle"	60.66—68.83 —77.0 (±11.87%)	opposite	remarkably finer	a little finer towards the tip	slightly curved course until short of the margin: nervatio camptodroma (pure camptodromous); along the course slight narrowing	loops among the secondaries short of the margin

Table 4 (Continued)

No.	Species	Angle of emergence	Position	Rel. thickness to the primary	Rel. thickness among each other	Direction and course	Termination
11	Chrysophyllum auratum "Caimito morado"	50.24—56.58 —62.92 (±11.21%)	opposite, towards the base more alternate	finer	finer towards the tip	straight to slightly curved course until short of the margin: nervatio camptodroma (pure to dentated camptodromous); along the course slight narrowing	delicate only at the tip but distinctive loops among the secondaries; in the middle and at the base very delicate loops, which can hardly be distinguished from the margin
12	Chrysophyllum sp. "Chupón" (9 m high)	52.33—56.25 —60.17 (±6.97%)	at the tip alternate, in the middle at the base slightly alternate opposite	finer	the strongest ones in the middle, finer towards the tip and the base finer	straight — especially at the tip — to slightly curved course until short of the margin: nervatio camptodroma (pure camptodromous); along the course slight narrowing	no prominent loops among the secondaries; curved endpiece runs parallel to the margin, joins it and disperses
13	"Pendarito"	61.02—65.75 —70.48 (±7.19%)	opposite	remarkably finer	alternating regularly from thicker to finer ones	straight, towards the tip slightly curved course until short of the margin: nervatio brochidodroma (pure brochidodromous); along the course no narrowing	distinctive anastomoses and loops among the stronger and finer secondaries
14	"Capurillo negro"	46.89—52.67 —58.45 (±10.97%)	slightly alternate opposite, partly opposite	finer	the strongest ones in the middle, finer towards the tip and the base	slightly curved course until short of the margin: nervatio camptodroma (pure to dentated camptodromous); along the course slight narrowing	loops among the secondaries short to the margin
15	"Purguillo"	65.77—71.58 —77.39 (±8.12%)	opposite, partly slightly alternate	conspicuously finer	alternating regularly from thicker to finer ones	straight, towards the tip slightly curved course until short of the margin: nervatio brochidodroma (curved brochidodromous); along the course no narrowing	distinctive anastomoses and loops among the stronger and finer secondaries

Table 5. Patterns of the veins of higher orders.

No.	Species	Tertiary veins	Quaternary veins	Quinternary veins	Free terminations[1)]	Islet pattern
1	Pouteria aff. anibaefolia "Caimito blanco"	+	+	+	+ 5 and higher orders	reticulate; irregular polygonal (generally penta- to hexagonal); easy to differentiate *type*: A 1, B 3, C 3, D 4, E 6, F 2, G 1
2	Pouteria egregia "Purguillo amarillo"	+	+	+	+ 5 and higher orders	reticulate; irregular polygonal (generally quadrangular to pentagonal); easy to differentiate *type*: A 1, B 3, C 3, D 4, E 3, F 2, G 1
3	Pouteria cf. trilocularis "Rosado"	+	+	+	+ 6	irregular polygonal (generally quadrangular to pentagonal); easy to differentiate *type*: A 3, B 3, C 4, D 5, E 6, F 3, G 2
4	Pouteria venosa "Bámpara"	+	+	+	+ 6	reticulate; irregular polygonal (generally quadrangular to pentagonal); easy to differentiate *type*: A 1, B 3, C 3, D 4, E 3, F 1, G 2
5	Pouteria sp. "Caimito negro"	+	+	+	+ 6 and higher orders	reticulate; irregular polygonal (generally penta- to hexagonal); easy to differentiate *type*: A 3, B 3, C 4, D 4, E 6, F 2, G 1
6	Manilkara sp. "Pendare"	+	+	+	+ 5 and higher orders	irregular polygonal (generally quadrangular to pentagonal); easy to differentiate *type*: A 1, B 3, C 3, D 2, E 3, F 2, G 2
7	Manilkara bidentata "Purguo"	+	+	+	+ 5 and higher orders	irregular polygonal (generally quadrangular to pentagonal); easy to differentiate *type*: leaf of the adult tree: A 1, B 3, C 2, D 2, E 3, F 1, G 1 leaf of the young tree: A 1, B 1, C 1, D 2, E 7, F 3, G 1
8	Manilkara bidentata "Purguo morado"	+	+	+	+ 5 and higher orders	irregular polygonal (generally quadrangular to pentagonal); easy to differentiate *type*: A 1, B 3, C 2, D 2, E 3, F 1, G 2
9	"Purguo blanco"	+	+	+	+ 5 and higher orders	irregular polygonal (generally quadrangular to pentagonal); easy to differentiate *type*: A 1, B 3, C 2, D 2, E 3, F 1, G 2
10	Ecclinusa guianensis "Chicle"	+	+	+	+ 6 and higher orders	irregular polygonal (generally quadrangular to pentagonal); squares partly almost vertical; easy to differentiate *type*: leaf of the adult tree: A 1, B 3, C 4, D 2, E 6, F 2, G 2 leaf of the young tree: A 2, B 3, C 2, D 5, E 3, F 2, G 3
11	Chrysophyllum auratum "Caimito morado"	+	+	+	+ partly 7 and higher orders	irregular longish polygonal (generally quadrangular to pentagonal); hard to differentiate *type*: leaf of the adult and young tree: A 2, B 5, C 3, D 6, E 1, F 1, G 1
12	Chrysophyllum sp. "Chupón" (9 m high)	+	+	+	+ sporadic; 6	reticulate; relatively regular polygons trapezoids and irregular polygonal forms (generally penta- to hexagonal); very easy to differentiate; "As seen in surface view, the vascular bundles form a pattern of squares around the 'air chambers'". (Roth, 1984, p. 392), *type*: A 1, B 4, C 2, D 4, E 1, F 2, G 3
13	"Pendarito"	+	+	+	+ sporadic;	reticulate; irregular longish polygonal (generally

Table 5 (Continued)

No.	Species	Tertiary veins	Quaternary veins	Quinternary veins	Free terminations[1]	Islet pattern
					3 and higher orders	quadrangular to pentagonal); hard to differentiate *type*: A 2, B 1, C 1, D 2, E 6, F 2, G 3
14	"Capurillo negro"	+	+	+	+ 6 and higher orders	irregular polygonal (generally quadrangular to hexagonal); hard to differentiate *type*: A 2, B 5, C 4, D 6, E 4, F 2, G 1
15	"Purguillo"	+	+	+	+ 6 and higher orders	irregular polygonal (generally quadrangular to pentagonal); easy to differentiate *type*: A 3, B 3, C 4, D 6, E 4, F 2, G 1

[1] The number indicates the degree of ramification.

(65.83°), "Pendarito" (65.75°), *Manilkara sp.* (62.25°) and *Pouteria cf. trilocularis* (60.0°). *Chrysophyllum auratum* (56.58°), *Chrysophyllum sp.* (56.25°) and "Capurillo negro" (52.67°) show values which lie clearly below 60°.

On the whole, I have determined that the widest angles of emergence occur in the genera *Manilkara* and *Pouteria,* but with remarkable variations among the species concerned. Extremely small differences appear in *Chrysophyllum.* (Genera of which only sample leaves of one species were available and species whose taxonomic origin has not been ascertained, were not taken into consideration in this survey.)

The position of the secondary veins in relation to each other ranges within the Sapotaceae from opposite through slightly alternate opposite to mixtures of opposite and alternate. A completely uniform picture is shown by the genus *Manilkara,* where only opposite secondaries appear. *Pouteria* is characterized by mixed forms.

With regard to the thickness of the secondaries in comparison to the primary vein the *Pouteria* species and the two representatives of *Chrysophyllum* possess finer secondaries, while the secondaries of *Manilkara* are much finer.

A comparison of the secondaries with each other shows that in *Pouteria* and *Chrysophyllum* the secondary veins of the middle are typically the thickest, becoming finer towards the base and the tip, or only towards the tip. The same holds true for "Capurillo negro" and *Ecclinusa guianensis.* The secondaries of *Manilkara,* "Pendarito" and "Purguillo" alternate regularly from thicker to finer veins.

All *Pouteria* species possess camptodromous nervation, as well as *Chrysophyllum auratum* and *Chrysophyllum sp., Ecclinusa guianensi* and "Capurillo negro". The remaining species show the brochidodromous venation type. The "nervatio camptodroma" appears either as the pure form (*Pouteria cf. trilocularis, P. venosa, Ecclinusa guianensis, Chrysophyllum sp.*) or as a mixture between pure and dentate camptodromous (*Pouteria aff. anibaefolia, P. egregia, P. sp., Chrysophyllum auratum,* "Capurillo negro"). The loops or the joinings of the secondaries, branching off from the medianus, develop shortly before the leaf margin. In the pure camptodromous venation the connecting arches are mostly difficult to differentiate from the leaf margin, as is the case, too, in the pure to dentate camptodromous course of vascular bundles of *Chrysophyllum auratum.*

The "nervatio brochidodroma" is also divided into two subgroups: "Pendarito" has the pure brochidodromous type, whereas the *Manilkara* species and "Purguillo" possess curved brochidodromous nervation. In the brochidodromous venation pattern loops and anastomoses are formed short of the leaf margin among the thicker and finer secondaries, branching off above each other from the medianus.

Although J.-A. Mouton does not provide any description beyond that of the tertiary veins —

"[. . .] la nervation tertiaire se raccorde à la secondaire. Il est pratiquement inutile d'aller au-delà." (1967, p. 166) — I consider it indispensable from a taxonomical as well as an ecological viewpoint to take the lateral veins of higher orders into consideration. Often the quaternary, quinternary, sesternary veins, or even those of further degrees of ramification, develop vein islets and free veinlet terminations. The "minor venation", in particular (Wylie, 1943), plays a vital role in the supply of nutrients in dicotyledonous plants.

The 15 species of the Sapotaceae possess ramifications to the degree of quinternaries at least. In *Pouteria aff. anibaefolia*, *P. egregia* and *Manilkara* these ramifications and/or those of higher orders make up the free terminations. The vein endings of the remaining species consist of sesternary, septernary or higher ramifications which are somewhat difficult to differentiate (*Chrysophyllum auratum*). *Manilkara, Ecclinusa guianensis, C. auratum*, "Pendarito" and "Capurillo negro" have free terminations, which are partly reduced to sclereids.

The vein islet patterns of the species studied belong mostly to the irregular polygonal type. *Chrysophyllum sp.* stands out from the other species due to the regular order of the vein islets. While in *C. sp.* rectangles and polygons are observed, which are clearly limited (cf. Roth 1984, p. 392: "[. . .] the vascular bundles form a pattern of squares around the 'air chambers'"), the vein network of *Chrysophyllum auratum* (leaf of adult and young tree), made up of irregular longish polygons, is hard to differentiate due to the sclereids. This difficulty within the genus is very noticeable. Both islet types are typical of the species concerned.

In the *Manilkara* species and *Ecclinusa guianensis* the vein islets, which appear in two ranks, and the specific dense network of sclereids are typical.

The leaf of the young tree of *Manilkara bidentata*/"Purguo" and *Ecclinusa guianensis* have very conspicuous veinlet networks. The islet pattern of the leaves of the young plants does not yet show the final pattern of the leaf of the adult tree. It is finer and wider meshed, but is clearly identifiable as a developing stage of the differentiated islet pattern.

The vascular bundle network of "Capurillo negro", which is abundantly provided with sclereids, is also typical of the species and is noninterchangable.

Only few of the venation characters investigated are suitable for the differentiation of Sapotaceous species. The genus *Manilkara* is distinguishable by the regular alternation of thicker and finer secondary veins. A much higher selectivity than the relative strength of the secondaries among each other is ensured by the pattern of the vein islets. For the determination of the *Manilkara* species the islet patterns together with cross-sections of the medianus are appropriate as definitive classification criteria.

Ecological significance of the anatomical investigations

Table 6. Measurements of the venation (leaves of adult trees).

Species	Leaf area(cm^2)	Number of secondaries/ leaf	Area[1] index (cm^2)	Mean distance of secondaries (cm)	Vein[2] islets/cm^2	Ramifications/islet	Veinlet terminations/cm^2	Vein length/cm^2 (cm)
Pouteria egregia	24.19—32.5 —40.81	26.37—29.0 —31.63	0.84—1.12 —1.4	0.53—0.65 —0.77	1059—1210 —1361	1.37—2.0 —2.63	1023—1340 —1657	90.84—94.19 —97.54
"Purguillo amarillo" (A)	(±25.57%)	(±9.07%)	(±25.0%)	(±18.46%)	(±12.48%)	(±31.5%)	(±23.66%)	(±3.56%)
Pouteria	38.56—54.36	12.53—14.5	2.92—3.73	1.4—1.51	1040—1160	1.2—1.9	940—1010	75.14—79.13

Table 6 (Continued)

Species	Leaf area(cm²)	Number of secondaries/ leaf	Area[1)] index (cm²)	Mean distance of secondaries (cm)	Vein[2)] islets/cm²	Ramifications/islet	Veinlet terminations/cm²	Vein length/cm² (cm)
venosa	−70.16	−16.47	−4.54	−1.62	−1280	−2.6	−1080	−83.12
"Bámpara" (A)	(±29.07%)	(±13.59%)	(±21.72%)	(±7.28%)	(±10.34%)	(±36.84%)	(±6.93%)	(±5.04%)
Manilkara spec.	51.36—67.22 −83.08	58.6—64.6 −70.6	0.84—1.03 −1.22	0.33—0.38 −0.43	236—275 −314	1.11—1.6 −2.09	506—570 −634	60.48—63.26 −66.04
"Pendare" (A)	(±23.59%)	(±9.29%)	(±18.45%)	(±13.16%)	(±14.09%)	(±30.63%)	(±11.23%)	(±4.39%)
Manilkara bidentata	34.21—47.64 −61.07	59.69—71.27 −82.85	0.54—0.66 −0.78	0.28—0.37 −0.46	453—590 −727	0.94—1.6 −2.26	1258—1360 −1462	45.39—48.66 −51.93
"Purguo" (A)	(±28.19%)	(±16.25%)	(±18.18%)	(±24.32%)	(±23.22%)	(±41.25%)	(±7.5%)	(±6.72%)
Manilkara bidentata	34.57—51.1 −67.63	57.27—65.1 −72.93	0.59—0.76 −0.93	0.37—0.4 −0.43	703—830 −957	0.84—1.3 −1.76	511—600 −689	52.66—58.09 −63.52
"Purguo morado" (A)	(±32.35%)	(±12.03%)	(±22.37%)	(±7.5%)	(±15.3%)	(±35.38%)	(±14.83%)	(±9.35%)
"Purguo blanco" (A)	53.68—66.73 −79.78	67.4—72.18 −76.96	0.78—0.92 −1.06	0.38—0.44 −0.5	488—630 −772	0.84—1.3 −1.76	1029—1270 −1511	52.1—53.57 −55.04
	(±19.56%)	(±6.62%)	(±15.22%)	(±13.64%)	(±22.54%)	(±35.38%)	(±18.98%)	(±2.74%)
Ecclinusa guianensis	8.54—24.61 −40.68	36.01—40.3 −44.59	0.42—0.59 −0.76	0.4—0.45 −0.5	223—268 −313	1.06—2.1 −3.14	580—710 −840	58.14—60.61 −63.08
"Chicle" (A)	(±65.3%)	(±10.65%)	(±28.81%)	(±11.11%)	(±16.73%)	(±49.52%)	(±18.31%)	(±4.08%)
Chrysophyllum auratum	72.71—134.43 −196.15	31.13—35.5 −39.87	2.36—3.69 −5.02	0.95—1.07 −1.19	848—940 −1032	cannot be differentiated	512—640 −768	39.9—42.27 −44.64
"Caimito morado" (A)	(±45.91%)	(±12.31%)	(±36.04%)	(±11.21%)	(±9.79%)		(±20.0%)	(±5.61%)
"Pendarito" (A)	23.61—29.8 −35.99 (±20.77%)	41.53—49.0 −56.47 (±15.24%)	0.47—0.58 −0.69 (±18.97%)	0.36—0.41 −0.46 (±12.2%)	56—68 −80 (±17.04%)	1—2	cannot be differentiated	19.50—21.38 −23.26 (±8.79%)
"Capurillo negro" (A?)	21.5—26.86 −32.22	16.59—18.0 −19.41	1.13—1.51 −1.89	0.74—0.9 −1.06	84—100 −116	0.8—1.7 −2.6	cannot be differentiated	37.38—39.81 −42.24
	(±19.96%)	(±7.83%)	(±25.17%)	(±17.78%)	(±15.75%)	(±52.94%)		(±6.1%)
"Purguillo" (A)	14.58—20.64 −26.7 (±29.36%)	49.59—52.09 −54.59 (±4.8%)	0.28—0.39 −0.5 (±28.21%)	0.27—0.34 −0.41 (±20.59%)	229—340 −451 (±32.65%)	1.58—2.5 −3.42 (±36.8%)	1240—1330 −1420 (±6.77%)	65.2—69.64 −74.08 (±6.38%)
Pouteria aff. anibaefolia	12.74—18.67 −24.6	22.33—25.33 28.33	0.55—0.73 −0.91	0.44—0.5 0.56	1000—1150 −1300	1.48—2.5 −3.52	927—1100 −1273	89.04—96.25 −103.46

Table 6 (Continued)

Species	Leaf area(cm²)	Number of secondaries/ leaf	Area[1] index (cm²)	Mean distance of secondaries (cm)	Vein[2] islets/cm²	Ramifications/islet	Veinlet terminations/cm²	Vein length/cm² (cm)
"Caimito blanco" (a)	(±31.76%)	(±11.84%)	(±24.66%)	(±12.0%)	(±13.04%)	(±40.8%)	(±15.73%)	(±7.49%)
Pouteria cf. trilocularis	38.85—63.5 —88.15	20.22—22.14 —24.06	1.85—2.84 —3.83	1.03—1.31 —1.59	352—430 —508	2.12—3.2 —4.28	855—980 —1105	46.34—47.86 —49.38
"Rosado" (a)	(±38.82%)	(±8.67%)	(±34.86%)	(±21.37%)	(±18.14%)	(±33.75%)	(±12.76%)	(±3.18%)
Pouteria spec.	15.42—20.23 —25.04	23.81—26.46 —29.11	0.61—0.76 —0.91	0.49—0.59 —0.69	395—450 —505	1.6—2.2 —2.8	966—1100 —1234	48.49—49.86 —51.23
"Caimito negro" (a)	(±23.78%)	(±10.02%)	(±19.74%)	(±16.95%)	(±12.17%)	(±27.27%)	(±12.18%)	(±2.75%)

[1] Quotient leaf-area: number of secondaries
[2] The vein islet number is based on round figures, while the percentage refers to the mathematical averages.

Table 7. Position of the species within the criteria investigated (leaves of adult trees).[1]

Position	Leaf area(cm²)	Number of secondaries/ leaf	Area index (cm²)	Mean distance of secondaries (cm)	Vein islets/ cm²	Ramifications/ islet	Veinlet terminations/cm²	Vein length/cm² (cm)
1	C. auratum	"Purguo blanco"	P. venosa	P. venosa	P. egregia	P. cf. trilocularis	"Purguo"	P. aff. anibaefolia
2	M. spec. "Purguo blanco	"Purguo"	C. auratum	P. cf. trilocularis	P. venosa	P. aff. anibaefolia "Purguillo"	P. egregia	P. egregia
3	P. cf. trilocularis	M. spec. "Purgro morado"	P. cf. trilocularis	C. auratum	P. aff. anibaefolia	P. spec.	"Purguillo"	P. venosa
4	P. venosa	"Purguillo"	"Capurillo negro"	"Capurillo negro"	C. auratum	E. guianensis	"Purguo blanco"	"Purguillo"
5	"Purguo morado"	"Pendarito"	P. egregia	P. egregia	"Purguo morado"	P. egregia	P. spec. P. aff. anibaefolia	M. spec.
6	"Purguo"	E. guianensis	M. spec.	P. spec.	"Purguo blanco"	P. venosa	P. venosa	E. guianensis
7	P. egregia	C. auratum	"Purguo blanco"	P. aff. anibaefolia	"Purguo"	"Capurillo negro"	P. cf. trilocularis	"Purguo morado"
8	"Pendarito"	P. egregia	"Purguo morado" P. spec.	E. guianensis	P. spec.	M. spec. "Purguo	E. guianensis	"Purguo blanco"
9	"Capurillo negro"	P. spec.	P. aff. anibaefolia	"Purguo blanco"	P. cf. trilocularis	"Purguo morado" "Purguo blanco"	C. auratum	P. spec.

[1] Decreasing values from 1—14.

Table 7 (Continued)

Position	Leaf area(cm²)	Number of secondaries/ leaf	Area index (cm²)	Mean distance of secondaries (cm)	Vein islets/ cm²	Ramifications/ islet	Veinlet terminations/cm²	Vein length/cm² (cm)
10	E. guianensis	P. aff. anibaefolia	"Purguo"	"Pendarito"	"Purguillo"	"Pendarito"	"Purguo morado"	"Purguo"
11	"Purguillo"	P. cf. trilocularis	E. guianensis	"Purguo morado"	M. spec.		M. spec.	P. cf. trilocularis
12	P. spec.	"Capurillo negro"	"Pendarito"	M. spec.	E. guianensis			C. auratum
13	P. aff. anibaefolia	P. venosa	"Purguillo"	"Purguo"	"Capurillo negro"			"Capurillo negro"
14				"Purguillo"	"Pendarito"			"Pendarito"

Approximate total vein lengths and conductive volumes/cm² (leaves of adult trees)

The approximate total vein length (average of the size of the prepared leaf of the adult tree × vein length/cm² of the lamina) produces the following data:

Table 8. Approximate total vein length.

Species	Total vein length (cm)
Pouteria egregia	3502.46
Pouteria venosa	2888.25
Manilkara spec.	5351.80
Manilkara bidentata/"Purguo"	3844.14
Manilkara bidentata/"Purguo morado"	2788.32
Manilkara bidentata/"Purguo blanco"	4017.75
Ecclinusa guianensis	1103.10
Chrysophyllum auratum	6551.85
"Pendarito"	598.64
"Capurillo negro"	1114.68
"Purguillo"	1845.46
Pouteria aff. anibaefolia	2839.38
Pouteria cf. trilocularis	2440.86
Pouteria spec.	1096.92

Results of the measurements of the Sapotaceous venation (leaves of adult trees)

The leaves of the largest average size, drawing on those collected samples available to me, are found in the genus *Chrysophyllum* (*C. auratum* with ca. 134 cm²). *Manilkara sp.* and *Manilkara bidentata*/"Purguo blanco" (both ca. 67 cm²), *Pouteria cf. trilocularis* (ca. 64 cm²), *Pouteria venosa* (ca. 54 cm²) as well as *M. bidentata*/"Purguo morado" (ca. 51 cm²) are sized between 70 and 50 cm². For *M. bidentata*/"Purguo" the average of ca. 48 cm² lies slightly lower. The leaf sizes of the remaining species range from ca. 33 cm² (*Pouteria egregia*) to ca. 19 cm² (*Pouteria aff. anibaefolia*). Between the enumerated species are "Pendarito" (ca. 30 cm²), "Capurillo negro" (ca. 27 cm²), *Ecclinusa guianensis* (ca. 25 cm²), "Purguillo" (ca. 21 cm²) and *Pouteria sp.* (ca. 20 cm²). It is noticeable that within the genus *Pouteria* greater vacillations (from ca. 19 cm² up to 64 cm²) are apparent than in *Manilkara*, where the values differ from ca. 48 cm² to 67 cm².

Irrespective of the leaf-blade, which is supplied by a single secondary vein, the species of the genus *Manilkara* possess the greatest number of secondary veins. The data range from ca. 72 (*M. bidentata*/"Purguo blanco") more than ca. 71 (*M. bidentata*/"Purguo") to ca. 65 (*M. sp.* and *M. bidentata*/"Purguo morado"). The deviation of seven secondaries per lamina is low in comparison to *Pouteria* (see below). The species "Purguillo" and "Pendarito" come next with ca. 52 and 49 secondary veins respectively. *Ecclinusa guianensis* follows with ca. 40 secondaries. For *Chrysophyllum auratum* ca. 36 ramifications were counted. The genus *Pouteria* makes up the last group with comparatively few secondary veins (*P. egregia* 29, *P. sp.* ca. 26, *P. aff. anibaefolia* ca. 25, *P. cf. trilocularis* ca. 22), finally "Capurillo negro" (18) and *P. venosa* (ca. 15). Within the genus of

Table 9. Approximate conductive volume per cm^2 of the lamina ($\pi r^2 \times$ vein length/cm^2 cf. Section 3.1).

Species	Gross conductive vol. (cm^3)	Net conductive vol. (cm^3)[1]
Pouteria egregia	0.0013	
Pouteria venosa	0.0016	0.0003 (=18.75%)
Manilkara spec.	0.0017	
Manilkara bidentata/"Purguo"	0.0009	
Manilkara bidentata/"Purguo morado"	0.0016	
Manilkara bidentata/"Purguo blanco"	0.0011	
Ecclinusa guianensis	0.0013	0.0003 (=23.08%)
Chrysophyllum auratum	0.0008	0.0001 (=12.5%)
"Pendarito"	0.0006	
"Capurillo negro"	0.0009	0.0001 (=11.11%)
"Purguillo"	0.0016	0.0005 (=31.25%)
Pouteria aff. anibaefolia	0.0017	
Pouteria cf. trilocularis	0.0005	
Pouteria spec.	0.0006	0.0001 (=16.67%)
	0.0008—0.0012—0.0016 (±33.33%)	0.0001—0.0002—0.0003 (±50%)

[1] After deduction of the estimated nonconductive tissues (cf. Section 3.1).

Pouteria the deviation of 14 secondaries is much higher than that of *Manilkara*.

The best innervation in relation to the leaf-size — documented as "area-index" — is measured for "Purguillo"; 0.39 cm^2 of the lamina are supplied by one secondary vein. "Pendarito" (0.58 cm^2) and *Ecclinusa guianensis* (0.59 cm^2) follow. Next on the list are the samples from *Manilkara*, together with two *Pouteria* species: *M. bidentata*/"Purguo" (0.66 cm^2), *P. aff. anibaefolia* (0.73 cm^2), *M. bidentata*/"Purguo morado" (0.76 cm^2), *P. sp.* (0.76 cm^2), *M. bidentata*/"Purguo blanco" (0.92 cm^2) and *M. sp.* (1.03 cm^2).

The distribution of the number of secondaries is reflected by their mean distance. The smallest distances were measured for "Purguillo" (0.34 cm) and especially *Manilkara* (*M. bidentata*/"Purguo" (0.37 cm), *M. sp.* (0.38 cm), *M. bidentata*/"Purguo morado" and "Purguo blanco" with 0.4 and 0.44 cm respectively). "Pendarito" (0.41 cm) is placed between the last two species of *Manilkara*. *Ecclinusa guianensis* takes the following position with 0.45 cm. Three representatives of *Pouteria* follow with 0.5 cm (*P. aff. anibaefolia*), 0.59 cm (*P. sp.*), and 0.65 cm (*P. egregia*). The remaining species of this genus occupy the last positions (*P. cf. trilocularis* 1.31 cm, *P. venosa* 1.51 cm) behind "Capurillo negro" (0.9 cm) and *Chrysophyllum auratum* (1.07 cm). In an analogous way to previous investigations the genus *Pouteria* represents a rather heterogeneous group; the distances

Cross-sections of the vascular bundles (schematized)
- xylem dotted -

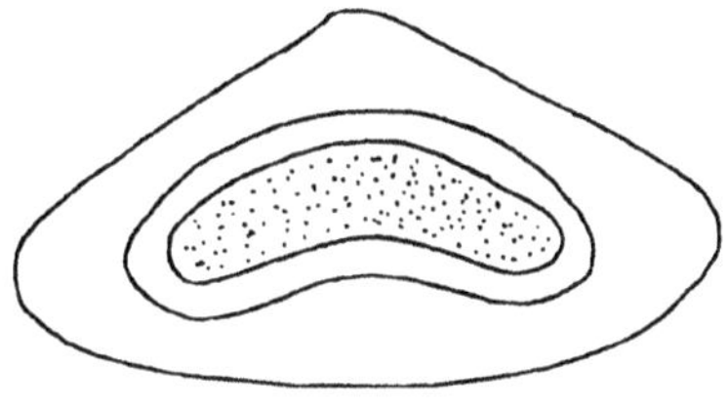

Fig. 20. Pouteria venosa. ca. 19% xylem

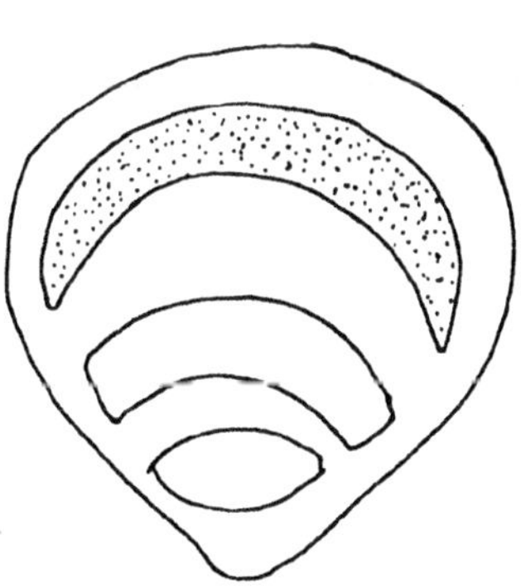

Fig. 21. Pouteria sp. ca. 18% xylem

Fig. 22. *Ecclinusa guianensis* (leaf of the adult tree).

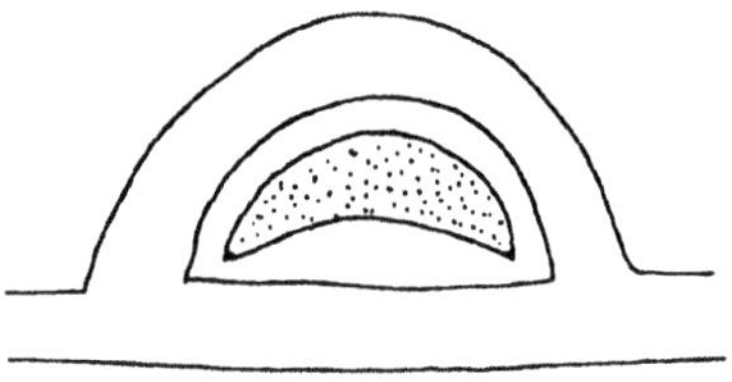

Fig. 23. *Chrysophyllum auratum* (leaf of the adult tree).

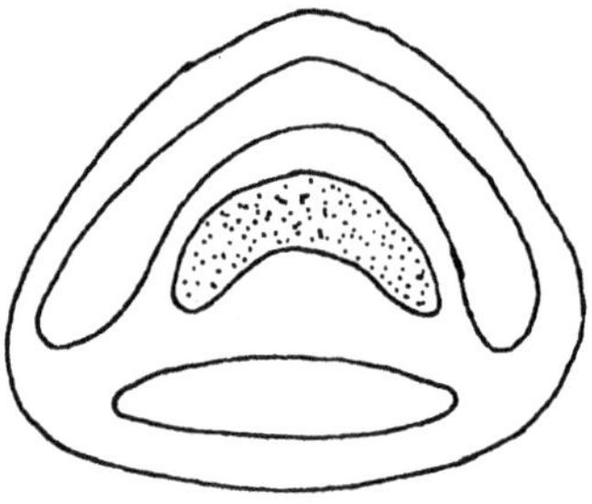

Fig. 24. "Capurillo negro".

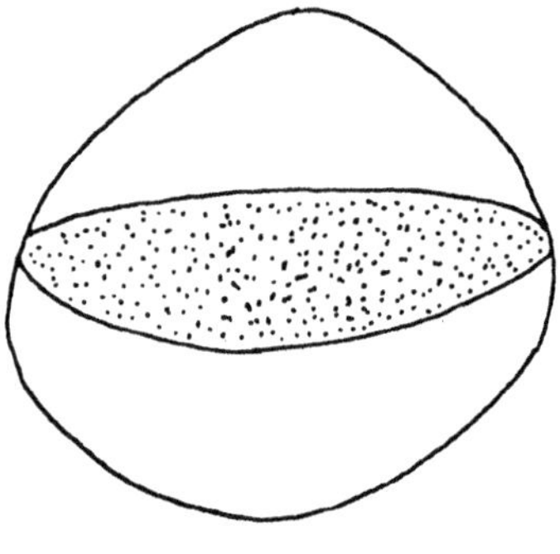

Fig. 25. "Purguillo".

of the secondary veins lie between 0.5 and 1.51 cm. For *Manilkara* only values from 0.37 to 0.44 cm were measured.

Of the 166 samples 125 (75.3%) leaves have the widest distances between the secondaries, at least in one half of the leaf, in the middle of the lamina.

The densest network of vein islets is found among the three representatives of *Pouteria: P. egregia* (1210 islets per cm^2 of the leaf blade), *P. venosa* (1160) and *P. aff. anibaefolia* (1150). *Chrysophyllum auratum* follows with 940 vein islets. The range of the vein islet number for *Manilkara* reaches from 830 (*M. bidentata*/"Purguo morado") to 275 (*M. sp.*). Between them are located *M. bidentata*/"Purguo blanco" (630), *M. bidentata*/"Purguo" (590) as well as *Pouteria sp.* (450), *Pouteria cf. trilocularis* (430) and "Purguillo" (340). *Ecclinusa guianensis* has fewer than 300 vein islets (268). Lastly, "Capurillo negro" (100) and "Pendarito" (68) have much the lowest number of vein islets.

Remarkable differences are apparent in the visible discrimination and distinctiveness of the vein islet patterns. The meshes of the *Pouteria* and *Manilkara* species are readily distinguishable from each other, as well as those of *Ecclinusa guianensis* and "Purguillo".

As far as *Chrysophyllum auratum* is concerned, one cannot clearly determine which vascular bundles branch off into one islet. It was not possible to identify the free veinlet terminations of "Pendarito" accurately because of the sclereids.

The enumeration of the ramifications, which run into one vein islet, resulted in values of 3—1 ramifications (mathematical averages: 3.2—1.3). *Pouteria cf. trilocularis* holds the highest value with 3 ramifications per islet (Ø = 3.2). With 2—3 ramifications (Ø = 2.5) *Pouteria aff. anibaefolia* and "Purguillo" take second place, followed by *Pouteria sp.* (2/2.2) and *Ecclinusa guianensis* (2/2.1). *Pouteria egregia* (2/2.0), *Pouterias venosa* (2/1.9) and "Capurillo negro" (2/1.7) too, can be allocated to the category of 2 branches. Between 2 and 1 ramification(s), inclining more towards one single ramification, are the following species: the *Manilkara* species — *M. sp.* and *M. bidentata*/"Purguo" (2/1.6), *M. bidentata*/"Purguo morado" and "Purguo blanco" (1/1.3) — and "Pendarito" (1—2 ramifications).

The results of the veinlet terminations/cm² show great differences within the Sapotaceous sample collection. *Manilkara bidentata*/"Purguo" (1360), *Pouteria egregia* (1340), "Purguillo" (1330) and *M. bidentata*/"Purguo blanco" (1270) show the greatest number of veinlet terminations. *Pouteria sp.* (1100), *Pouteria aff. anibaefolia* (1100) and *Pouteria venosa* (1010) follow. The value for *Pouteria cf. trilocularis* (980) is slightly lower. The *Pouteria* species — apart from *P. egregia* — fluctuate around 1100. They thus form a comparatively homogeneous group. *Ecclinusa guianensis* shows 710 and *Chrysophyllum auratum* 640 terminations per cm² of the lamina. A mean value of 600 was counted for *Manilkara bidentata*/"Purguo morado". Finally, with 570 veinlet terminations, *Manilkara sp.* is only slightly below this number. The veinlet terminations of "Pendarito" and "Capurillo negro" could not be discriminated. (For *Manilkara sp.* and *M. bidentata* the results only refer to those terminations which were clearly identifiable.)

The best vein supply per cm² of the leaf blade was measured for *Pouteria aff. anibaefolia* (96.25 cm) and *Pouteria egregia* (94.19 cm). The third *Pouteria* species also possesses a dense venation system (79.13 cm/cm²). "Purguillo" (69.64 cm), *Manilkara sp.* (63.26 cm) and *Ecclinusa guianensis* (60.16 cm) show lengths from 70 to 60 cm. *Manilkara bidentata*/"Purguo morado" (58.09 cm) is in the same category. The second species, classified as *M. bidentata*, "Purguo blanco" has a somewhat smaller vein length per cm² of the blade than "Purguo morado": 53.57 cm. The mean innervation of *Pouteria sp.* (49.86 cm) and *M. bidentata*/"Purguo" (48.66) lies slightly below 50 cm/cm². The next group comprises *Pouteria cf. trilocularis* (47.86 cm), *Chrysophyllum auratum* (42.27 cm) and "Capurillo negro" (39.81 cm). An extremely weak supply of veins was established for "Pendarito" (21.38 cm/cm²).

The genus *Pouteria* is noticeable for its heterogeneity with regard to the vein length, while the range of the four representatives of *Manilkara* is not as great (*Pouteria*: from 94.19 to 47.86 cm/cm², *Manilkara*: 63.26 to 48.66 cm/cm²):

With reference to Table 9, we can state that the study of the cross-sections of the vascular bundles of six Sapotaceous species (Figures 20 to 25) shows that the cross-sections have an oval to roundish form. The xylem of the vascular bundles investigated is kidney-shaped with a transversal stretching. It can be concluded that within the family of the Sapotaceae the error in measurement, caused by the formula πr^2 (calculation of the vascular bundle cross-section) is almost on the same scale. Therefore, the values of the gross and net conductive volumes of the Sapotaceous representatives can be compared with each other.

From the figures it can be concluded that the xylem makes up percentages of ca. 10.5% to ca. 22.5% of the vascular bundle cross-section. The net conductive volumes, which were calculated only on the basis of the water-conducting tissues, comprise ca. 11% to ca. 31% of the gross conductive volume.

Approximate total vein lengths and conductive volumes/cm² (leaves of young trees)

The approximate total vein length (average of the size of the prepared leaves of young trees × vein length/cm²) produces the following data:

Table 10. Approximate total vein length.

Species	Total vein length (cm)
Chrysophyllum spec.	4613.87
Chrysophyllum auratum	9491.02
Manilkara bidentata/"Purguo"	3668.62
Ecclinusa guianensis	1684.62

Table 11. Approximate conductive volume per cm² of the lamina (cf. Section 3.1).

Species	Gross conductive volume (cm³)
Chrysophyllum spec.	0.0009
Chrysophyllum auratum	0.0007
Manilkara bidentata/"Purguo"	0.0010
Ecclinusa guianensis	0.0008
	0.0008–0.0009–0.0010 (±11.11%)

Table 12. Measurements of the venation (leaves of young trees).

Species	Leaf area (cm^2)	Number of secondaries/ leaf	Area[1] index (cm^2)	Mean distance of secondaries (cm)	Vein[2] islets/cm^2	Ramifications/ islet	Veinlet terminations/cm^2	Vein length/cm^2 (cm)
Chrysophyllum spec.	55.84—88.83 —121.82	33.76—37.25 —40.74	1.52—2.41 —3.3	0.84—1.04 —1.24	1134—1350 —1566	sporadically 1—3 ramifications	60—170 —280	60.61—67.75 —74.89
"Chupón" (9 m high)	(±37.14%)	(±9.37%)	(±36.93%)	(±19.23%)	(±16.0%)		(±64.71%)	(±10.54%)
Chrysophyllum auratum	136.36—159.09 —181.82	33.83—37.5 —41.17	3.48—4.21 —4.94	0.83—1.02 —1.21	890—1040 —1190	sporadically 1 ramification	40—130 —220	68.09—72.23 —76.37
"Caimito morado" (7 m high)	(±14.29%)	(±9.79%)	(±17.34%)	(±18.63%)	(±14.42%)		(±69.23%)	(±5.73%)
Manilkara bidentata	100.4	83	1.21	0.44	89—100 —111	0.91—1.4 —1.89	cannot be differentiated	35.25—36.54 —37.83
"Purguo" (6 m high)					(±11.25%)	(±35.0%)		(±3.53%)
Ecclinusa guianensis	52.7	53	0.99	0.42	144—173 —202	0.05—0.8 —1.55	dto.	31.22—31.96 —32.7
"Chicle" (1.5 m high)					(±16.52%)	(±93.75%)		(±2.32%)

[1] Quotient leaf area: number of secondaries.
[2] The vein-islet number is based on round figures, while the percentage refers to the mathematical averages.

Table 13. Position of the species within the criteria investigated (leaves of young trees).[1]

Position	Leaf area (cm^2)	Number of secondaries/ leaf	Area index (cm^2)	Mean distance of secondaries (cm)	Vein islets/ cm^2	Ramifications/ islet	Veinlet terminations/cm^2	Vein length/cm^2 (cm)
1	C. auratum	"Purguo"	C. auratum	C. spec.	C. spec.	"Purguo"	C. spec.	C. auratum
2	"Purguo"	E. guianensis	C. spec.	C. auratum	C. auratum	E. guianensis	C. auratum	C. spec.
3	C. spec.	C. auratum	"Purguo"	"Purguo"	E. guianensis	C. spec.		"Purguo"
4	E. guianensis	C. spec.	E. guianensis	E. guianensis	"Purguo"	C. auratum		E. guianensis

[1] Decreasing values from 1—4.

Results of the measurements of the Sapotaceous venation (comparison between leaves of young and adult trees)[1]

From the Sapotaceous species *Manilkara bidentata*/"Purguo", *Ecclinusa guianensis* and *Chrysophyllum auratum* adult leaves of young trees were available, as well as those from adult plants. The species *Chrysophyllum sp.* was only represented by leaves of young trees. Although I had seven and ten samples$_{YT}$ of *C. auratum* and *C. sp.* at my disposal, respectively, I could only study one single sample of the other species mentioned.

[1] In this section the name of a species refers without further explanation to the leaves of young trees.
Abbreviations: leaf$_{YT}$ = adult leaf of the young tree
leaf$_{AT}$ = adult leaf of the adult tree.

The largest leaves (ca. 160 cm^2 on average) belong to *Chrysophyllum auratum* and *Manilkara bidentata*/"Purguo" (ca. 100 cm^2). The samples of *Chrysophyllum sp.* had a mean size of ca. 89 cm^2. The smallest leaf in my collection, of ca. 53 cm^2, originated from the 1.5-m-high "Chicle" tree (*Ecclinusa guianensis*). A comparison of the size was convenient only for the leaves of young and adult plants of *C. auratum*, because sufficient leaf-material was available only for this species. The mean values (ca. 160 cm^2) for the $leaf_{YT}$, ca. 134 cm^2 for the $leaf_{AT}$ differ only slightly. In order to give a complete survey, the data for *Manilkara bidentata*/"Purguo" and *E. guianensis* are also mentioned: the $leaf_{YT}$ of "Purguo" is larger by ca. 53 cm^2 and that of *E. guianensis* is larger by ca. 28 cm^2 than the $leaf_{AT}$.

The $leaf_{YT}$ of *Manilkara bidentata*/"Purguo" possesses on average 12 secondary veins more than the leaf of the adult tree (ca. 71). For *Ecclinusa guianensis* a distinctive difference is revealed, amounting to ca. 13 in favour of the $leaf_{YT}$ (53 secondaries). The $leaf_{YT}$ of *Chrysophyllum auratum* has only two secondaries more than the $leaf_{AT}$: ca. 38. *Chrysophyllum sp.* also yields a mean number of ca. 37 secondary veins.

Just as with the adult plants, I calculated the smallest area index for *Ecclinusa guianensis* (0.99 cm^2). This means an increase of 0.4 cm^2 in comparison to the $leaf_{AT}$, where only 0.59 cm^2 of the leaf blade is supplied by one secondary. A higher index of 0.55 cm^2 (1.21 cm^2) was calculated for *Manilkara bidentata*/"Purguo" in relation to the $leaf_{AT}$ (0.66 cm^2). For the $leaf_{YT}$ of *Chrysophyllum auratum* there was a difference of merely +0.52 (4.21 to 3.69 cm^2 for the $leaf_{AT}$). *Chrysophyllum sp.*, finally, had an index of 2.41 cm^2.

The narrowest distances between the secondary veins were measured for *Ecclinusa guianensis* (0.42); i.e. −0.03 cm in comparison with the $leaf_{AT}$. The distances of "Purguo" are 0.07 cm wider than for the $leaf_{AT}$, i.e. 0.44 cm. *Chrysophyllum auratum* shows a still smaller deviation: −0.05 cm (1.02 cm for the $leaf_{YT}$ to 1.07 cm of the $leaf_{AT}$). The second species of *Chrysophyllum, C. sp.*, yields a value (1.04 cm), comparable to that of *C. auratum*.

Fifteen (= 68.18%) of the 22 leaves of young plants investigated reveal, at least in one half of the leaf, the greatest distances in the middle of the lamina.

Chrysophyllum sp. and *Chrysophyllum auratum* with a mean vein islet number of 1350 and 1040, respectively, have the densest networks of vascular bundles. In relation to the $leaf_{AT}$ *C. auratum* possesses on average 100 islets more per cm^2 of the blade. *Ecclinusa guianensis* follows (173 islets); with 95 the $leaf_{YT}$ lies below the number of the $leaf_{AT}$. With the help of a 70-fold magnification of the venation pattern of the $leaf_{YT}$ I succeeded in counting the immensely finely-meshed network of sclereids, which pass through the intercostal fields. On 1 cm^2 of the blade I counted 20967 minute areas of tissue, enclosed by sclereids. The length of the sclereids per cm^2 of the lamina reaches almost 274 cm.

The vascular bundle network of *Manilkara bidentata*/"Purguo" (6 m high), which is very complicated due to the sclereids, consists of 100 vein islets/cm^2 of the blade. That corresponds to −490 meshes in relation to the $leaf_{AT}$ (590 islets).

While the $leaf_{YT}$ of the "Purguo" tree shows one ramification on average (exact value 1.4), the number of the $leaf_{AT}$ displays two ramifications (exact value 1.6). The comparison of the numbers to the first significant figure reveals only an insignificant deviation. The comparison of the degree of ramification for *Ecclinusa guianensis* is therefore more distinctive. In the leaves of the young tree one branch (0.8) appears on average, whereas the $leaves_{AT}$ possess two (2.1). In the islets of *Chrysophyllum sp.* 1—3 ramifications are occasionally presents. The $leaf_{YT}$ of *Chrysophyllum auratum* also occasionally reveals one ramification.

Within the group of $leaves_{YT}$ free veinlet terminations are only distinguishable in the two *Chrysophyllum* species. They possess almost the same number of terminations, i.e. comparatively few: *C. sp.* 170 and *C. auratum* 130 (640 for the $leaf_{AT}$).

Within the sample collection of the young

Table 14. Direct comparison: Measurements of the venation of leaves of adult and young trees.

Species	Leaf area (cm^2)	Number of secondaries/ leaf	Area index (cm^2)	Mean distance of secondaries (cm)	Vein islets/ cm^2	Ramifications/ islet	Veinlet termina- tions/cm^2	Vein length/cm^2 (cm)
Manilkara bidentata								
"Purguo" (A)	47.64	71.27	0.66	0.37	590	1.6	1360	48.66
"Purguo" (6 m high)	/100.4/	/83.0/	/1.21/	/0.44/	100	1.4	cannot be differentiated	36.54
Difference	52.76	11.73	0.55	0.07	490	0.2		12.12
leaf_{AT}	−110.75%	−16.46%	−83.33%	−18.92%	+83.05%	+12.5%		+24.91%
leaf_{YT}	+52.55%	+14.13%	+45.45%	+15.91%	−490%	−14.29%		−33.17%
Factory	2.11×	1.16×	1.83×	1.19×	16.95%	87.5%		75.09%
Ecclinusa guianensis								
"Chicle" (A)	24.61	40.3	0.59	0.45	268	2.1	710	60.61
"Chicle" (1.5 m high)	/52.7/	/53.0/	/0.99/	/0.42/	173	0.8	cannot be differentiated	31.96
Difference	28.09	12.7	0.4	0.03	95	1.3		28.65
leaf_{AT}	−114.14%	−31.51%	−67.8%	+6.67%	+35.45%	+61.9%		+47.27%
leaf_{YT}	+53.3%	+23.96%	+40.4%	−7.14%	−54.91%	−162.5%		−89.64%
Factor	2.14×	1.32×	1.68×	93.33%	64.55%	38.1%		52.73%
Chryso- phyllum auratum								
"Caimito morado" (A)	134.43	35.5	3.69	1.07	940	cannot be differentiated	640	42.27
"Caimito morado" (7 m high)	159.09	37.5	4.21	1.02	1040	sporadically 1 ramification	130	72.23
Difference	24.66	2.0	0.52	0.05	100		510	29.96
leaf_{AT}	−18.34%	−5.63%	−14.09%	+4.67%	−10.64%		+79.69%	−70.88%
leaf_{YT}	+15.5%	+5.33%	+12.35%	−4.9%	+9.62%		−392.31%	+41.48%
Factor	1.18×	1.06×	1.14×	95.33%	1.11×		20.31%	1.71×

The numerical value '/. . ./' does not indicate the average, as only one single sample leaf was available.
Difference: "leaf_{AT}" = difference between the leaf of the adult tree and the leaf of the young tree.
"leaf_{YT}" = difference between the leaf of the young tree and the leaf of the adult tree.
"Factor" means the relation between the leaf of the young tree and that of the adult tree.

plants the mean vein length per cm^2 of the leaf blade of *Chrysophyllum* is distinctly higher than for *Manilkara bidentata*/"Purguo" and *Ecclinusa guianensis*. The leaf$_{YT}$ of *Chrysophyllum auratum* has the greatest count with 72.23 cm vein length per cm^2 of the lamina (ca. 30 cm more than the leaf$_{AT}$). The vein length of *Chrysophyllum sp.* shows a somewhat lower value: 67.75 cm/cm^2. Roughly half the previous length was measured on "Purguo" (36.54 cm, −12.12 in relation to the leaf$_{AT}$) and *E. guianensis* (31.96 cm, −28.65 in relation to the leaf$_{AT}$).

Summary. The leaves of young plants of the Sapotaceae are on average larger than the leaves of the adult trees, and they have more secondary veins per leaf. Consequently, their area indexes are higher. The ramification number per vein islet, however, is smaller in the sample collection of the leaves$_{YT}$. The remaining four investigational criteria fail to reveal any uniform tendencies with regard to the leaves of the young plants.

3.4. Lauraceae

Taxonomical utilization of the characteristics compared with leaf anatomy

Taxonomical characteristics of the Lauraceous venation

The 11 Lauraceous species studied possess a medianus as well as closed venation. Apart from *Aniba riparia, Aniba excelsa* and "Laurel canelo", whose venation type is a brochidodromous one, the "nervatio camptodroma" dominates. *Ocotea duotincta* and *Endlicheria cocuirey* show a venation type which occupies a mean position between the two forms mentioned.

The transitional forms ranging from brochidodromous to actinodromous patterns, observed by Kim and Kim (1984) in Korean specimens of the Lauraceae are not present among the 11 species from the tropical Venezuelan Guyana.

The medianus of *Ocotea nicaraguensis* and *Aniba excelsa* is extremely well developed compared with the thickness of the secondary veins. The remaining species reveal a comparatively thick medianus.

This continuous, uninterrupted primary vein, present in all 11 species, divides the blade into two equal parts, except in *Ocotea martiana* and *Beilschmiedia curviramea*, where nearly equally large areas are present.

A common characteristic of the medianus of the investigated Lauraceae is its regular narrowing from the base of the leaf to the tip and its termination there. Apart from *Ocotea martiana* and "Laurel canelo" the medianus displays a slight curvature either along its course or near the tip of the leaf.

The secondary veins are generally characterized by a usually slightly alternate arrangement, partly in conjunction with an opposite position.

Without exception the secondaries of all Lauraceous species studied are thinner in comparison to the medianus.

When comparing the secondaries the thickest secondary veins are found in the middle of the leaf, becoming thinner towards the base and the tip (*Ocotea duotincta, O. martiana, O. nicaraguensis, Aniba, Nectandra grandis, Beilschmiedia curviramea*, "Laurel rastrojero"). With *Ocotea aff. subalveolata* and *Endlicheria cocuirey* the thickness decreases from the base to the tip. The secondaries of "Laurel canelo" have the same width throughout the whole lamina.

The two nervation types found in the present species — "nervatio camptodroma" and "brochidodroma" — appear as pure or dentate camptodromous (*Ocotea martiana, O. nicaraguensis, Nectandra grandis*, "Laurel rastrojero" and *Ocotea aff. subalveolata, Beilschmiedia curviramea*), respectively. *Aniba* and "Laurel canelo" show the curved brochidodromous type. The leaves of *Ocotea duotincta* and *Endlicheria cocuirey* (only a single sample leaf) have the characteristics of the camptodromous as well as of the brochidodromous nervation. The venation pattern, classified as pure camptodromous, is characterized by secondary veins which reach the leaf margin in a remarkably steep manner. The secondaries end in

Table 15. Leaf material investigated.

Family	Genus	Species	Vernacular name	Tree height	Individuals/ number of the leaves studied	No.[1]
Lauraceae	Ocotea	duotincta C. K. Allen	Laurel verde	a	207/1	X – 46 (337)
			Laurel verde	5 m	5	X – 46 (337)
	Ocotea	martiana (Nees) Mez	Laurel baboso	A	7/5	D – 379 (326)
	Ocotea	nicaraguensis Mez	Laurel blanco	A	43/5	D – 336 (328)
			Laurel blanco	11 m	1	D – 336 (328)
	Ocotea	aff. subalveolata C. K. Allen	Laurel paraguito	a	5	(339)
	Aniba	riparia (Nees) Mez	Laurel amarillo	A	34/5	D – 359 (325)
	Aniba	excelsa Kosterm.	Laurel Rollet	A	134/5	X – 117 (335)
	Nectandra	grandis (Mez) Kosterm.	Laurel	A	20/5	L – 4 (324)
	Beilschmiedia	curviramea (Meissn.) Mez	Aguacatillo moises	a	139/9	D – 275 (5)
			Aguacatillo moises	10 m	1	D – 275 (5)
	Endlicheria	cocuirey Kosterm.	Laurel negro	?	1	X – 1 (331)
			Laurel canelo	A?	5	?
			Laurel rastrojero	?	5	X – 183 (334)

[1] cf. Table 2

Table 16. Venation type, patterns of the primary veins.

No.	Species	Venation pattern: Name	Venation pattern: Formation of the margin	Patterns of the primaries: Relative thickness	Patterns of the primaries: Leaf symmetry	Course	Termination
1	Ocotea duotincta "Laurel verde"	mixture of campto- and brochidodromous	closed	thick	equally large areas	slightly curved; regular narrowing	ending at the tip
2	Ocotea martiana	camptodromous	"	"	almost equally large areas	straight; regular narrowing	"

Table 16 (Continued)

		Venation pattern		Patterns of the primaries			
No.	Species	Name	Formation of the margin	Relative thickness	Leaf symmetry	Course	Termination
	"Laurel baboso"						
3	Ocotea nicaraguensis "Laurel blanco"	"	"	extremely thick	equally large areas	slightly curved towards the tip; regular narrowing	"
4	Ocotea aff. subalveolata "Laurel paraguito"	"	"	thick	"	"	"
5	Aniba riparia "Laurel amarillo"	brochidodromous	"	"	"	"	"
6	Aniba excelsa "Laurel Rollet"	"	"	extremely thick	"	straight to slightly curved; regular narrowing	"
7	Nectandra grandis "Laurel"	camptodromous	"	thick	"	"	"
8	Beilschmiedia curviramea "Aguacatillo moises"	"	"	"	almost equally large areas	"	"
9	Endlicheria cocuirey "Laurel negro"	mixture of campto- and brochidodromous	"	"	equally large areas	slightly curved; regular narrowing	"
10	"Laurel canelo"	brochidodromous	"	"	"	straight; regular narrowing	"
11	"Laurel rastrojero"	camptodromous	"	"	"	"	"

the ramification above, forming anastomoses. In the case of the dentate camptodromous nervation the secondary veins anastomose clearly short of the leaf margin. The most striking character of the curved "nervatio brochidodroma" is the formation of arches of vascular bundles that decrease in size above the loops between the secondaries.

All of the species investigated possess at least some degree of ramification up to the quinternary vein stage, as well as freely ending veinlets. These species include at least quinternary (*Ocotea duotincta, Aniba riparia, Beilschmiedia curviramea*, "Laurel canelo", "Laurel rastrojero") or sesternary veins (*Ocotea aff. subalveolata, Aniba excelsa,*

Table 17. Patterns of the secondary veins.

No.	Species	Position	Rel. thickness to the primary	Rel. thickness among each other	Direction and course	Termination
1	Ocotea duotincta "Laurel verde"	slightly alternate opposite	finer	the thickest ones in the middle, thinner towards the tip and the base	straight to slightly curved course until short of the margin: nervatio camptodroma (pure camptodromous); along the course slight narrowing left half of the leaf with a tendency to the brochidodromous type	loops among the secondaries short of the margin as a rather steep, curved joining of one secondary into the secondary above
2	Ocotea martiana "Laurel baboso"	slightly alternate opposite to alternate	"	"	slightly curved course until short of the margin: nervatio camptodroma (pure camptodromous); along the course slight narrowing	"
3	Ocotea nicaraguensis "Laurel blanco"	opposite, towards the tip partly alternate	"	"	"	"
4	Ocotea aff. subalveolata "Laurel paraguito"	towards the base slightly alternate opposite, towards the middle and the tip more alternate	"	diminishing thicker from the base to the tip	slightly curved course until short of the margin: nervatio camptodroma (dentate camptodromous); along the course slight narrowing	loops and anastomoses among the secondaries short of the margin
5	Aniba riparia "Laurel amarillo"	slightly alternate opposite, partly alternate	finer	the thickest ones in the middle, towards the tip and the base thinner	slightly curved course towards the margin: nervatio brochidodroma (curved brochidodromous); along the course slight narrowing	curled fusion of the secondaries short of the leaf margin; lateral veins of higher orders form loops above those of the secondaries
6	Aniba excelsa "Laurel Rollet"	slightly alternate opposite	"	"	"	"
7	Nectandra grandis "Laurel"	"	"	"	straight to slightly curved course until short of the margin: nervatio camptodroma (pure camptodromous); along the course slight narrowing	loops among the secondaries short of the margin as a rather steep curved joining of one secondary into the secondary above
8	Beilschmiedia curviramea	"	"	"	slightly curved course until short of the	loops and anastomoses among the secondaries

Table 17 (Continued)

No.	Species	Position	Rel. thickness to the primary	Rel. thickness among each other	Direction and course	Termination
	"Aguacatillo moises"				margin: nervatio camptodroma (dentate camptodromous); during the course slight narrowing	short of the margin
9	Endlicheria cocuirey "Laurel negro"	slightly alternate opposite	finer	diminishing thickness from the base to the tip	slightly curved course until short of the margin: nervatio camptodroma (pure camptodromous); along the course slight narrowing; left half of the single sample leaf with a tendency to the brochidodromous type	loops among the secondaries just short of the margin
10	"Laurel canelo"	"	"	equal thickness	straight course towards the margin: nervatio brochidodroma (curved brochidodromous); not narrowing along the course	curled fusion of the secondaries before the leaf margin; lateral veins of higher orders form loops above those of the secondaires
11	"Laurel rastrojero"	"	"	the thickest ones in the middle, thinner towards the base and the tip	slightly curved course until short of the margin: nervatio camptodroma (pure camptodromous); along the course slight narrowing	loops among the secondaries short of the margin as a rather steep curved joining of one secondary into the secondary above

Table 18. Patterns of the veins of higher orders.

No.	Species	Tertiary veins	Quaternary veins	Quinternary veins	Free terminations [1)]	Islet pattern
1	Ocotea duotincta "Laurel verde"	+	+	+	+ 5 and higher orders	reticulate; ± regular polygonal (generally quadrangular to pentagonal); islets partly rectangular; easy to differentiate; division of the larger rectangles into 2 equally sized rectangles; in the middle of the dividing vein mostly ramification of another lateral vein, which ends freely *type*: leaf of the adult tree: A 1, B 4, C 2, D 4, E 3, F 1, G 3 leaf of the young tree: A 1, B 4, C 1, D 4, E 4, F 3, G 2
2	Ocotea martiana "Laurel baboso"	+	+	+	+ 7 and higher orders	reticulate; irregular polygonal (generally penta- to hexagonal); "cocktread-like" islet-pattern; easy to differentiate *type*: A 2, B 3, C 1, D 4, E 5, F 2, G 1
3	Ocotea nicaraguensis	+	+	+	+ 7 and higher	reticulate; irregular polygonal (generally quadrangular to pentagonal); "cocktread-

Table 18 (Continued)

No.	Species	Tertiary veins	Quaternary veins	Quinternary veins	Free terminations [1]	Islet-pattern
	"Laurel blanco"				orders	like" islet-pattern; easy to differentiate *type*: leaf of the adult tree: A 1, B 3, C 2, D 4, E 5, F 1, G 1 leaf of the young tree: A 2, B 3, C 4, D 4, E 6, F 3, G 2
4	Ocotea aff. subalveolata "Laurel paraguito"	+	+	+	+ 6 and higher orders	reticulate; irregular polygonal (generally quadrangular to pentagonal); easy to differentiate *type*: A 1, B 3, C 2, D 4, E 6, F 1, G 1
5	Aniba riparia "Laurel amarillo"	+	+	+	+ 5 and higher orders	reticulate; irregular polygonal (generally penta- to hexagonal); easy to differentiate *type*: A 2, B 5, C 4, D 6, E 1, F 1, G 3
6	Aniba excelsa "Laurel Rollet"	+	+	+	+ 6 and higher orders	reticulate; irregular polygonal (generally quadrangular to pentagonal); ± easy to differentiate *type*: A 1, B 4, C 2, D 4, E 3?, F 2, G 3
7	Nectandra grandis "Laurel"	+	+	+	+ 7 and higher orders	irregular polygonal (generally quadrangular to hexagonal); "cock-tread-like" islet-pattern; easy to differentiate *type*: A 2, B 5, C 4, D 6, E 5, F 2, G 1
8	Beilschmiedia curviramea "Aguacatillo moises"	+	+	+	+ 5 and higher orders	irregular polygonal (generally penta- to hexagonal); easy to differentiate *type*: leaf of the adult tree: A 2, B 3, C 4, D 5, E 4, F 1, G 2 leaf of the young tree: A 2, B 3, C 4, D 5, E 4, F 2, G 1
9	Endlicheria cocuirey "Laurel negro"	+	+	+	+ 6 and higher orders	reticulate; irregular polygonal (generally quadrangular to pentagonal); ± easy to differentiate *type*: A 3, B 3, C 4, D 4, E 6, F 3, G 2
10	"Laurel canelo"	+	+	+	+ 5 and higher orders	reticulate; irregular polygonal (generally quadrangular to pentagonal); easy to differentiate; frequent transversal division of larger rectangular islets; from the dividing vein mostly ramification of a lateral vein, which ends freely *type*: A 1, B 4, C 4, D 4, E 3, F 1, G 2
11	"Laurel rastrojero"	+	+	+	+ 5 and higher orders	reticulate; irregular polygonal (generally quadrangular to pentagonal); easy to differentiate; frequent transversal division of larger rectangular islets; from the dividing vein mostly ramification of a lateral vein, which ends freely *type*: A 1, B 4, C 1, D 4, E 3, F 1, G 3

[1] The number indicates the degree of ramification.

Endlicheria cocuirey). *Ocotea martiana, O. nicaraguensis* and *Nectandra grandis* even have septernaries as freely ending terminations.

Although the Lauraceous species in general have a reticulate venation, consisting of irregular polygonal islets, the patterns differ considerably. The meshes, especially of *Ocotea duotincta*, "Laurel canelo" and "Laurel rastrojero", are easy to differentiate. The larger rectangular areas are often divided into two smaller squares. A remarkable venation pattern can be found in *Ocotea martiana, O. nicaraguensis* and *Nectandra grandis*: the ramifications and terminations, penetrating into the islets, resemble the footprint of a cock ("cocktread" in the tables).

The comparison of the leaf of the adult and young tree of *Ocotea duotincta* for both kinds of leaf revealed an almost similar pattern of "squares" (cf. Roth, 1984, p. 392), which is larger and has more ramifications in the leaf of the young tree. I interpret this structure of the islets of the leaves of the adult tree as being the result of a retarded surface growth of the intercostal fields, since the leaf area of the adult tree is on average half as large again in comparison with the corresponding leaf of the young tree (cf. Figures 95, 96 and Table 27).

The venation pattern of the leaf of the adult and young tree of *Ocotea nicaraguensis* is basically similar, too. The vein islets and ramifications of the leaf of the juvenile plant are on the whole finer and less symmetrical. The "cocktread" ramification has not yet developed so clearly.

The only essential difference between the venation pattern of the leaf of the adult and young plant of *Beilschmiedia curviramea* likewise lies in the more intensive development of islets and ramifications of the leaves of the adult tree.

The vascular bundles forming the islets of *Aniba riparia* are comparatively fine. The meshes themselves appear to be "edged" with few ramifications.

In the differentiation of the Lauraceous species, the pattern of the vein islets is also the most important criterion for classification.

Ecological significance of the anatomical investigations

Table 19. Measurements of the venation (leaves of adult trees).

Species	Leaf area (cm^2)	Number of secondaries/ leaf	Area[1] index (cm^2)	Mean distance of secondaries (cm)	Vein[2] islets/cm^2	Ramifications/ islet	Veinlet terminations/cm^2	Vein length/cm^2 (cm)
Ocotea martiana	46.28–61.8 –77.32	14.16–16.2 –18.24	2.99–3.8 –4.61	1.59–1.72 –1.85	187–235 –283	0.92–1.7 –2.48	1626–1910 –2194	62.17–64.2 –66.23
"Laurel baboso" (A)	(±25.11%)	(±12.59%)	(±21.32%)	(±7.56%)	(±20.32%)	(±45.88%)	(±14.87%)	(±3.16%)
Ocotea nicaraguensis	34.28–64.6 –94.92	17.59–19.0 –20.41	1.97–3.31 –4.65	0.92–1.25 –1.58	340–385 –430	1.0	654–820 –986	44.41–48.58 –52.75
"Laurel blanco" (A)	(±46.93%)	(±7.42%)	(±40.48%)	(±26.4%)	(±11.69%)		(±20.24%)	(±8.58%)
Aniba riparia	43.34–66.8 –90.26	20.55–22.4 –24.25	2.08–2.92 –3.76	1.07–1.24 –1.41	169–200 –231	0.11–0.6 –1.09	30–130 –230	27.88–28.67 –29.46

Table 19 (Continued)

Species	Leaf area (cm^2)	Number of secondaries/ leaf	Area[1] index (cm^2)	Mean distance of secondaries (cm)	Vein[2] islets/cm^2	Ramifications/ islet	Veinlet terminations/cm^2	Vein length/cm^2 (cm)
"Laurel amarillo" (A)	(±35.12%)	(±8.26%)	(±28.77%)	(±13.71%)	(±15.75%)	(±81.67%)	(±76.92%)	(±2.76%)
Aniba excelsa	118.77—168.2 —217.63	38.76—40.8 —42.84	2.81—4.15 —5.49	1.15—1.44 —1.73	95—115 —135	1.0	cannot be differentiated	21.68—23.6 —25.52
"Laurel Rollet" (A)	(±29.39%)	(±5.0%)	(±32.29%)	(±20.14%)	(±17.39%)			(±8.14%)
Nectandra grandis	25.49—38.0 —50.51	10.6—11.4 —12.2	2.29—3.33 —4.37	2.06—2.32 —2.58	212—255 —298	1.06—1.7 —2.34	845—990 —1135	51.4—55.34 —59.28
"Laurel" (A)	(±32.92%)	(±7.02%)	(±31.23%)	(±11.21%)	(±16.86%)	(±37.65%)	(±14.65%)	(±7.12%)
"Laurel canelo" (A?)	8.2—10.12 —12.04	19.21—21.0 —22.79	0.39—0.48 —0.57	0.44—0.49 —0.54	843—970 —1097	0.6—0.9 —1.2	585—770 —955	62.48—64.83 —67.18
	(±18.97%)	(±8.52%)	(±18.75%)	(±10.2%)	(±13.09%)	(±33.33%)	(±24.03%)	(±3.62%)
Ocotea duotincta	49.3	25	1.97	1.08	342—420 —498	0.11—0.6 —1.09	540—710 —880	31.5—37.59 —43.68
"Laurel verde" (a)					(±18.57%)	(±81.67%)	(±23.94%)	(±16.2%)
Ocotea aff. subalveolata	10.64—25.6 —40.56	11.47—12.8 —14.13	1.76—2.4 —3.04	1.1—1.36 —1.62	507—590 —673	0.83—1.5 —2.17	1022—1150 —1278	50.83—52.57 —54.31
"Laurel paraguito" (a)	(±58.44%)	(±10.39%)	(±26.67%)	(±19.12%)	(±14.07%)	(±44.67%)	(±11.13%)	(±3.31%)
Beilschmiedia miedia curviramea	22.83—35.46 —48.09	37.96—42.33 —46.7	0.6—0.83 —1.06	0.5—0.6 —0.7	123—150 —177	1.2—1.8 —2.4	476—590 —704	43.82—45.46 —47.1
"Aguacatillo moises" (a)	(±35.62%)	(±10.32%)	(±27.71%)	(±16.67%)	(±18.33%)	(±33.33%)	(±19.32%)	(±3.61%)
"Laurel rastrojero" (?)	48.48—58.2 —67.92	20.54—24.0 —27.46	1.98—2.46 —2.94	1.02—1.17 —1.32	279—315 —351	0.11—0.6 —1.09	147—280 —413	36.77—40.11 —43.45
Endlicheria cocuirey	(±16.7%)	(±14.42%)	(±19.51%)	(±12.82%)	(±11.35%)	(±81.67%)	(±47.5%)	(±8.33%)
"Laurel negro" (?)	31.1	19	1.64	1.08	144—160 —176	0.84—1.3 —1.76	588—710 —832	47.98—48.22 —48.46
					(±10.31%)	(±35.38%)	(±17.18%)	(±0.5%)

[1] Quotient leaf area: number of secondaries

[2] The vein islet number is based on round figures, while the percentage refers to the mathematical averages.

Table 20. Position of the species within the criteria investigated (leaves of adult trees).[1)]

Position	Leaf area (cm²)	Number of secondaries/ leaf	Area index (cm²)	Mean distance of secondaries (cm)	Vein islets/cm²	Ramifications/ islet	Veinlet terminations/cm²	Vein length/cm² (cm)
1.	A. excelsa	B. curviramea	A. excelsa	N. grandis	"Laurel canelo"	B. curviramea	O. martiana	"Laurel canelo"
2	A. riparia	A. excelsa	O. martiana	O. martiana	O. aff. subalveolata	N. grandis O. martiana	O. aff. subalveolata	O. martiana
3	O. nicaraguensis	O. duotincta	N. grandis	A. excelsa	O. duotincta	O. aff. subalveolata	N. grandis	N. grandis
4	O. martiana	"Laurel rastrojero"	O. nicaraguensis	O. aff. subalveolata	O. nicaraguensis	E. cocuirey	O. nicaraguensis	O. aff. subalveolata
5	"Laurel rastrojero"	A. riparia	A. riparia	O. nicaraguensis	"Laurel rastrojero"	A. excelsa O. nicaraguensis	"Laurel canelo"	O. nicaraguensis
6	O. duotincta	"Laurel canelo"	"Laurel rastrojero"	A. riparia	N. grandis	"Laurel canelo"	O. duotincta E. cocuirey	E. cocuirey
7	N. grandis	O. nicaraguensis E. cocuirey	O. aff. subalveolata	"Laurel rastrojero"	O. martiana	"Laurel rastrojero" A. riparia O. duotincta	B. curviramea	B. curviramea
8	B. curviramea	O. martiana	O. duotincta	O. duotincta E. cocuirey	A. riparia		"Laurel rastrojero"	"Laurel rastrojero"
9	E. cocuirey	O. aff. subalveolata	E. cocuirey	B. curviramea	E. coruirey		A. riparia	O. duotincta
10	O. aff. subalveolata	N. grandis	B. curviramea	"Laurel canelo"	B. curviramea			A. riparia
11	"Laurel canelo"		"Laurel canelo"		A. excelsa			A. excelsa

[1)] Decreasing values from 1—11.

Approximate total vein lengths and conductive volumes/cm² (leaves of adult trees)

The approximate total vein length (average of the size of the prepared leaf of the adult tree × vein length/cm² of the lamina) results in the following data:

Table 21. Approximate total vein length.

Species	Total vein length (cm)
Ocotea martiana	5521.20
Ocotea nicaraguensis	4420.78
Aniba riparia	2379.61
Aniba excelsa	4609.08
Nectandra grandis	3209.72
"Laurel canelo"	790.93
Ocotea duotincta	1853.19
Ocotea aff. subalveolata	2050.23
Beilschmiedia curviramea	2245.72
"Laurel rastrojero"	2887.92
Endlicheria cocuirey	1499.64

Results of the measurements of the Lauraceous venation (leaves of adult trees)

The range of the average leaf size extends from ca. 168 cm² (*Aniba excelsa*) to ca. 10 cm² ("Laurel canelo"). The second *Aniba* species (*A. riparia*) takes second place with a mean leaf size of ca. 67 cm², followed by *Ocotea nicaraguensis* (ca. 65 cm²), *O. martiana* (ca. 62 cm²), "Laurel rastrojero" (ca. 58 cm²) and *O. duotincta* (ca. 50 cm² — value from the single sample leaf). The leaves of adult trees of *Nectandra grandis, Beilschmiedia curviramea* and *Endlicheria cocuirey* have blades ranging between 40 and 30 cm²: 38 cm² (*N. grandis*), ca. 35 cm² (*B. curviramea*), ca. 31 cm² (the single sample leaf of *E. cocuirey*). With an area of ca. 26 cm² the leaves of *Ocotea aff. subalveolata* are remarkably smaller than the remaining ones of the genus. On average the smallest leaves are found in "Laurel canelo" (ca. 10 cm²).

Table 22. Approximate conductive volume per cm^2 of the lamina (cf. Section 3.1).

Species	Gross conductive vol. (cm^3)	Net conductive vol. (cm^3)[1)]
Ocotea martiana	0.0005	0.00004 (= 8.0%)
Ocotea nicaraguensis	0.0017	0.0002 (= 11.76%)
Aniba riparia	0.0004	0.00003 (= 7.5%)
Aniba excelsa	0.0008	0.0001 (= 12.5%)
Nectandra grandis	0.0032	
"Laurel canelo"	0.0010	
Ocotea duotincta	0.0012	0.0002 (= 16.67%)
Ocotea aff. subalveolata	0.0012	
Beilschmiedia curviramea	0.0023	0.0001 (= 4.35%)
"Laurel rastrojero"	0.0007	0.0001 (= 14.29%)
Endlicheria cocuirey	0.0021	
	0.0006—0.0014—0.0022 ($\pm$57.14%)	0.00004—0.0001—0.00016 ($\pm$60%)

[1)] After deduction of the estimated non-conductive tissues (cf. Section 3.1).

Cross-sections of vascular bundles (schematized)
- xylem dotted -

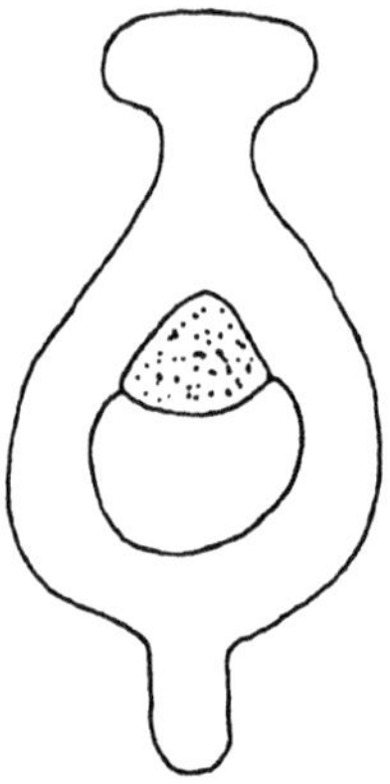

ca. 7.5% xylem

Fig. 26. Ocotea martiana.

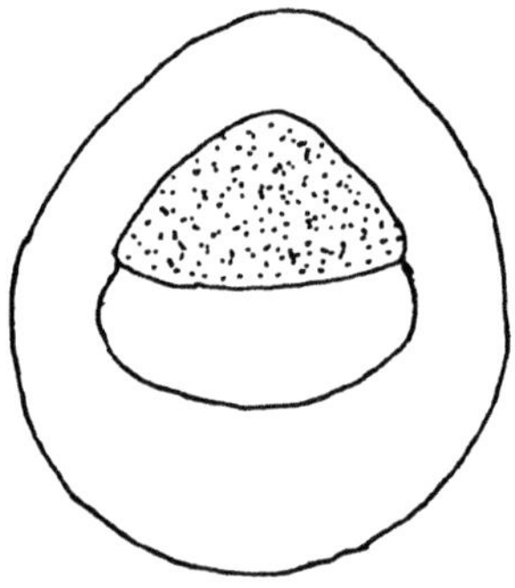

ca. 16.5% xylem

Fig. 28. Ocotea duotincta (leaf of the adult tree).

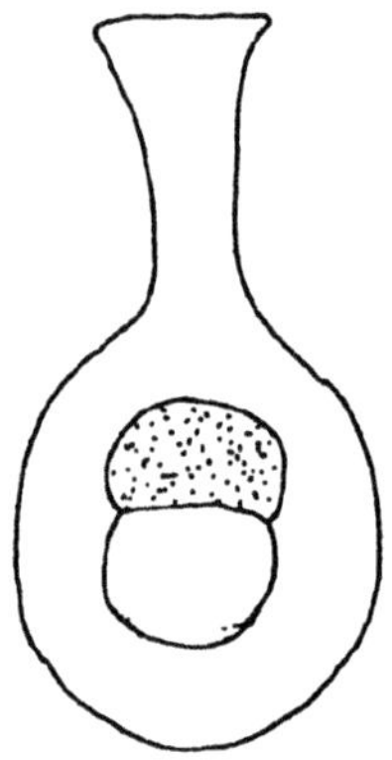

ca. 10% xylem

Fig. 27. Ocotea nicaraguensis (leaf of the adult tree).

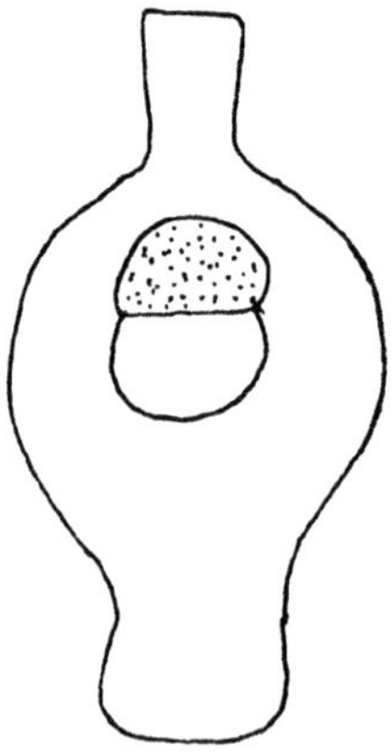

ca. 7% xylem

Fig. 29. Aniba riparia.

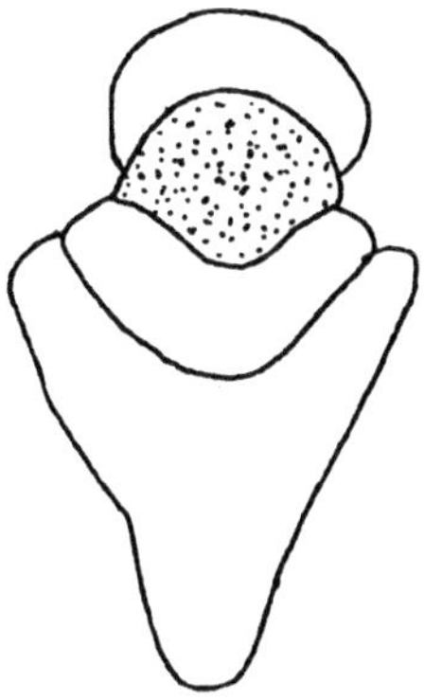

Fig. 30. Aniba excelsa.

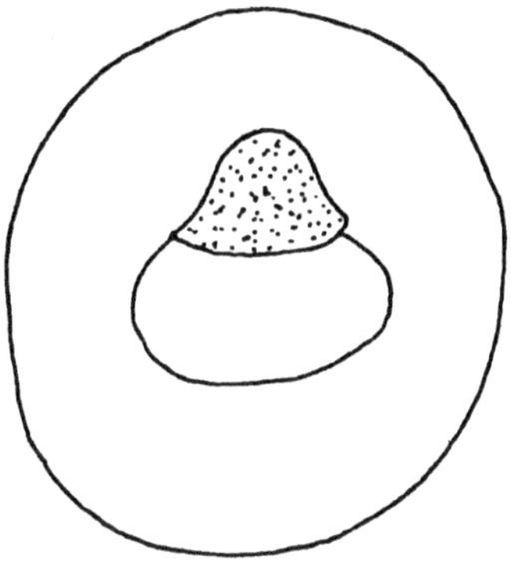

Fig. 31. Beilschmiedia curviramea (leaf of the adult tree).

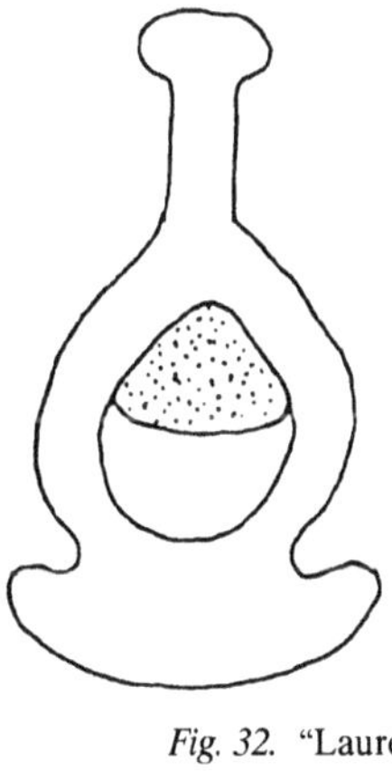

Fig. 32. "Laurel rastrojero".

In relation to the other nine species measured, *Beilschmiedia curviramea* (ca. 42 secondaries) and *Aniba excelsa* (ca. 41) have a high number of secondary veins. One of the four *Ocotea*-species (*O. duotincta*) follows with 25 secondaries. Other species in the 30 to 20 secondary ramifications grouping are: "Laurel rastrojero" (24), *Aniba riparia* (ca. 22) as well as "Laurel canelo" (21). The last group is made up of *Ocotea nicaraguensis* and *Endlicheria cocuirey* (both 19), *Ocotea martiana* (ca. 16), *O. aff. subalveolata* (ca. 13) and *Nectandra grandis* (ca. 11). These numerical values can be included in the 20 to 10 secondaries range.

The smallest area of the lamina supplied by one secondary vein was measured for "Laurel canelo" and *Beilschmiedia curviramea*, which with 0.48 cm^2 and 0.83 cm^2 are clearly under 1 cm^2. *Endlicheria cocuirey* follows with an index of 1.64 cm^2. One secondary of *Ocotea duotincta* supplies almost 2 cm^2 (exact value 1.97 cm^2). *Ocotea aff. subalveolata* (2.4 cm^2), "Laurel rastrojero" 2.46 cm^2) and *Aniba riparia* (2.92 cm^2) range between 2 and 3 cm^2. *O. nicaraguensis* (3.31 cm^2), *Nectandra grandis* (3.33 cm^2) and *O. martiana* (3.8 cm^2) form the next group. Finally, the leaves of adult trees of *Aniba excelsa* possess the smallest number of secondaries in relation to the leaf blade (index = 4.15 cm^2). Within the specimens of the genus *Ocotea* as well as in *Aniba* the averages of the sample species differ remarkably.

As with the smallest area-indexes, "Laurel canelo" and *Beilschmiedia curviramea* (0.49 cm, 0.6 cm) possess the smallest mean distances between the secondary veins. *Ocotea duotincta*, "Laurel negro" (both 1.08 cm) and "Laurel rastrojero" (1.17 cm) also belong in the same range. The next group comprises *Aniba riparia, Ocotea nicaraguensis* as well as *O. aff. subalveolata* (distances of 1.24 cm, 1.25 cm and 1.36 cm). The next higher values are found in *Aniba excelsa* (1.44 cm) and *Ocotea martiana* (1.72 cm). Lastly, *Nectandra grandis* possesses the greatest mean distances between the secondaries (2.32 cm).

Of the 50 leaves studies — at least in one half of the blade — 82% (41 leaves) have the relatively largest distances between the secondaries in the middle of the leaf.

The densest veinlet network is recorded for "Laurel canelo" (970 vein islets per cm^2 of the lamina). *Ocotea aff. subalveolata* (590), *O. duotincta* (420) and *O. nicaraguensis* (385) join. After "Laurel rastrojero" (315) and *Nectandra grandis*

(255) the fourth *Ocotea* species (*O. martiana*) follows with 235 islets/cm^2. The mean value of *Aniba riparia* is 200 vein islets. *Endlicheria cocuirey* (160) and *Beilschmiedia curviramea* (150) respectively possess 40 and 50 islets fewer. *Aniba excelsa* (115 islets) takes the last place. Apart from *O. martiana*, the high vein islet number, according to the results, is an ecologic-morphologic character of the leaves of adult trees of the genus *Ocotea*, which were collected in Venezuelan Guyana.

The mathematical mean values of the ramifications per vein islet range from 2 to 1 single ramification (1.8 and 0.6). While *Beilschmiedia curviramea, Nectandra grandis, Ocotea martiana* and *O. aff. subalveolata* possess an average of 2 ramifications (correct data: 1.8—1.7—1.7—1.5), the vein islets of *Endlicheria cocuirey* show 1, partly 2 ramifications (Ø = 1.3). *Ocotea nicaraguensis* and *Aniba excelsa* have generally one branch (Ø = 1). For "Laurel canelo" a rate of ramification of 90% — on the basis of 1 ramification/islet — was counted, giving an average of 0.9. Nearly 50% of the meshes of *Ocotea duotincta, Aniba riparia* and "Laurel rastrojero" possess one branch (Ø = 0.6).

Although the vein islets of *Nectandra grandis* and *Ocotea martiana* contain on the average 1.7 ramifications, the veinlet termination number of *O. martiana* (1910 terminations/cm^2 of the lamina) is remarkably higher than that of *N. grandis* (990). The 1910 veinlet terminations also represent the maximum for the group including the leaves of adult trees. A second specimen of *Ocotea* (*O. aff. subalveolata*) takes next place (1150 terminations). *N. grandis* (990) precedes *O. nicaraguensis* (820) in the hierarchy. "Laurel canelo" (770) is followed by *Ocotea duotincta* together with *Endlicheria cocuirey* (both 710). With a relatively small distance the 590 terminations/cm^2 of *Beilschmiedia curviramea* take the seventh of nine positions. The difference is much greater between the last two species "Laurel rastrojero" (280 terminations) and *Aniba riparia* (130). For this investigation criterion there was no reliable result for *Aniba excelsa*.

With a mean vein length of 64.83 and 64.2 cm per cm^2 of the leaf blade "Laurel canelo" and *Ocotea martiana* have the greatest supply of veins. The following group comprises *Nectandra grandis* and *Ocotea aff. subalveolata*, whose values range between 50 and 60 cm/cm^2: 55.34 (*N. grandis*) and 52.57 (*O. aff subalveolata*). Almost the same vein lengths are measured for *Ocotea nicaraguensis* — 48.58 cm — and *Endlicheria cocuirey* — 48.22 cm. *Beilschmiedia curviramea* follows with 45.46 cm/cm^2. The mean vein length of "Laurel rastrojero" (40.11 cm) and *Ocotea duotincta* (37.59) belong to the 30 to ca. 40 cm category. With a vein length of 28.67 cm/cm^2 and 23.60 cm, respectively *Aniba riparia* and *Aniba excelsa* possess the weakest vein supply among the leaves of adult Lauraceous trees.

While the form of the cross-sections of the vascular bundles (Figures 26—32) of the seven Lauraceous specimens differs considerably from longish stretched to almost roundish, the xylems show a uniform semicircular to semioval area. Their percentage of the total cross section lies between 6.5% and 16.5%.

The calculation of the net conductive volume including only the xylem leads to net volumes, which make up ca. 4% to ca. 17% of the gross conductive volume.

Approximate total vein lengths and conductive volumes/cm^2 (leaves of young trees)

The approximate total vein length (average of the size of the prepared leaves of young trees × vein length/cm^2 of the lamina) provides the following data:

Table 23. Approximate total vein length.

Species	Total vein length (cm)
Ocotea nicaraguensis	15024.59
Beilschmiedia curviramea	2375.33
Ocotea duotincta	3347.04

Table 24. Approximate conductive volume per cm^2 of the lamina (cf. Section 3.1).

Species	Gross conductive volume (cm^3)
Ocotea nicaraguensis	0.0014
Beilschmiedia curviramea	0.0010
Ocotea duotincta	0.0009
	0.0009—0.0011—0.0013 (± 18.18%)

Table 25. Measurements of the venation (leaves of young trees).

Species	Leaf area (cm²)	Number of secondaries/ leaf	Area[1] index (cm²)	Mean distance of secondaries (cm)	Vein[2] islets/cm²	Ramifications/ islet	Veinlet terminations/cm²	Vein length/cm² (cm)
Ocotea nicaraguensis	271.3	24	11.3	1.82	303—335 —367	1.06—1.7 —2.34	640—780 —920	52.65—55.38 —58.11
"Laurel blanco" (11 m high)					(±9.55%)	(±37.65%)	(±17.95%)	(±4.93%)
Beilschmiedia curviramea	40.5	32	1.27	0.71	103—133 —163	1.0—1.5 —2.0	745—820 —895	58.32—58.65 —58.98
"Aguacatillo moises" (10 m high)					(±22.45%)	(±33.33%)	(±9.15%)	(±0.56%)
Ocotea duotincta	98.2—111.1 —124.0	14.06—16.2 —18.34	6.25—6.87 —7.49	2.26—2.81 —3.36	227—243 —259	0.8—1.1 —1.4	930—1050 —1170	32.90—34.47 —36.04
"Laurel verde" (5 m high)	(±11.61%)	(±13.21%)	(±9.02%)	(±19.57%)	(±6.6%)	(±27.27%)	(±11.43%)	(±4.55%)

[1] Quotient leaf area: number of secondaries
[2] The vein islet number is based on round figures, while the percentage refers to the mathematical averages.

Table 26. Position of the species within the criteria investigated (leaves of young trees).[1]

Position	Leaf area (cm²)	Number of secondaries/ leaf	Area index (cm²)	Mean distance of secondaries (cm)	Vein islets/cm²	Ramifications/ islet	Veinlet terminations/cm²	Vein length/cm² (cm)
1	O. nicaraguensis	B. curviramea	O. nicaraguensis	O. duotincta	O. nicaraguensis	O. nicaraguensis	O. duotincta	B. curviramea
2	O. duotincta	O. nicaraguensis	O. duotincta	O. nicaraguensis	O. duotincta	B. curviramea	B. curviramea	O. nicaraguensis
3	B. curviramea	O. duotincta	B. curviramea	B. curviramea	B. curviramea	O. duotincta	O. nicaraguensis	O. duotincta

[1] Decreasing values from 1—3.

Results of the measurements of the Lauraceous venation (comparison between leaves of young and adult trees)[1]

The data of the 5-m-high *Ocotea duotincta* refer to mean values in the 5 sample leaves investigated, while the other two leaves of young plants (*Ocotea nicaraguensis* and *Beilschmiedia curviramea*) were only available as single sample leaves.

With ca. 271 cm² *Ocotea nicaraguensis* displays the largest leaf, compared with the ca. 65 cm² of the blade of the $leaf_{AT}$. The laminas of the $leaves_{YT}$ of *Ocotea duotincta* cover a mean area of 111 cm² (as opposed to ca. 49 cm² of the $leaf_{AT}$). The blade of *Beilschmiedia curviramea* is ca. 41 cm² large; here, the difference with the $leaf_{AT}$ (ca. 35 cm²) is comparatively small.

[1] In this section the name of a species refers without further explanation to the leaves of young trees.
Abbreviations: $leaf_{YT}$ = adult leaf of the young tree
$leaf_{AT}$ = adult leaf of the adult tree

The smallest leaf (*Beilschmiedia curviramea*) exhibits the greatest number of secondary veins, having 32 of these ramifications, whereas only 24 secondaries supply the comparatively largest blade of *Ocotea nicaraguensis*. The lamina of *Ocotea duotincta* is netted with nearly 16 branches of the medianus. For *B. curviramea* the difference of the $leaf_{AT}$ amounts to 10 secondary veins. The $leaf_{YT}$ of *O. nicaraguensis* has 5 secondaries more and *O. duotincta* has on average ca. 9 secondary veins fewer than the leaf of the adult tree.

With reference to the relation of the leaf-blade to the number of secondaries, the lowest area index results for *Beilschmiedia curviramea* (1.27 cm^2) compared to 0.83 cm^2 of the $leaf_{AT}$. For the $leaf_{YT}$ of *Ocotea nicaraguensis* I calculated the maximum within the group of $leaves_{YT}$ (11.3 cm^2). *Ocotea duotincta* appears in the middle (6.87 cm^2). While the area index of the $leaf_{YT}$ of *O. duotincta* is 4.9 cm^2 greater than that of its $leaf_{AT}$ (1.97 cm^2), the corresponding values of *O. nicaraguensis* vary by 7.99 cm^2 (11.3 cm^2 for the $leaf_{YT}$ in relation to 3.31 cm^2 for the $leaf_{AT}$). Thus, both specimens of the genus *Ocotea* show remarkable differences between the area indices of leaves of young and adult trees, with higher values registered for each $leaf_{YT}$.

The sequence of the number of secondaries is reflected in the mean distance of the secondary

Table 27. Direct comparison: Measurements of the venation of leaves of adult and young trees.

Species	Leaf area (cm^2)	Number of secondaries/ leaf	Area index (cm^2)	Mean distance of secondaries (cm)	Vein islets/cm^2	Ramifications/ islet	Veinlet terminations/cm^2	Vein length/cm^2 (cm)
Ocotea duotincta								
"Laurel verde" (a)	/49.3/	/25.0/	/1.97/	/1.08/	420	0.6	710	37.59
"Laurel verde" (5 m high)	111.1	16.2	6.87	2.81	243	1.1	1050	34.47
Difference	61.8	8.8	4.9	1.73	177	0.5	340	3.12
$leaf_{AT}$	−125.35%	+35.2%	−248.73%	−160.19%	+42.14%	−83.33%	−47.89%	+8.30%
$leaf_{YT}$	+55.63%	−54.32%	+71.32%	+61.57%	−72.84%	+45.45%	+32.38%	−9.05%
Factor	2.25×	64.8%	3.49×	2.6×	57.86%	1.83×	1.48×	91.70%
Ocotea nicaraguensis								
"Laurel blanco" (A)	64.6	19	3.31	1.25	385	1.0	820	48.58
"Laurel blanco" (11 m high)	/271.3/	/24/	/11.3/	/1.82/	335	1.7	780	55.38
Difference	206.7	5	7.99	0.57	50	0.7	40	6.8
$leaf_{AT}$	−319.97%	−26.32%	−241.39%	−45.6%	+12.99%	−70.0%	+4.88%	−14.0%
$leaf_{YT}$	+76.19%	+20.83%	+70.71%	+31.32%	−14.93%	+41.18%	−5.13%	+12.28%

Table 27 (Continued)

Species	Leaf area (cm^2)	Number of secondaries/ leaf	Area index (cm^2)	Mean distance of secondaries (cm)	Vein islets/cm^2	Ramifications/ islet	Veinlet terminations/cm^2	Vein length/cm^2 (cm)
Factor	4.2×	1.26×	3.41×	1.46×	87.01%	1.7×	95.12%	1.14×
Beilschmiedia curviramea								
"Aguacatillo moises" (a)	35.46	42.33	0.83	0.6	150	1.8	590	45.46
"Aguacatillo moises" (10 m high)	/40.5/	/32.0/	/1.27/	/0.71/	133	1.5	820	58.65
Difference	5.04	10.33	0.44	0.11	17	0.3	230	13.19
$leaf_{AT}$	−14.21%	+24.4%	−53.01%	−18.33%	+11.33%	+16.67%	−38.98%	−29.01%
$leaf_{YT}$	+12.44%	−32.28%	+34.65%	+15.49%	−12.78%	−20.0%	+28.05%	+22.49%
Factor	1.14×	75.6%	1.53×	1.18×	88.67%	83.33%	1.39×	1.29×

The numerical value '/. . ./' does not indicate the average, as only one single sample leaf was available.
Difference: "$leaf_{AT}$" = difference between the leaf of the adult tree and the leaf of the young tree.
"$leaf_{YT}$" = difference between the leaf of the young tree and the leaf of the adult tree.
"Factor" means the relation between the leaf of the young tree and that of the adult tree.

veins: here too, *Beilschmiedia curviramea* (0.71 cm) lies above of *Ocotea nicaraguensis* (1.82 cm) and *Ocotea duotincta* (2.81 cm). In this category too, the *Ocotea* species show greater differences with reference to the $leaves_{AT}$ than *B. curviramea*. The $leaf_{YT}$ of *B. curviramea* has mean distances, which are 0.11 cm wider. For *O. nicaraguensis* this difference amounts to +0.57 cm ($leaf_{YT}$ 1.82 cm in relation to 1.25 cm of the $leaf_{AT}$), whereas for *O. duotincta* a greater distance of 1.73 cm was measured ($leaf_{YT}$ 2.81 cm, $leaf_{AT}$ 1.08 cm).

For the seven leaves of young plants investigated, it may be stated that the secondaries of the middle of the leaf are the farthest from each other, compared to the secondary veins of the base and the tip.

The enumeration of the vein islets per cm^2 of the lamina resulted in the highest vein islet numbers for the largest leaves: *Ocotea nicaraguensis* 335 and *Ocotea duotincta* 243 meshes/cm^2. The difference to the corresponding $leaf_{AT}$ for *O. nicaraguensis* (−50) is slight. The $leaf_{YT}$ of *O. duotincta* however, on 1 cm^2 of the leaf-blade, possesses on average 177 islets fewer. Within the group of leaves of young trees *Beilschmiedia curviramea* (133 islets) is last in the group. The difference from the 150 vein islets of the $leaf_{AT}$ (−17) proves to be relatively insignificant.

In one vein islet of *Ocotea nicaraguensis* (average = 1.7) 1—3 ramifications appear in comparison to one branch of the $leaf_{AT}$. *Beilschmiedia curviramea* displays 1—2 ramifications per islet in the relation 1 : 1, i.e. 1.5 as mean value ($leaf_{AT}$ 1.8). From the vein islets of *Ocotea duotincta* generally one single ramification emerges; 10% of the areas of the leaves studied show two branches (mathematical average 1.1 in relation to 0.6 for the $leaf_{AT}$).

The data for the veinlet terminations per cm^2 of the blade differ only slightly. *Ocotea duotincta* has 1050 terminations on average (+340 compared with the $leaf_{AT}$). *Beilschmiedia curviramea* follows with 820 (+230), and the $leaf_{YT}$ of *Ocotea nicaraguensis* with 780 (−40) has developed the least number of veinlet-terminations in its group. This last number is remarkable, because of the comparatively small difference in number to the leaf of the adult plant (820 veinlet terminations/ cm^2).

With a vein length of 58.65 cm per cm^2 of the lamina the foliage leaf of the 10 m high *Beilschmiedia curviramea* produces the densest vascular system among the leaves of young trees. In comparison to the $leaf_{AT}$ the vascular bundles are longer by 13.19 cm/cm^2. The $leaf_{YT}$ of *Ocotea nicaraguensis* is equally endowed with its vein length amounting to 55.38—6.8 cm more than the $leaf_{AT}$. Considerably smaller than the data already noted is the vein length of the $leaf_{YT}$ of *Ocotea duotincta*. With a mean value of 34.47 cm/cm^2 the vein length not only takes the last position within the group of young plants, but indicates at the same time a negative difference of -3.12 cm/cm^2 in relation to the leaf of the adult tree.

Summary. The leaves of young trees of the three species investigated possess larger leaf blades than the leaves of adult plants. They have higher area indexes and show wider distances between the secondary veins. The vein islet number of the lamina for all $leaves_{YT}$ is below the numbers of the corresponding $leaf_{AT}$. The remaining four investigational criteria fail to reveal any uniform tendencies with regard to the group of young plants.

The differences ascertained for the pairs of leaves of young and adult trees are attributed to the reduced growth of the laminas of the $leaves_{AT}$.

3.5. Euphorbiaceae

Taxonomical utilization of the characteristics compared with leaf anatomy

Taxonomical characters of the Euphorbiaceous venation

Only two of the 13 species studied belong to the camptodromous venation type. The remaining 11 species have a brochidodromous or (*Conceveiba guyanensis*) a craspedodromous venation pattern.

In contrast to the Sapotaceae and Lauraceae, the closed termination of the veins at the leaf margin is no common anatomical characteristic of the Euphorbiaceous specimens. There is a dissimilarity between the 10 species with closed venation at the margin, and *Conceveiba guyanensis* as well as *Pausandra flagellorhachis*, whose secondaries and ramifications respectively end directly in the margin. At the join the laminas form an indentation in the coarsely dentate leaf margin. This kind of margin can thus be described as a mixture between open and closed nervation or even a transitional form, mostly characterized by closed venation. According to Troll (1938) these characteristics of a "sekundär offenen Nervatur" (Troll, 1938, p. 1092) can be caused by a retardation in the development of the vein islets in the margin. (Although the leaf margin of *Sapium sp.*, too, exhibits a slight dentation, there is no such transition, but rather closed venation.) In contrast, the veins of *Pera schomburgkiana* end completely free at the margin with typical islets and strong terminations.

Apart from the partly open form of the leaf margin, *Conceveiba guyanensis* is also conspicuous because of the small, freely terminating veinlets of higher orders within the Euphorbiaceae studied. *Mabea piriri* also shows similar fine freely ending ramifications at the leaf margin.

The medianus runs from tip to end or occasionally in a camptodromous way and has a regular narrowing from the base to the tip. Apart from *Croton matourensis* which has an extremely thick primary vein, the medianus, in comparison to the secondaries of the other Euphorbiaceous species is thick. The medianus of *Drypetes variabilis, Piranhea longepedunculata* and "Kerosén blanco" divides the blade into two almost equally large areas, whereas the lamina of the remaining species is divided into completely equal parts.

In general, the medianus ends directly at the tip of the leaf. The sample leaves of *Drypetes variabilis* display either a primary vein, terminating at the tip or a ramified subtilization of the same. Finally, in "Kerosén blanco" the medianus ends at the tip or has the ramified form.

Except for three species, the secondaries are slightly opposite to each other. In *Sapium sp.* this arrangement is replaced by a more alternating position towards the base of the leaf. *Piranhea longepedunculata* displays opposite ramifications of the first order and *Conceveiba guyanensis* has a combination of both types.

In comparison to the primary vein the secondaries are finer, those of *Pera schomburgkiana* even much finer.

With regard to the relative thickness of the secondary veins among each other, I found four different classes and a mixed form. *Sapium sp., Mabea piriri* and *M. taquiri* exhibit a regular changing of stronger and finer secondaries, while the strongest secondary veins of *Pogonophora sagotii, Drypetes variabilis* and "Kerosén blanco" appear in the middle of the leaf, becoming finer along their course to the base and tip. A mixture of both types described appears in *Piranhea longepedunculata, Chaetocarpus schomburgkianus, Conceveiba guyanensis* and *Pausandra flagellorhachis*, where the thickest ramifications run in the middle and the finer ones at the base and the tip. In some sample leaves there is a finer vein lying between the thicker ones. The thickness of

Table 28. Leaf material investigated.

Family	Genus	Species	Vernacular name	Tree height	Individuals/ number of the leaves studied	No.[1]
Euphorbiaceae	Sapium	spec. P. Br.	Lechero blanco	A	12/1	D – 9 (341)
	Mabea	piriri Aubl.	Pata de pauji	6 m	1	D – 3 (439)
	Mabea	taquiri Aubl.	Pata de pauji lacreado	?	8	X – 180 (440)
	Pogonophora	sagotii ?	Flor de mayo	A	190/8	X – 40 (215)
	Pera	schomburgkiana (Benth.) Muell. Arg.	Pilón rosado	A	95/8	X – 45 (457)
	Drypetes	variabilis Uittien	Kerosén	A	423/6	D – 119 (318)
			Kerosén	5 m	1	D – 119 (318)
	Hieronyma	laxifolia (Tul.)	Aguacatillo	A	43/8	X – 128 (4)
		Muell. Arg.	Aguacatillo	seedling (1 m)	1	X – 128 (4)
	Piranhea	longepedunculata Jabl.	Caramacate	A (1 m Ø)	33/10	(95)
			Caramacate	A (3 m)	1	(95)
	Chaetocarpus	schomburgkianus (O. Ktze.) Pax & K. Hoffm.	Cacho	A	3679/8	X – 124 (64)
	Conceveiba	guyanensis Aubl.	Nicolás	a	14/11	D – 380 (406)
	Pausandra	flagellorhachis Lanj.	Manglillo	aa	7	(364)
	Croton	matourensis Aubl.	Canelo	A bush (10 cms Ø)	8	(84)
			Kerosén blanco	?	6	?

[1] cf. Table 2

Table 29. Venation type, patterns of the primary veins.

		Venation pattern		Patterns of the primaries			
No.	Species	Name	Formation of the margin	Relative thickness	Leaf symmetry	Course	Termination
1	Sapium sp. "Lechero blanco"	Camptodromous	closed; leaf margin coarsely dentate	thick	equally large areas	slightly curved; regular narrowing	not ascertainable from the single sample leaf
2	Mabea piriri "Pata de pauji" (6 m high)	brochidodromous	closed	〃	〃	straight; regular narrowing	ending at the tip
3	Mabea taquiri "Pata de pauji lacreado"	〃	〃	〃	〃	〃	〃
4	Pogonophora sagotii "Flor de mayo"	camptodromous	〃	〃	〃	slightly curved; regular narrowing	〃
5	Pera schomburgkiana "Pilón rosado"	brochidodromous	open	〃	〃	straight to slight narrowing	〃
6	Drypetes variabilis "Kerosén"	〃	closed	〃	almost equally large areas	〃	ending at the tip or ramified subtilization
7	Hieronyma laxifolia "Aguacatillo"	〃	〃	〃	equally large areas	〃	ending at the tip
8	Piranhea longepedunculata "Caramacate"	〃	〃	〃	almost equally large areas	〃	〃
9	Chaetocarpus schomburgkianus "Cacho"	〃	〃	〃	equally large areas	〃	〃
10	Conceveiba guyanensis "Nicolás"	craspedodromous	closed; transition to open venation where the secondaries run directly into the leaf margin and the margin shows crenatures	〃	〃	〃	〃
11	Pausandra	brochidodromous	〃	〃	〃	〃	〃

Table 29 (Continued)

No.	Species	Venation pattern: Name	Venation pattern: Formation of the margin	Patterns of the primaries: Relative thickness	Patterns of the primaries: Leaf symmetry	Course	Termination
	flagellorhachis "Manglillo"						
12	Croton matourensis "Canelo" (10 cms Ø)	"	closed	extremely thick	"	"	"
13	"Kerosén blanco"	"	"	thick	almost equally large areas	"	ending at the tip or ramified subtilization

Table 30. Patterns of the secondary veins.

No.	Species	Position	Rel. thickness to the primary	Rel. thickness among each other	Direction and course	Termination
1	Sapium sp. "Lechero blanco"	slightly alternate opposite, more alternate towards the base	finer	regular alternating from thicker to thinner ones	curved course until short of the margin: nervatio camptodroma (pure camptodromous); along the course slight narrowing	loops among the secondaries short of the margin and termination in the margin? (not ascertainable from the single sample leaf)
2	Mabea piriri "Pata de pauji" (6 m high)	slightly alternate opposite	"	"	straight course towards the margin; greater approach to the margin towards the base and the tip: nervatio brochidodroma (dentate brochidodromous); along the course slight narrowing	distinctive loops among the secondaries comparatively far from the margin; further clear anastomoses above the loops of the secondaries
3	Mabea taquiri "Pata de pauji lacreado"	"	"	"	"	"
4	Pogonophora sagotii "Flor de mayo"	"	"	the thickest ones in the middle, towards the base and the tip finer	slightly curved course until short of the margin: nervatio camptodroma (pure camptodromous); along the course slight narrowing	loops among the secondaries short of the margin

Table 30 (Continued)

No.	Species	Position	Rel. thickness to the primary	Rel. thickness among each other	Direction and course	Termination
5	Pera schomburgkiana "Pilón rosado"	″	remarkably finer	equal thickness	straight course towards the margin; greater approach to the margin towards the base and the tip: nervatio brochidodroma (dentate brochidodromous); not narrowing along the course	distinctive loops among the secondaries comparatively far from the margin: further clear anastomoses above the loops of the secondaries
6	Drypetes variabilis "Kerosén"	″	finer	the thickest ones in the middle, towards the base and the tip finer	slightly curved course towards the margin; greatest approach to the margin towards the base and the tip: nervatio brochidodroma (dentate brochidodromous); along the course slight narrowing	distinctive loops among the secondaries (in the middle comparatively further from the margin than at the tip and the base); further clear anastomoses — partly 2 fields or ranks slightly dislocated above each other — above the loops of the secondaries
7	Hieronyma laxifolia "Aguacatillo"	″	″	decreasing thickness from the base to the tip	straight, towards the tip slightly curved course until short of the margin: nervatio brochidodroma (pure brochidodromous); along the course slight narrowing	distinctive loops among the secondaries short of the margin
8	Piranhea longepedunculata "Caramacate"	opposite	finer	the thickest ones in the middle, towards the base and the tip finer; occasionally one finer secondary between 2 stronger ones	slightly curved course towards the margin; greater approach to the margin towards the base and the tip: nervatio brochidodroma (dentate brochidodromous); along the course slight narrowing	distinctive loops among the secondaries comparatively far from the margin, further clear anastomoses above the loops of the secondaries
9	Chaetocarpus schomburgkianus "Cacho"	″	″	″	straight course until short of the margin: nervatio brochidodroma (pure brochidodromous); not narrowing along the course	distinctive loops among the secondaries, towards the tip short of the margin, towards the base further from the margin

Table 30 (Continued)

No.	Species	Position	Rel. thickness to the primary	Rel. thickness among each other	Direction and course	Termination
10	Conceveiba guyanensis "Nicolás"	at the base opposite, towards the middle and the tip slightly alternate opposite to alternate	〃	〃	straight course into the margin: nervatio craspedodroma; not narrowing along the course	termination of all, or at least of the stronger secondaries or their ramifications in the margin; leaf margin coarsely dentate
11	Pausandra flagellorhachis "Manglillo"	slightly alternate opposite	〃	〃	straight, towards the tip curved course until almost directly at the margin: nervatio brochidodroma (pure brochidodromous); along the course slight narrowing	distinctive loops among the secondaries short of the margin, leaf margin coarsely dentate
12	Croton matourensis "Canelo" (10 m Ø)	〃	〃	diminishing thickness from the base to the tip	straight course until short of the margin: nervatio brochidodroma (pure brochidodromous); along the course slight narrowing	distinctive loops among the secondaries short of the margin
13	"Kerosén blanco"	〃	〃	the thickest ones in the middle, towards the base and the tip finer	slight, at the tip more curved course towards the margin: greatest approach to the margin towards the base and the tip: nervatio brochidodroma (dentate brochidodromous); along the course slight narrowing	distinctive loops among the secondaries (in the middle comparatively further from the margin than at the tip and the base); further clear anastomoses — occasionally 2 fields or ranks slightly dislocated above each other — above the loops of the secondaries

Table 31. Patterns of the veins of higher orders.

No.	Species	Tertiary veins	Quaternary veins	Quinternary veins	Free terminations [1)]	Islet pattern
1	Sapium sp. "Lechero blanco"	+	+	+	+ 7 and higher orders	irregular, relatively large polygons, which are passed through by finest, very crooked ramifications, partly further divided; thus small quadrangular islets develop; ± easy to differentiate *type*: A 2, B 3, C 4, D 6, E 6, F 2, G 1

Table 31 (Continued)

No.	Species	Tertiary veins	Quaternary veins	Quinternary veins	Free terminations [1]	Islet pattern
2	Mabea piriri "Pata de pauji" (6 m high)	+	+	+	+ 7 and higher orders	irregular polygons (generally penta- to hexagonal) of very different form and size; partly parallel to the direction of the secondaries; easy to differentiate *type*: A 2, B 5, C 1, D 6, E 6, F 2, G 1
3	Mabea taquiri "Pata de pauji lacreado"	+	+	+	+ 7 and higher orders	irregular polygons (generally penta- to hexagonal) of very different form and size; easy to differentiate *type*: A 2, B 3, C 4, D 5, E 6, F 2, G 1
4	Pogonophora sagotii "Flor de mayo"	+	+	+	+ 5 and higher orders	irregular polygonal (quadrangular to pentagonal); mostly 2 ranks of islets between the tertiaries, which run vertically to the secondaries; easy to differentiate *type*: A 3, B 3, C 2, D 2, E 3, F 2, G 2
5	Pera schomburgkiana "Pilón rosado"	+	+	+	+ 5 and higher orders	relatively large, irregular polygons (generally pentagonal); "cocktread-like" ramifications in the islets; easy to differentiate; pattern suggests a tangle of broken branches *type*: A 2, B 5, C 1, D 6, E 5, F 2, G 1
6	Drypetes variabilis "Kerosén"	+	+	+	+ 5 and higher orders	irregular polygons (generally quadrangular to pentagonal) of very different form and size; easy to differentiate *type*: leaf of the adult tree: A 2, B 3, C 4, D 5, E 4, F 2, G 2; leaf of the young tree: A 2, B 5, C 4, D 6, E 5, F 2, G 2
7	Hieronyma laxifolia "Aguacatillo"	+	+	+	+ 7 and higher orders	reticulate; irregular polygons (generally quadrangular to pentagonal); easy to differentiate; "cocktread-like" ramifications in the islets of the leaves of the adult trees *type*: leaf of the adult tree: A 2, B 3, C 3, D 4, E 5, F 3, G 1; leaf of the young tree: A 2, B 5, C 1, D 6, E 6, F 1, G 3
8	Piranhea longepedunculata "Caramacate"	+	+	+	+ sporadic; degree?	very small-meshed, relatively regular vascular bundle network; small irregular polygons (generally quadrangular to pentagonal); very easy to differentiate; islet network suggests dry fissured clay soil *type*: leaves of the adult and young tree: A 1, B 4, C 3, D 4, E 1, F 3, G 3
9	Chaetocarpus schomburgkianus "Cacho"	+	+	+	+ 3 and higher orders; partly direct ramifications of the secondaries	irregular polygonal islets; pentagonal and more angular larger units of islets without principle of order *type*: A 2, B 5, C 1, D 6, E 7, F 3, G 1
10	Conceveiba guyanensis	+	+	+	+ 5 and higher	irregular polygonal (generally quadrangular to pentagonal); mostly 2—3 ranks of islets between the

Table 31 (Continued)

No.	Species	Tertiary veins	Quaternary veins	Quinternary veins	Free terminations [1]	Islet pattern
	"Nicolás"				orders	tertiaries, which run vertically to the secondaries; easy to differentiate *type*: A 3, B 3, C 4, D 2, E 6, F 2, G 1
11	Pausandra flagellorhachis "Manglillo"	+	no differentiation of the veins of higher orders because of the sclereids			irregular, noticeably large polygons; because of the very dense network of sclereids very hard to differentiate *type*: A 2, B 5, C 1, D 6, E 3, F 2, G 3
12	Croton matourensis "Canelo" (10 cm Ø)	+	+	+	+ 3 and higher orders; partly direct ramifications of the secondaries	finest vascular bundles as threadlike, superposing fibres, which form small irregular polygonal islets (generally quadrangular to pentagonal); hard to differentiate because of the stellate hairs on the lower leaf-surface *type*: A 1, B 5, C 4, D 6, E 6, F 2, G 1
13	"Kerosén blanco"	+	+	+	+ 7 and higher	irregular relatively large polygons (generally quadrangular to pentagonal) with dendroid ramifications; easy to differentiate *type*: A 2, B 3, C 2, D 4, E 6, F 3, G 2

[1] The number means the degree of ramification.

the secondaries of *Pera schombirgkiana* remains unchanged throughout the blade, while a decreasing thickness from the base to the tip is observed in the leaves of *Hieronyma laxifolia* and *Croton matourensis.*

The main venation types in the prepared Euphorbiaceous species are the camptodromous and brochidodromous nervation, but with the craspedodromous form in *Conceveiba guyanensis.* The "nervatio craspedodroma" is characterized by a direct termination of the secondaries or their ramifications in the leaf margin. At the joining the margin of *C. guyanensis* shows a coarse dentation. *Sapium sp.* (as well as *Mabea piriri* as a single sample leaf) and *Pogonophora sagotii* reveal a camptodromous course of the secondary veins which anastomose just short of the margin. Those species belonging to the pure brochidodromous type are *Hieronyma laxifolia, Chaetocarpus schomburgkianus, Pausandra flagellorhachis* and *Croton matourensis.* The net formation of anastomoses among the secondaries takes place more or less distant from the leaf margin. The similarly coarsely dentate margin of *P. flagellorhachis* must be emphasized. As the dentation occurs together with the "nervatio brochidodroma" too, this anatomical leaf characteristic cannot be interpreted as an indication or criterion for craspedodromous venation, as could have been concluded from the formation of the margin and the venation of *Conceveiba guyanensis.* Rather it is a feature of the transitional type of closed and open nervation. The venation patterns of the remaining six species can be classified as dentate brochidodromous. The characteristics of this kind of course and termination of the secondaries are to be found in their net anastomoses close to the leaf margin, or a little further from it, together with further anastomoses above these loops.

A really impressive vein network, compared with the remaining species, is present in *Pausandra flagellorhachis.* The thread-like convoluted sclereids are fine and dense in such a manner that a detailed analysis of the degrees of ramification is not possible.

No less striking is the vein islet pattern of the leaf of the young tree of *Croton matourensis.* The hierarchy of veins up to the degree of quinternary ramifications is clearly definable. The free terminations originate to some degree as direct ramifications of the secondary veins.

In *Chaetocarpus schomburgkianus* too, the free terminations are partly formed by direct runners from the secondaries. A network of sclereids composed of extremely fine meshes is typical of this plant.

Freely terminating veinlets occur in all sample species. In *Pogonophora sagotii, Pera schomburgkiana, Drypetes variabilis* and *Conceveiba guyanensis* they can be classified as quinternaries or veins of higher orders. *Sapium sp., Mabea, Hieronyma laxifolia* as well as "Kerosén blanco" display veinlet terminations which consist of ramifications with septernary veins or higher. *Piranhea longepedunculata* is noticeable for the infrequent appearance of free terminations. The net of very small islets is made up of clearly distinct polygons which are remarkable in comparison with the other species of the Euphorbiaceae because of their regular arrangement.

Within the remaining species polygonal types of islets can be found varying considerably in arrangement, form and size together with the corresponding degree of ramification which mostly constitute a typical characteristic of the species.

Typical of *Mabea* are the predominantly dendroid ramifications of the vein islets. The leaf of the young tree of *M. piriri* shows very irregular large meshes. The species *Pera schomburgkiana* too, displays these wide mesh islets but in conjunction with the ramifications reminiscent of cocktreads. The widespread meshes in the leaf of the young tree of *Drypetes variabilis* with the characteristic cocktread-like form of ramification prove to be typical of the species. The leaf of the adult tree has fewer typical vein islets and less pronounced ramifications. The vein network of the seedling of *Hieronyma laxifolia* is also more wide meshed and is supplied with finer veins than the leaf of the adult plant. It has many fewer and also weaker ramifications. The veinlet network on the whole seems to be fragile compared with the leaf of an adult tree.

The most curious venation patterns of the whole sample collection are apparent in *Chaetocarpus schomburgkianus, Pausandra flagellorhachis* and *Croton matourensis* (leaf of the young tree). The veinlet network of *Chaetocarpus schomburgkianus* achieves its specific characteristic by means of a dense net of sclereids, strengthening the leaf texture. The sclereids, observed in *P. flagellorhachis*, which run parallel to the surface of the leaf cause difficulty in distinguishing the veinlet network because of their fine, threadlike structure. The meshes themselves are outstandingly large in comparison to those of the other species. The very irregular islets of *Croton matourensis* together with the stellate hairs of the supper surface form the characteristics that are typical of the species. On the preparation these trichomes penetrate visibly into the islet network.

Summarizing statements are not possible with regard to the classification of the vein islet patterns of the Euphorbiaceae, since my studies have demonstrated that within this family many diverse patterns exist which are especially typical of their species. In comparison with the Sapotaceae and Lauraceae the tropical species of the Euphorbiaceae are the most heterogeneous group (cf. Metcalfe and Chalk, 1965, vol. II, pp. 1207 f.).

Ecological significance of the anatomical investigations (see Tables 32 and 33)

Approximate total vein lengths and conductive volumes/cm² (leaves of adult trees)

The approximate total vein length (average of the size of the prepared leaf of the adult tree × vein length/cm² of the lamina) results in the data given in Tables 34 and 35.

Results of the measurements of the Euphorbaceous venation (leaves of adult trees)

The species *Pausandra flagellorhachis* has on average the largest foliage leaves with ca. 213 cm². With leaf sizes between 100 cm² and 50 cm² *Conceveiba guyanensis* (ca. 90 cm²), *Pogonophora sagotii* (ca. 76 cm²), *Hieronyma laxifolia* (ca. 67 cm²) and "Kerosén blanco" (ca. 57 cm²) follow. *Chaetocarpus schomburgkianus* as well as *Piran-*

Table 32. Measurements of the venation (leaves of adult trees).

Species	Leaf area (cm[2]))	Number of secondaries/ leaf	Area[1]) index (cm²)	Mean distance of secondaries (cm)	Vein[2]) islets/cm²	Ramifications/ islet	Veinlet terminations/cm²	Vein length/cm² (cm)
Sapium spec.	25	51	0.49	0.39	253—275 −297	0.45—1.2 −1.95	cannot be differentiated	54.0—57.65 −61.3
"Lechero blanco" (A)					(±8.09%)	(±62.5%)		(±6.33%)
Pogonophora sagotii	48.19—76.13 −104.07	20.08—22.63 −25.18	2.32—3.31 −4.3	1.09—1.28 −1.47	315—380 −445	0.6—1.7 −2.8	698—890 −1082	68.91—69.89 −70.87
"Flor de mayo" (A)	(±36.7%)	(±11.27%)	(±29.91%)	(±14.84%)	(±17.11%)	(±64.71%)	(±21.57%)	(±1.4%)
Pera schomburgkiana	11.24—17.33 −23.42	17.23—22.25 −27.27	0.64—0.77 −0.9	0.46—0.53 −0.6	89—120 −151	4.74—6.0 −7.26	1125—1370 −1615	62.09—63.24 −64.39
"Pilón rosado" (A)	(±35.14%)	(±22.65%)	(±16.88%)	(±13.21%)	(±26.04%)	(±21.0%)	(±17.88%)	(±1.82%)
Drypetes variabilis	19.25—28.47 −37.69	19.69—22.5 −25.31	0.93—1.24 −1.55	0.73—0.8 −0.87	166—195 −224	0.6—1.4 −2.2	370—470 −570	43.63—46.07 −48.51
"Kerosén" (A)	(±32.38%)	(±12.49%)	(±25.0%)	(±8.75%)	(±15.0%)	(±57.14%)	(±21.28%)	(±5.3%)
Hieronyma laxifolia	41.36—67.11 −92.86	17.62—19.88 −22.14	2.29—3.32 −4.35	1.04—1.21 −1.38	323—345 −367	1.06—1.7 −2.34	1521—1700 −1879	76.43—78.83 −81.23
"Aguacatillo" catillo" (A)	(±38.37%)	(±11.37%)	(±31.02%)	(±14.05%)	(±6.38%)	(±37.65%)	(±10.53%)	(±3.04%)
Piranhea longepedunculata	19.61—31.23 −42.85	30.11—34.4 −38.69	0.61—0.93 −1.25	0.45—0.54 −0.63	3058—3430 −3802	0—1	229—340 −451	97.76—102.82 −107.88
"Caramacate" (1 m Ø) (A)	(±37.21%)	(±12.47%)	(±34.41%)	(±16.67%)	(±10.85%)		(±32.65%)	(±4.92%)
Chaetocarpus schomburgkianus	24.64—35.36 −46.08	32.27—34.0 −35.73	0.74—1.05 −1.36	0.58—0.69 −0.8	56—75 −94	1.22—2.2 −3.18	cannot be differentiated	30.28—31.57 −32.86
"Cacho" (A)	(±30.32%)	(±5.09%)	(±29.52%)	(±15.94%)	(±25.67%)	(±44.55%)		(±4.09%)
Conceveiba guyanensis	56.15—89.62 −123.09	20.75—22.36 −23.97	2.57—4.01 −5.45	1.16—1.5 −1.84	322—373 −424	2.75—4.6 −6.45	1340—1460 −1580	53.17—55.16 −57.15
"Nicolás" (a)	(±37.35%)	(±7.2%)	(±35.91%)	(±22.67%)	(±13.56%)	(±40.22%)	(±8.22%)	(±3.61%)
Pausandra flagellorhachis	127.87—213.29 −298.71	32.74—35.29 −37.84	3.4—6.17 −8.94	1.14—1.54 −1.94	12—14 −16	0—2	cannot be differentiated	15.56—17.52 −19.48
"Manglillo" (aa)	(±40.05%)	(±7.23%)	(±44.89%)	(±25.97%)	(±14.29%)			(±11.19%)
Mabea taquiri	11.06—16.16 −21.26	30.83—34.13 −37.43	0.36—0.46 −0.56	0.36—0.39 −0.42	383—423 −463	1.6—2.9 −4.2	1089—1310 −1531	68.88—71.24 −73.6

Table 32 (Continued)

Species	Leaf area (cm²⁾)	Number of secondaries/ leaf	Area[1] index (cm²)	Mean distance of secondaries (cm)	Vein[2] islets/cm²	Ramifications/ islet	Veinlet terminations/cm²	Vein length/cm² (cm)
"Pata de pauji lacreado" (?)	(±31.56%)	(±9.67%)	(±21.74%)	(±7.69%)	(±9.35%)	(±44.83%)	(±16.87%)	(±3.31%)
"Kerosén blanco" (?)	39.1—57.33 —75.56 (±31.8%)	16.46—19.5 —22.54 (±15.59%)	2.04—2.96 —3.88 (±31.08%)	0.97—1.14 —1.31 (±14.91%)	229—253 —277 (±9.31%)	0.33—1.2 —2.07 (±72.5%)	369—450 —531 (±18.0%)	42.27—43.88 —45.49 (±3.67%)

[1] Quotient leaf area: number of secondaries
[2] The vein islet number is based on round figures, while the percentage refers to the mathematical averages.

Table 34. Approximate total vein length.

Species	Total vein length (cm)
Sapium spec.	1441.25
Pogonophora sagotii	4542.85
Pera schomburgkiana	1188.91
Drypetes variabilis	1547.95
Hieronyma laxifolia	3697.13
Piranhea longepedunculata	3269.68
Chaetocarpus schomburgkianus	1041.81
Conceveiba guyanensis	6111.73
Pausandra flagellorhachis	2960.88
Mabea taquiri	1353.65
"Kerosén blanco"	2391.46

hea longepedunculata lie between 40 and 30 cm² with ca. 35 cm² and ca. 31 cm². *Drypetes variabilis* (ca. 18 cm²) and the single sample leaf of *Sapium sp.* (25 cm²) can be classified under the 30—20 cm² range. With ca. 17 cm² and ca. 16 cm² *Pera schomburgkiana* and *Mabea taquiri* have the smallest blades of the leaves of the adult plants.

With 51 secondary veins *Sapium sp.* takes first place with regard to the second criterion of Table 32. I counted ca. 35 and 34 ramifications of first order, respectively for the leaves of *Pausandra flagellorhachis* and *Piranhea longepedunculata, Mabea taquiri, Chaetocarpus schomburgkianus.* The next group includes *Pogonophora sagotii,*

Table 35. Approximate conductive volume per cm² of the lamina (cf. Section 3.1).

Species	Gross conductive vol. (cm³)	Net conductive vol. (cm³)[1]
Sapium spec.	0.0012	
Pogonophora sagotii	0.0009	0.0002 (= 22.22%)
Pera schomburgkiana	0.0008	0.0003 (= 37.5%)
Drypetes variabilis	0.0009	0.0002 (= 22.22%)
Hieronyma laxifolia	0.0009	
Piranhea longepedunculata	0.0007	
Chaetocarpus schomburgkianus	0.0007	0.0001 (= 14.29%)
Conceveiba guyanensis	0.0009	0.0001 (= 11.11%)
Pausandra flagellorhachis	0.0012	
Mabea taquiri	0.0015	
"Kerosén blanco"	0.0013	
	0.0008—0.0010—0.0012 (±20.0%)	0.00013—0.0002—0.00027 (±35.0%)

[1] After deduction of the estimates non-conductive tissues (cf. Section 3.1).

Table 33. Position of the species within the criteria investigated (leaves of adult trees).[1]

Position	Leaf area (cm^2)	Number of secondaries/ leaf	Area index (cm^2)	Mean distance of secondaries (cm)	Vein islets/cm^2	Ramifications/ islet	Veinlet terminations/cm^2	Vein length/cm^2 (cm)
1	P. flagellorhachis	S. spec.	P. flagellorhachis	P. flagellorhachis	P. longepedunculata	P. schomburgkiana	H. laxifolia	P. longepedunculata
2	C. guyanensis	P. flagellorhachis	C. guyanensis	C. guyanensis	M. taquiri	C. guyanensis	C. guyanensis	H. laxifolia
3	P. sagotii	P. longepedunculata M. taquiri C. schomburgkianus	H. laxifolia	P. sagotii	P. sagotii	M. taquiri	P. schomburgkiana	M. taquiri
4	H. laxifolia	P. sagotii D. variabilis	P. sagotii	H. laxifolia	C. guyanensis	C. schomburgkianus	M. taquiri	P. sagotii
5	"Kerosén blanco"	C. guyanensis P. schomburgkiana	"Kerosén blanco"	"Kerosén blanco"	H. laxifolia	H. laxifolia P. sagotii	P. sagotii	P. schomburgkiana
6	C. schomburgkianus	H. laxifolia "Kerosén blanco"	D. variabilis	D. variabilis	S. spec.	D. variabilis	D. variabilis	S. spec.
7	P. longepedunculata		C. schomburgkianus	C. schomburgkianus	"Kerosén blanco"	"Kerosén blanco" S. spec.	"Kerosén blanco"	C. guyanensis
8	D. variabilis		P. longepedunculata	P. longepedunculata	D. variabilis	P. flagellorhachis	P. longepedunculata	D. variabilis
9	S. spec.		P. schomburgkiana	P. schomburgkiana	P. schomburgkiana	P. longepedunculata		"Kerosén blanco"
10	P. schomburgkiana		S. spec.	S. spec. M. taquiri	C. schomburgkianus			C. schomburgkianus
11	M. taquiri		M. taquiri		P. flagellorhachis			P. flagellorhachis

[1] Decreasing values from 1—11.

Cross-sections of vascular bundles (schematized)
- xylem dotted -

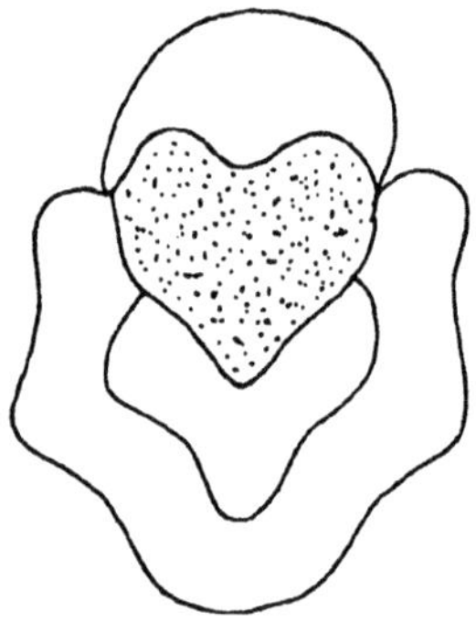

ca. 23% xylem

Fig. 33. Pogonophora sagotii.

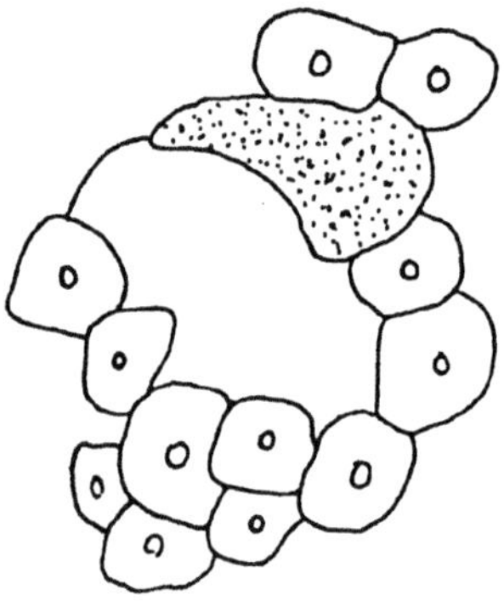

ca. 32.5% xylem

Fig. 34. Pera schomburgkiana.

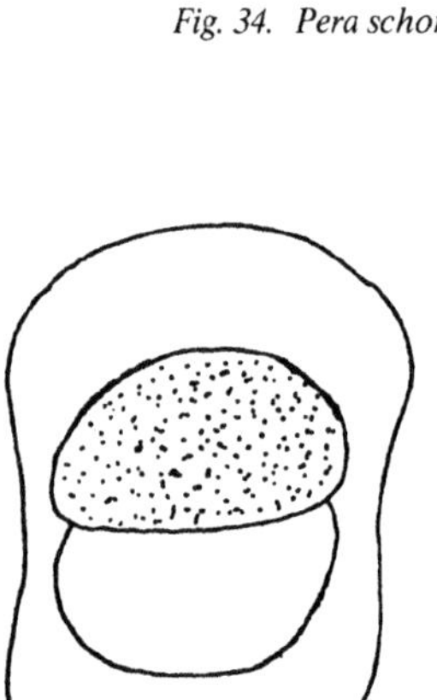

ca. 22.5% xylem

Fig. 35. Drypetes variabilis (leaf of the adult tree).

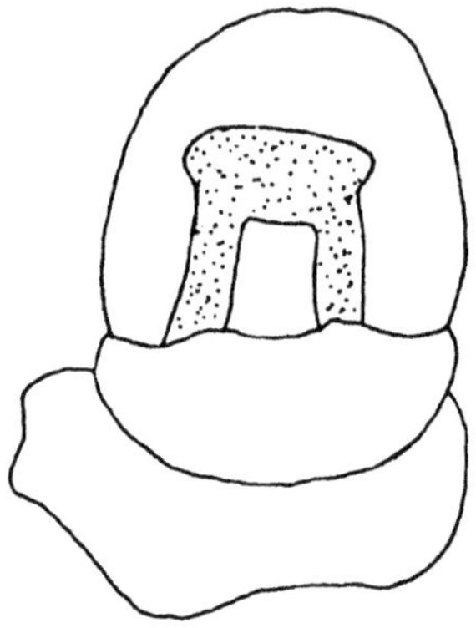

ca. 18.5% xylem

Fig. 36. Chaetocarpus schomburgkianus.

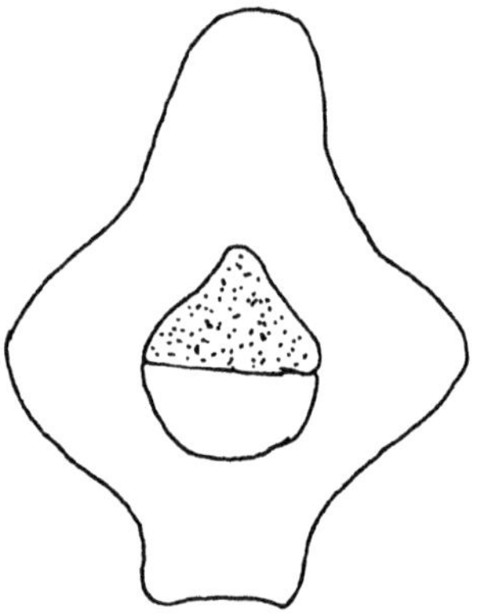

ca. 8.5% xylem

Fig. 37. Conceveeiba guyanensis.

Drypetes variabilis (both ca. 23 secondaries) and *Conceveiba guyanensis, Pera schomburgkiana* (both ca. 22 secondary veins). The remaining species *Hieronyma laxifolia* and "Kerosén blanco" possess ca. 20 ramifications of the medianus.

Because of its relatively low area index of less than 0.5 cm^2 *Mabea taquiri* (0.46 cm^2) and *Sapium sp.* (0.49 cm^2) stand out clearly from the other species. Nearly 0.5—1 cm^2 of the lamina are supplied by one secondary in *Pera schomburgkiana* (0.77 cm^2) and *Piranhea longepedunculata* (0.93 cm^2). At least 1 cm^2 and 1.25 cm^2 are innerved by one secondary vein in *Chaetocarpus schomburgkianus* (1.05 cm^2) and *Drypetes variabilis* (1.24 cm^2). "Kerosén blanco" follows (2.96 cm^2) at a distance. *Pogonophora sagotii* and

Hieronyma laxifolia possess per 3.31 cm^2 and 3.32 cm^2 of the leaf area one secondary ramification, respectively. The comparatively largest leaves of *Conceveiba guyanensis* and *Pausandra flagellorhachis* also show the highest area indices (4.01 and 6.17 cm^2).

Corresponding to the small area index the mean distance of the secondaries in *Mabea taquiri* and *Sapium sp.* (both 0.39 cm) is also the lowest. *Pera schomburgkiana* (0.53 cm), *Piranhea longepedunculata* (0.54 cm), *Chaetocarpus schomburgkianus* (0.69 cm), *Drypetes variabilis* (0.8 cm) and "Kerosén blanco" (1.14 cm) follow according to the index order. With regard to *Hieronyma laxifolia* (1.21 cm) and *Pogonophora sagotii* (1.28 cm) when compared with the area index, the order is reversed. Also analogously to the index, the ramifications of first order of *Conceveiba guyanensis* (1.5 cm) and *Pausandra flagellorhachis* (1.54 cm) have the mean largest distances between them.

Of the 71 leaves of adult trees studied, 65 examples (= 91.55%), on at least one half of the leaf, have the greatest distances in the middle compared with the base and the tip.

Due to the complicated branching of the sclereids in *Chaetocarpus schomburgkianus* and *Pausandra flagellorhachis* it was not possible to discriminate the islet network clearly enough to include them in the following result of my investigation.

By far the finest vein network, or rather the one with the finest meshes which were still countable, was found in *Piranhea longepedunculata* (3430 vein islets/cm^2 of the lamina). The second higher values of 423 (*Mabea taquiri*), 380 (*Pogonophora sagotii*), 373 (*Conceveiba guyanensis*) and 345 (*Hieronyma laxifolia*) appear minimal compared with the first. *Sapium sp.*, "Kerosén blanco" and *Drypetes variabilis* rank lower, with 275, 253 and 195 meshes. Extremely few vein islets per cm^2 of the lamina were counted for *Pera schomburgkiana* and *Chaetocarpus schomburgkianus*: the corresponding mean value showing 120 and 75 meshes/cm^2, respectively.

The most widely meshed vein network of the whole sample collection is found in *Pausandra flagellorhachis*, where only 14 islets are present per cm^2 of the blade.

Piranhea longepedunculata also has a distinctive position as far as the vein islet number is concerned: there is only one single ramification, appearing occasionally in an islet. Only 7 ramifications with a free ending could be discriminated among 100 meshes, counted with the help of a photograph (50-fold magnification). With on average 6 ramifications the vein islets of *Pera schomburgkiana* exhibit the highest number. In *Conceveiba guyanensis* each islet produces 4.6 lateral veins of higher order. *Mabea taquiri* (Ø = 2.9) displays nearly 3 ramifications. The calculation for *Chaetocarpus schomburgkianus* (2.2), *Hieronyma laxifolia* and *Pogonophora sagotii* (both 1.7), indicates an approximate value of 2. A mean value of 1.4 is found for *Drypetes variabilis*. The vein islets of "Kerosén blanco" have no, or a maximum of 3 ramfications, giving an average of 1.2. For *Sapium sp.* too, I calculated a mean number of 1.2 ramifications per mesh. The range extends from 0 to maximally 2 ramifications, whereby only the veins were considered which can be clearly discriminated. Either no, or a maximum of 2 ramifications are present in the islets of *Pausandra flagellorhachis*.

The range of the veinlet termination number per cm^2 of the lamina ranges from 1700 (*Hieronyma laxifolia*) to 340 (*Piranhea longepedunculata*). With 1460, 1370 and 1310 terminations/cm^2 *Conceveiba guyanensis, Pera schomburgkiana* and *Mabea taquiri* approach the upper limit, whereas the 470 terminations of *Drypetes variabilis* and the 450 terminations of "Kerosén blanco" approach the lower limits of this range. *Pogonophora sagotii* takes a mean place with 890 veinlet terminations.

In comparison to the whole sample collection of the leaves of adult trees *Piranhea longepedunculata* has the highest innervation of 102.82 cm vein length per cm^2 of the lamina. As described above, this species is conspicuous for its smallest meshed venation system. The next smaller value is 78.83 cm (*Hieronyma laxifolia*). The distance between the following vein lengths are far less: 71.24 cm/cm^2 (*Mabea taquiri*), 69.89 cm (*Pogonophora sagotii*), 63.24 cm (*Pera schomburg-

kiana), 57.65 cm (*Sapium sp.*) and 55.16 cm (*Conceveiba guyanensis*). *Drypetes variabilis*, whose vein length is almost 10 cm shorter (46.07 cm), follows. The vein length of "Kerosén blanco" (43.88 cm) belongs to this range, too. In the next species, *Chaetocarpus schomburgkianus*, the difference consists of more than 10 cm, the mean vein length is 31.57 cm/cm^2. With comparatively the smallest supply of veins *Pausandra flagellorhachis* (17.52 cm) is the last specimen in the group of the leaves of adult plants.

Neither the cross-sections of the vascular bundles of the five Euphorbiaceae studied, nor their xylems show common structures with regard to the form (Figures 33—37). The xylem comprises between ca. 8.5% and 32.5% of the total cross section. Thus, taking only the xylem into consideration, the net conductive volume yields ca. 11% to ca. 38% of the gross volume.

Approximate total vein lengths and conductive volumes/cm^2 (leaves of young trees)

The approximate total vein length (average of the size of the prepared leaves of the young trees × vein length/cm^2 of the lamina) produces the following data:

Table 36. Approximate total vein length

Species	Total vein length (cm)
Mabea piriri	2668.17
Drypetes variabilis	5320.91
Piranhea longepedunculata	2040.95
Hieronyma laxifolia	?
Croton matourensis	5925.69

Table 37. Approximate conductive volume per cm^2 of the lamina (cf. Section 3.1).

Species	Gross conductive volume (cm^3)
Mabea piriri	0.0009
Drypetes variabilis	0.0015
Piranhea longepedunculata	0.0003
Hieronyma laxifolia	0.0002
Croton matourensis	0.0004
	0.0002—0.0007—0.0012 (± 71.43%)

Results of the measurements of the Euphorbiaceous venation (comparison between leaves of young and adult trees)[1)]

Apart from *Croton matourensis* I had only one leaf of the corresponding tree at my disposal. The leaf size of the seedling of *Hieronyma laxifolia* could not be measured because of missing parts of the blade and the fragility of the dried leaf material. Thus, the area index is missing.

The largest leaf$_{YT}$ (ca. 152 cm^2) originates from *Drypetes variabilis* — in comparison with the relatively smallest mean area of the leaf$_{AT}$ of 28 cm^2. *Croton matourensis* (ca. 127 cm^2), *Mabea piriri* (ca. 39 cm^2) and the 3 m-high tree of *Piranhea longepedunculata* (ca. 31 cm^2) follow. The leaf area of "Caramamacate" (1 m in diameter) amounts to ca. 31 cm^2, too.

The greatest number of secondary veins can be counted for *Croton matourensis* (ca. 49) and *Mabea piriri* (47). *Drypetes variabilis* and *Piranhea longepedunculata* follow with 34 and 25. The leaf of the seedling of *Hieronyma laxifolia* has an estimated number of 18 secondaries. The leaves$_{AT}$ of *Drypetes variabilis* and *Piranhea longepedunculata* exhibit almost the same number in reverse order: 22—23 (*D. variabilis*) and ca. 34 (*P. longepedunculata*). In contrast to the leaves$_{AT}$, there are differences of +11 to +12 (*D. variabilis*) and ca. −9 (*P. longepedunculata*), respectively. As for *Hieronyma laxifolia*, there is also — in analogy to the leaf size — only a slight difference to the leaf$_{AT}$ (ca. 20 secondaries).

Because of the small size of the leaf — in comparison with the remaining leaves — and the comparatively high number of secondary veins *Mabea piriri* has the lowest area index (0.84 cm^2). Thus, one secondary has only to supply a relative small part of the blade. *Piranhea longepedunculata* has an area of 1.22 cm^2, supplied by one secondary vein. The values 2.61 and 4.48 for *Croton matourensis* and *Drypetes variabilis* belong, therefore, to a higher category. They provide the highest data within the leaves of the young

[1)] In this chapter the name of a species refers without further explanation to the leaves of young trees.
Abbreviations: leaf$_{YT}$ = adult leaf of the young tree.
leaf$_{AT}$ = adult leaf of the adult tree.

Table 38. Measurements of the venation (leaves of young trees).

Species	Leaf area (cm^2)	Number of secondaries/ leaf	Area[1] index (cm^2)	Mean distance of secondaries (cm)	Vein[2] islets/cm^2	Ramifications/ islet	Veinlet termina-nations/cm^2	Vein length/cm^2 (cm)
Mabea piriri	39.4	47	0.84	0.38	324—348 —372	1.45—2.2 —2.95	1068—1230 —1392	64.35—67.72 —71.09
"Pata de pauji" (6 m high)					(±6.76%)	(±34.09%)	(±13.17%)	(±4.98%)
Drypetes variabilis	152.2	34	4.48	1.25	92—115 —138	1.78—3.9 —6.02	306—410 —514	31.84—34.96 —38.08
"Kerosén" 5 m high)					(±20.0%)	(±54.36%)	(±25.37%)	(±8.92%)
Piranhea longepedunculata	30.6	25	1.22	0.78	1304—1550 —1796	0—1	112—220 —328	63.07—66.05 —69.03
"Caramacate" (3 m high)					(± 15.87%)		(±49.09%)	(±4.51%)
Hieronyma laxifolia	?	18	?	2.09	140—163 —186	0.74—1.4 —2.06	237—300 —363	39.84—41.07 —42.3
"Aguacatillo" (seedling 1 m high)					(±14.15%)	(±47.14%)	(±21.0%)	(±2.99%)
Croton matourensis	93.05—127.43 —161.81	45.93—48.75 —51.57	1.91—2.61 —3.31	0.61—0.74 —0.87	158—190 —222	0.94—1.6 —2.26	881—1130 —1379	36.81—38.73 —40.65
"Canelo" (10 cm Ø) (A)	(±26.98%)	(±5.78%)	(±26.82%)	(±17.57%)	(±16.84%)	(±41.25)	(±22.04%)	(±4.96%)

[1] Quotient leaf area: number of secondaries.
[2] The vein islet number is based on round figures, while the percentage refers to the mathematical averages.

trees. The corresponding results of the $leaves_{AT}$ of *P. longepedunculata* (0.93 cm^2) and *D. variabilis* (1.24 cm^2) only reveal a net deviation in the case of *D. variabilis*. While the figure for *P. longepedunculata* is 0.29 cm^2, it is 3.24 cm^2 in favour of the $leaf_{AT}$ for *D. variabilis*.

The 47 lateral veins of first order of *Mabea piriri* are on average 0.38 cm distant from each other. For *Croton matourensis* the mean distance amounts to 0.74 cm, for *Piranhea longepedunculata* 0.78 cm (by comparison, the mean distance for the $leaf_{AT}$ is 0.54 cm). While the distance of 1.25 cm for *Drypetes variabilis* (0.8 cm in the $leaf_{AT}$) occupies a mean position, the secondaries of *Hieronyma laxifolia* with a distance of 2.09 cm (1.21 cm for the $leaf_{AT}$) are the farthest from each other. The corresponding measurements for the $leaves_{AT}$ of the species mentioned above, reveal generally greater distances for the leaves of the young plants: +0.24 cm (*P. longepedunculata*), +0.45 cm (*D. variabilis*) and + 0.88 cm (*H. laxifolia*).

Twenty leaves of the 22 leaves of young trees studied have, at least in one half of the lamina, the widest distances between the secondary veins in the middle of the leaf (= 90.91%).

The four investigational criteria, still pending, were rendered considerably more complicated in *Croton matourensis* because of the stellate hairs on the upper leaf surface (cf. Roth, 1984).

Within the group of the young trees, too, the vein network of *Piranhea longepedunculata* comprises a comparatively extremely high vein islet number: on average 1550 per cm^2. This numerical

Table 39. Position of the species within the criteria investigated (leaves of young trees)[1)]

Position	Leaf area (cm^2)	Number of secondaries/leaf	Area index (cm^2)	Mean distance of secondaries (cm)	Vein islets/cm^2	Ramifications/islet	Veinlet terminations/cm^2	Vein length/cm^2 (cm)
1	D. variabilis	C. matourensis	D. variabilis	H. laxifolia	P. longepedunculata	D. variabilis	M. piriri	M. piriri
2	C. matourensis	M. piriri	C. matourensis	D. variabilis	M. piriri	M. piriri	C. matourensis	P. longepedunculata
3	M. piriri	D. variabilis	P. longepedunculata	P. longepedunculata	C. matourensis	C. matourensis	D. variabilis	H. laxifolia
4	P. longepedunculata	P. longepedunculata	M. piriri	C. matourensis	H. laxifolia	H. laxifolia	H. laxifolia	C. matourenis
5		H. laxifolia		M. piriri	D. variabilis	P. longepedunculata	P. longepedunculata	D. variabilis

[1)] Decreasing values from 1—4.

value proves to be low compared to that of the $leaf_{AT}$ (3430 islets/cm^2). The remaining four species have islet numbers as follows: *Mabea piriri* 348, *Croton matourensis* 190, *Hieronyma laxifolia* 163 (i.e. −182 in comparison with the 345 meshes of the $leaf_{AT}$). *Drypetes variabilis*, finally, displays 115 vein islets (−80 in relation to 195 of the $leaf_{AT}$). Summing up, the vein islets of the leaves of the young trees reveal values which are slight above or below 50% of the number of the corresponding leaves of the adult plants.

Just as for the foliage leaves of the adult tree, only one single ramification per vein islet is occasionally present in *Piranhea longepedunculata* (3 m high). These are ca. 14 islets with a single ramification per 100 meshes, as a freely ending termination, just as for the $leaves_{AT}$. Compared to the leaves of the adult trees, the $leaf_{YT}$ has twice as many vein islets with ramifications.

The following mean data for the ramifications per islet stand at 3.9 ramifications (*Drypetes variabilis*), ca. 2 (average = 2.2 — *Mabea piriri* and 1.6 — *Croton matourensis*) and 1.4 ramifications (seedling of *Hieronyma laxifolia*). While the approximate mean value for the seedling amounts to 1 ramification per islet, the leaves of the adult tree show ca. 2 branches (exactly 1.7). With +2.5 the ramification number of the $leaf_{YT}$ of *Drypetes variabilis* lies considerably above that of the $leaf_{AT}$ (= 1.4).

Just as for the criterion of the veinlet terminations/cm^2 the $leaves_{YT}$ of *Mabea piriri* and *Croton matourensis* are noticeable because of their considerably greater number: 1230 and 1130. The remaining species have merely 410 (*Drypetes variabilis*, 470 for the $leaf_{AT}$), 300 (*Hieronyma*

Table 40. Direct comparison. Measurements of the venation of leaves of adult and young trees.

Species	Leaf area (cm^2)	Number of secondaries/ leaf	Area index (cm^2)	Mean distance of secondaries (cm)	Vein islets/cm^2	Ramifications/ islet	Veinlet terminations/cm^2	Vein length/cm^2 (cm)
Drypetes variabilis								
"Kerosén" (A)	28.47	22.5	1.24	0.8	195	1.4	470	46.07
"Kerosén" (5 m high)	/152.2/	/34.0/	/4.48/	/1.25/	115	3.9	410	34.96
Difference	123.73	11.5	3.24	0.45	80	2.5	60	11.11
$leaf_{AT}$	−434.6%	−51.11%	−261.29%	−56.25%	+41.03%	−178.57%	+12.77%	+24.12%
$leaf_{YT}$	+81.29%	+33.82%	+72.32%	+36.0%	−69.57%	+64.1%	−14.63%	−31.78%
Factor	5.35×	1.51×	3.61×	1.56×	58.97%	2.79%	87.23%	75.88%
Hieronyma laxifolia								
"Aguacatillo" (A)	67.11	19.88	3.32	1.21	345	1.7	1700	78.83
"Aguacatillo" (seedling)	?	/18.0/	?	/2.09/	163	1.4	300	41.07

Table 40 (Continued)

Species	Leaf area (cm²)	Number of secondaries/ leaf	Area index (cm²)	Mean distance of secondaries (cm)	Vein islets/cm²	Ramifications/ islet	Veinlet termina- tions/cm²	Vein length/cm² (cm)
Difference	?	1.88	?	0.88	182	0.3	1400	37.76
$leaf_{AT}$		+9.46%		−72.73%	+52.75%	+17.65%	+82.35%	+47.9%
$leaf_{YT}$		−10.44%		+42.11%	−111.66%	−21.43%	−466.67%	−91.94%
Factor		90.54%		1.73×	47.25%	82.35%	17.65%	52.1%
Piranhea longepedunculata								
"Caramacate" (1 m Ø) (A)	31.23	34.4	0.93	0.54	3430	0–1	340	102.82
"Caramacate" (3 m high)	/30.6/	/25.0/	/1.22/	/0.78/	1550	0–1	220	66.05
Difference	0.63	9.4	0.29	0.24	1880		120	36.77
$leaf_{AT}$	+2.02%	+27.33%	−31.18%	−44.44%	+54.81%		+35.29%	+35.76%
$leaf_{YT}$	−2.06%	−37.6%	+23.77%	+30.77%	−121.29%		−54.55%	−55.67%
Factor	97.98%	72.67%	1.31×	1.44×	45.19%		64.71%	64.24%

The numerical value '/. . ./' does not indicate the average, as only one single sample leaf was available.
Difference: "$leaf_{AT}$" = difference between the leaf of the adult tree and the leaf of the young tree
"$leaf_{YT}$" = difference between the leaf of the young tree and the leaf of the adult tree
"Factor" means the relation between the leaf of the young tree and that of the adult tree.

laxifolia, 1700 for the $leaf_{AT}$) and 220 terminations for *Piranhea longepedunculata* ($leaf_{AT}$: 340). The most distinct differences from the leaves of the adult trees appear in *H. laxifolia*: the $leaf_{YT}$ only has 17.65% of the veinlet terminations/cm² in comparison with its $leaf_{AT}$. In the case of *D. variabilis* this deviation is comparatively small: +60 (referring to the $leaf_{YT}$). *P. longepedunculata* takes a mean position in the range of the deviations from the leaf of the adult plant; it has 120 terminations/cm² fewer.

Within the collection of the leaves of the young trees, *Mabea piriri* (67.72 cm vein length per cm of the lamina) and *Piranhea longepedunculata* (66.05 cm/cm²) exhibit the densest innervation. For "Caramacate" the comparison with the $leaf_{AT}$ reveals a vein length which is smaller by 36.77 cm. *Hieronyma laxifolia* follows with a clearly shorter vein length of 41.07 cm, giving a decrease of −37.76 cm in relation to that of the $leaf_{AT}$. Of all $leaves_{YT}$ studied, the weakest supply of veins was registered in *Croton matourensis* (38.73 cm/cm²) and *Drypetes variabilis* (34.93 cm). The difference of the $leaf_{YT}$ of *D. variabilis* amounts to −11.11 cm compared with the leaf of the adult tree.

Summary. A comparison of the area indices and the mean distance of the secondary veins of leaves of young and adult trees produces higher numerical values for the group of the leaves of young trees. The $leaves_{YT}$ have considerably fewer vein islets per cm² of the lamina. A comparison of the veinlet terminations and vein length per cm² shows negative results for the leaves of the young plants. The remaining three investigational criteria fail to indicate any general tendencies with regard to the group of the young trees.

3.6. Summarizing comparison of the results for Sapotaceae, Lauraceae and Euphorbiaceae

Synopsis of the taxonomical utilization of the characteristics in comparison with the leaf anatomy

Within the family of the Sapotaceae the camptodromous venation prevails over the brochidodromous one. The pure campto- and brochidodromous type as well as a mixture of both appears in the Lauraceae. The Euphorbiaceae almost exclusively display the "nervatio brochidodroma". In addition to the two species with a camptodromous course of the vascular bundles, *Conceveiba guyanensis* alone affords a unique example of craspedodromous venation (Picture 38).

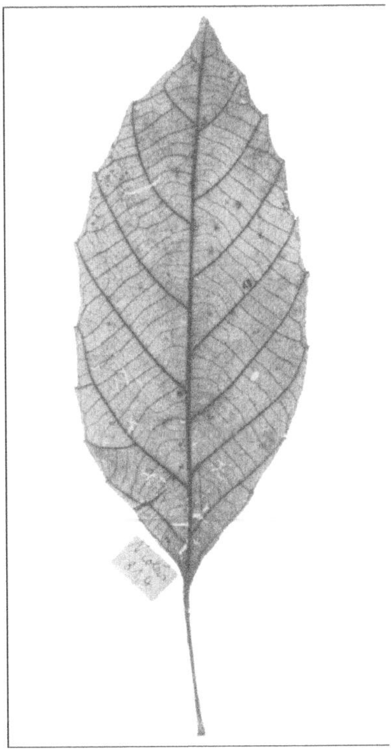

Fig. 38. Conceveiba guyanensis

81, 4 cm²

A common consideration of nervation-type and tree-height ("A", "a", "aa") together leads to interesting observations. The similar types of venation in the genera with several species (*Pouteria, Manilkara, Chrysophyllum, Ocotea, Aniba, Mabea*) lead first of all to the conclusion that the venation type is a characteristic typical of the family (i.e. of the sample collection).

While species with the camptodromous type appear in "A" as well as in "a", the brochidodromous venation exists only in leaves from trees of the height "A" (exception: *Pausandra flagellorhachis*). It could be concluded that in the sample collection the "nervatio brochidodroma" is predominantly present in individuals of the highest stratum. (A conclusion as to the asssociation of this venation type with the height "aa" would be precipitate, as I had only one example of "aa" at my disposal.)

A comparison of the measurements of the venation (cf. Tables 6, 19, 32) shows the following tendencies: the observation of the whole of the material investigated (leaves of adult trees) indicate that the leaves of the camptodromous type generally possess fewer secondary veins than those of the brochidodromous type (the exceptions being species with "nervatio camptodroma", which have a relatively high number of secondaries in relation to the remaining camptodromous leaves of adult trees: *Ecclinusa guianensis, Chrysophyllum auratum, Beilschmiedia curviramea, Sapium sp.*). The group of brochidodromous leaves does not show such a clear tendency: there is indeed a higher number of secondary veins present (especially in the *Manilkara* species and some other Sapotaceae), but in the Lauraceae and Euphorbiaceae there exist brochidodromous species with comparatively few secondaries. The tendencies (= "nervatio camptodroma" with relatively few and "nervatio brochidodroma" with in contrast, many secondary veins) apply mainly to the Sapotaceae.

No correlations were found between the remaining measurements and the venation type.

Whereas the venation of the Sapotaceae and Lauraceae is closed, three of the species of the

Euphorbiaceae show an open form of the leaf margin or a mixture of closed and open venation.

The relative thickness of the primary veins compared to the secondaries varies in the Sapotaceae between "thick" and "fine/thin". The primaries of the Lauraceae are predominantly "thick". in two species they are even "extremely thick". The thick medianus also dominates among the Euphorbiaceae; only *Croton matourensis* has an extremely thick primary vein (Picture 39).

In the three families the medianus divides the lamina of most of the species in two equal or almost equal parts. Only the Sapotaceous species *Pouteria cf. trilocularis* displays a distinctly unequally divided lamina (Picture 40).

In the leaves of all species of the Sapotaceae, Lauraceae and Euphorbiaceae a sole primary vein (medianus) runs from the base of the leaf to the tip.

In the majority of the foliar leaves this medianus runs straight, in the remaining ones it is slightly curved, and almost without exception it narrows

Fig. 39. Croton matourensis
(leaf of a young tree)

103 cm²

regularly (Sapotaceae), while in the specimens of the Lauraceae, there is a predominantly slightly curved course and a mixture between a straight and slightly curved course with regular narrowing. This mixture together with an equal decrease in thickness is characteristic of the primary veins of the Euphorbiaceae.

Apart from the Sapotaceous species "Capurillo negro", (Picture 41 — ramified subtilization of the termination), *Drypetes variabilis*, and also "Kerosén blanco" (Euphorbiaceae — both kinds of termination possible) the medianus in the leaves of the three families ends directly at the tip.

The position of the secondary veins varies from pure opposite to alternate, including a series of transitional stages. Oppositifolious species prevail in the samples of the Sapotaceae, Lauraceae and Euphorbiaceae.

In relation to their primary veins, the secondaries of the Lauraceous species are more finely developed. In the other two families one also finds considerably finer secondary veins in comparison to the medianus.

A description of the relative thickness of the secondaries among each other produces a varying result within the sample collection. In each family the thickest secondary veins occur in the middle, becoming finer towards the base and the tip. Secondaries which become finer from the base towards the tip or which remain unchanged throughout the whole leaf are only present in the Lauraceae and Euphorbiaceae. For the Sapotaceae and Euphorbiaceae the more or less regular combination of thicker and finer secondary veins is characteristic in some species (Picture 42).

The direction and the course of the secondary veins specify the types of venation described in the first part of the taxonomical synopsis.

The terminations of the secondaries in all three families are characterized by loop development among the secondary veins ending at the leaf margin (Figures 1, 2). These loops occur mainly short of the margin in the foliar leaves of the Sapotaceae, whereby the bows of vascular bundles are in part difficult to discriminate from the leaf margin (camptodromous venation, Picture 40). Very clear loops among the thicker and thinner

Chrysophy auratum
91,5 cm²
—— 1 cm

Pouteria venosa
90 cm²

Manilkara spec.
76,2 cm²

Pouteria cf. trilocularis
75 cm²

Fig. 40.

Manilkara bidentata / *"Purguo morado"*
41,9 cm²
—— 1 cm

Pouteria egregia
34 cm²

Ecclinusa guianensis
33, 8 cm²

"Capurillo negro"
33 cm²

Fig. 41.

Manilkara spec.

61,6 cm²
—— 1 cm

Chrysophyllum spec.
(leaf of a young tree)

56, 2 cm²

Pouteria venosa

49 cm²

Manilkara bidentata/
"Purguo blanco"

46 cm²

Fig. 42.

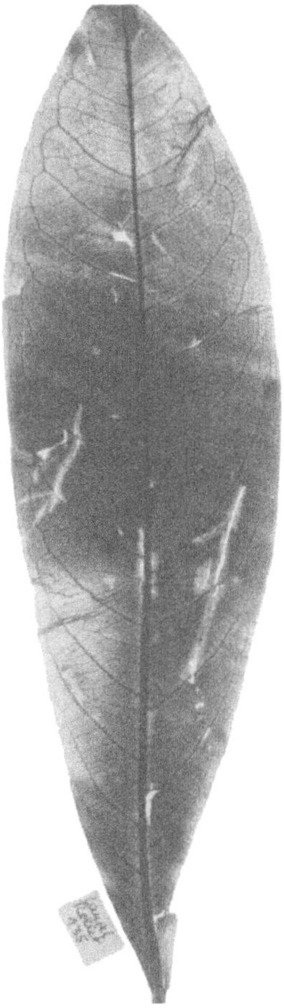

Fig. 43. Aniba excelsa

135 cm^2

—— 1 cm

secondary veins are to be found in the brochidodromous type of venation (Picture 42). In addition to such a loop-like fusion of these veins, in some Lauraceous leaves further curves of vascular bundles of higher orders are visible above the bows formed by the secondaries (Picture 43). The pure camptodromous venation in this family is indicated by the steep, bow-like joining of one secondary into the one above (Picture 44). Some examples of the Euphorbiaceae also show the simple formation of loops as well as a second row of anastomosing loops above those of the secondaries: in one representative of this venation type the leaf margin is coarsely dentate (Picture 45). The only species in the whole sample collection to display craspedodromous venation with a coarsely dentate margin too, is the Euphorbiacee *Conceveiba guyanensis* (Picture 38).

All leaves of the collection have lateral veins of higher orders which exceed the degree of ramification of quinternary veins.

The free veinlet terminations, also present in all species, consist generally of quinternaries or lateral veins of higher orders. Only "Pendarito", *Chaetocarpus schomburgkianus* (Picture 46) and *Croton matourensis* occasionally show freely ending tertiary veins.

The Sapotaceous species investigated have vein islets consisting of irregularly or reticulately formed polygons. A double-rowed islet pattern is typical of the genus *Manilkara* in this collection. The vein network of *Chrysophyllum sp.* (Picture 47 — leaf of a young tree), composed of "squares" (cf. Roth, 1984, p. 392), is almost atypical, since such forms appear repeatedly in the Lauraceae. Here, an investigation of a leaf of an adult tree of *C. sp.* would be necessary in order to come to a definitive conclusion. Patterns which are typical of the species are chiefly present in those species which possess sclereids: *Chrysophyllum auratum, Ecclinusa guianensis, Manilkara.* Leaves of young and adult trees differ only very little: the leaves of young plants clearly show the latter "adult pattern".

The irregular or reticulate polygonal type is also found among the specimens of the Lauraceae. Transversely divided "squares" are typical of *Ocotea doutincta, Aniba excelsa,* "Laurel canelo", and "Laurel rastrojero". Three species *Ocotea martiana, Ocotea nicaraguensis, Nectandra grandis* (Picture 48) are clearly distinguishable from the original pattern due to the "cocktread-like" ramifications of the islets. The comparison of leaves of adult and young trees of the same species revealed only slight differences; the venation patterns are essentially equal.

Extremely various and specific patterns of vascular bundles predominate in the investigated Euphorbiaceae. *Piranhea longepedunculata* (Picture 49) can be quoted as the only example of the "square"-type. The remaining species have specifically arranged polygonal islets or a particularly

Fig. 44. Ocotea duotincta
(leaf of a young tree)

100, 5 cm²

"Laurel rastrojero"

59 cm²

Fig. 44.

Fig. 45. Pausandra flagellorhachis

116 cm²
—— 1 cm

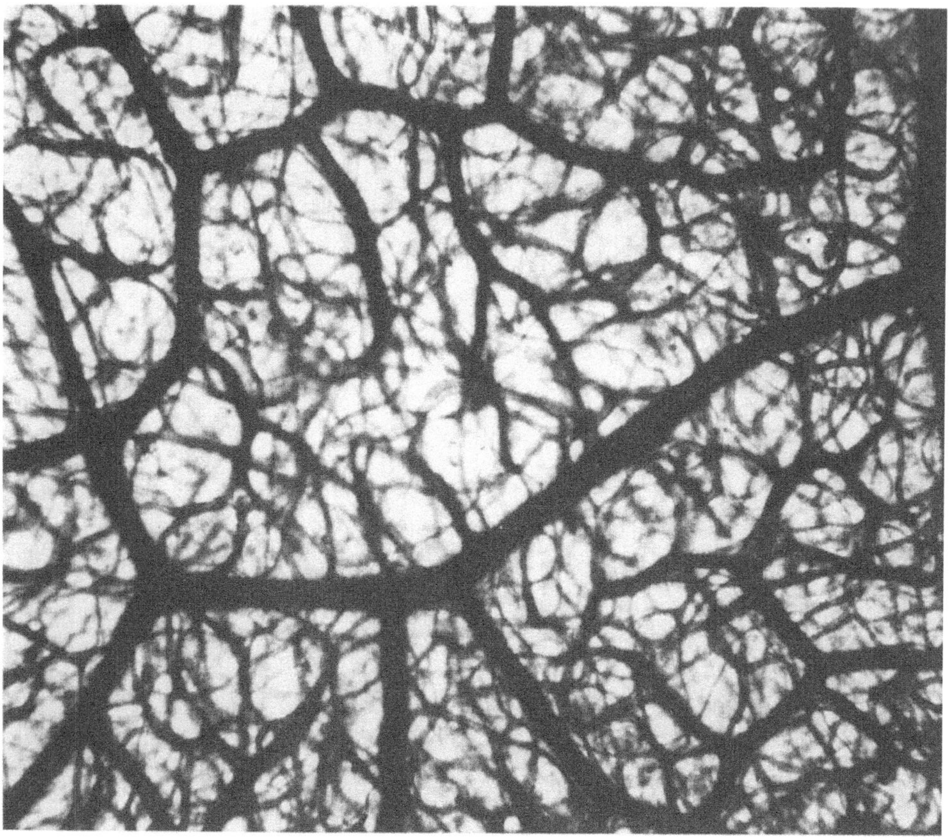

1 mm

Fig. 46. Chaetocarpus schomburgkianus — venation pattern
6, 3 × 4

distinctive form of ramification. Dense sclereids or stellate hairs on the lower side of the leaf cause the specific islet pattern of *Chaetocarpus schomburgkianus* (Picture 46), *Pausandra flagellorhachis* and *Croton matourensis* (leaf of a young tree — Picture 50). The islet pattern of those species where leaves of adult and young trees were available differ considerably.

In comparison with the Sapotaceae and Lauraceae, the Euphorbiaceae are the most heterogeneous group with regard to the variety and typical development of the islet pattern. In contrast, the Sapotaceae show hardly any outstanding formation types. This family is, as far as can be ascertained from the patterns of the species, investigated, very homogeneous.

Correlations between vein islet pattern and vein length/cm²

When searching for a relation between the venation pattern (cf. determination key) and the result of the measurements, it must be considered that only one criterion of description from A—G correlates with one of the results, because neither the characteristic of the vein islets nor that of the leaves of adult and young trees which were

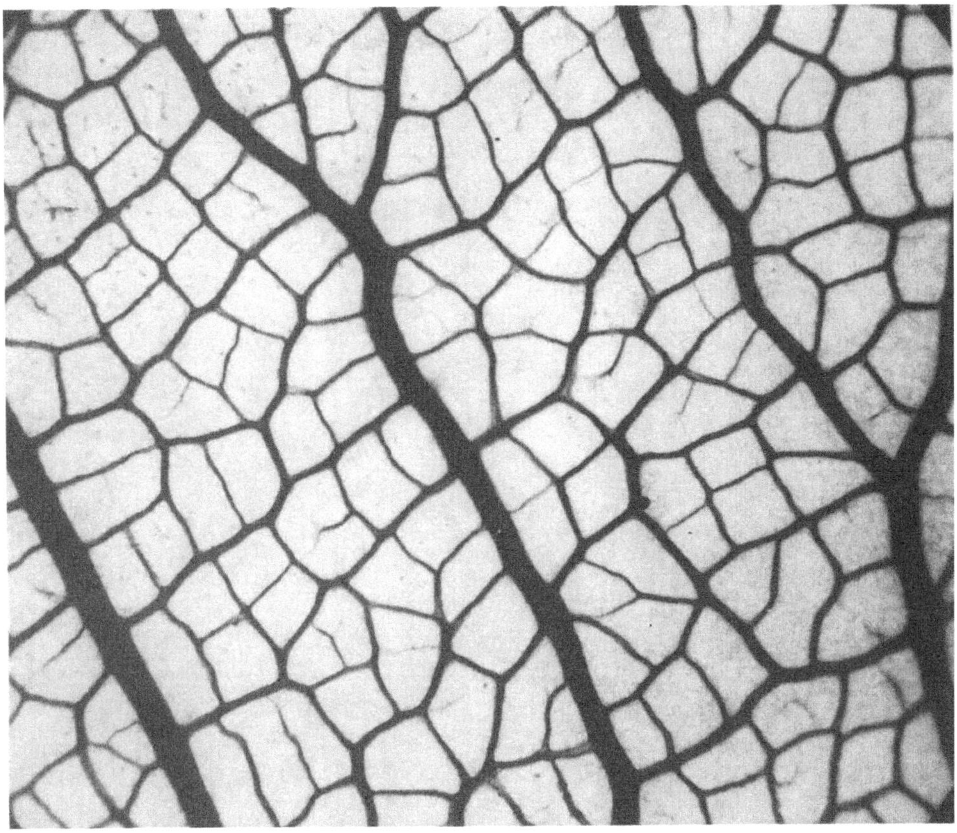

Fig. 47. Chrysophyllum spec. (leaf of a young tree) — venation pattern
6, 3 × 4

demonstrated by the venation measurements can be classified under one heading.

As the vein length per cm^2 is one of the most relevant ecological characteristics of the venation (cf. Zalenski, 1902, Wylie, 1943a, Plymale and Wylie, 1944, Shields, 1950, Geiger and Cataldo, 1969) I examined to what extent a correlation exists between the data and the different criteria of classification.

1. *A. Regularity in form and size — vein length/cm²*

Those species (leaves of adult trees) which have a very high vein length/cm^2 within the sample collection are distinguishable by the islet character A 1 ("regular"): *Pouteria aff. anibaefolia, P. egregia, P. venosa, Piranhea longepedunculata.* This conformity is extremely noticeable in the three *Pouteria* species. As the unnamed species of *Pouteria, P. cf. trilocularis*, and *P. sp.* have neither a high vein length nor the characteristic A 1, it cannot be said to be peculiar to the genus.

Contrasting examples, such as *Manilkara sp., M. bidentata*/"Purguo" (leaves of adult and young trees) *Ocotea duotincta* (leaves of adult and young trees), *O. nicaraguiensis* (leaf of the adult tree), *O. aff. subalveolata, Aniba excelsa*, "Laurel canelo" as well as "Laurel rastrojero" show that the islet character A 1 by no means gives rise to a high vein

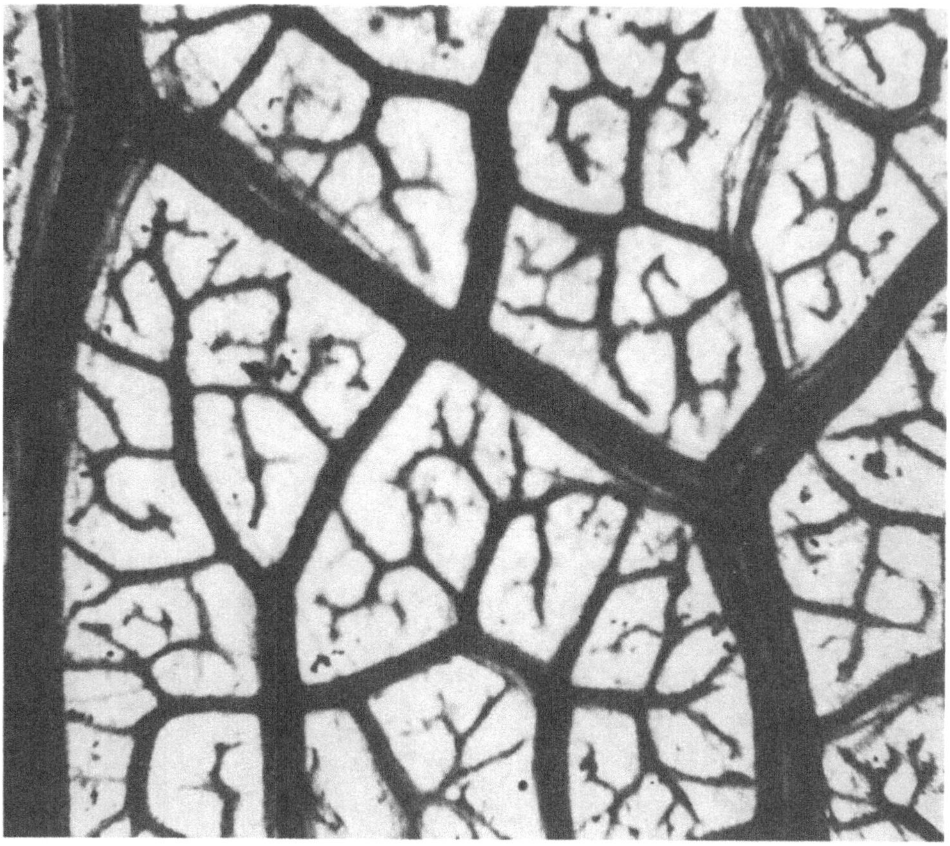

1 mm

Fig. 48. Nectandra grandis — venation pattern
6, 3 × 4

length. A conclusion can thus only be drawn as regards the (high) vein length to the characteristic "regular", and not vice versa.

2. *B. Form of the islets — vein length/cm²*

This comparison reveals no relation between the two criteria of investigation.

3. *C. Size of the vein islets — vein length/cm²*

The same correlation as under 1 is reflected in the relation of the vein length to the size of the meshes. The leaves of the adult trees with a high vein length have "comparatively small" islets (C 3) in common. However, it would be wrong to conclude that there is a strong innervation for C 3. This is demonstrated by the species *Manilkara sp.* and *Hieronyma laxifolia*. As the two *Pouteria* species which possess a relatively short vein length (*P. cf. trilocularis, P. sp.*) do not belong to this type of islet at all, it is confirmed that a high vein length correlates with small meshes, but the converse conclusion does not hold.

4. *D. Arrangement of the islets — vein length/cm²*

The leaves of adult trees with a strong innervation mentioned under 1 and 3, all have a "reticulate" arrangement of islets in common (D 4). But this

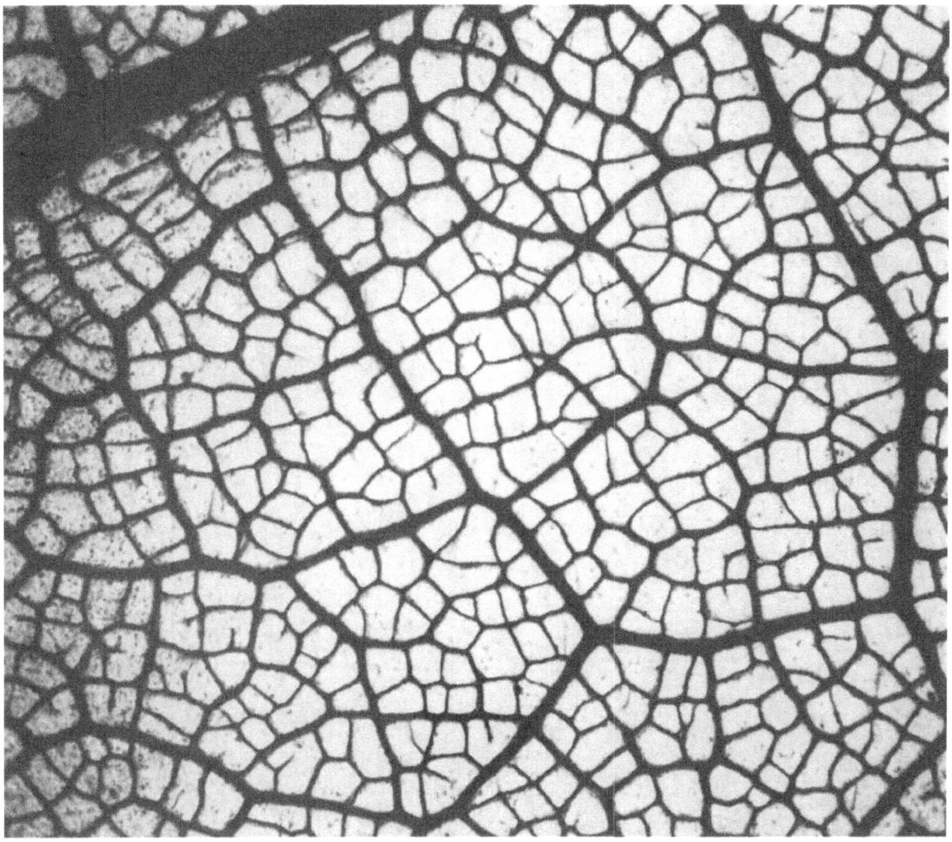

1 mm

Fig. 49. Piranhea longepedunculata – venation pattern 6, 3 × 4

generally descriptive type also appears in several other species, so that it correlates less with a high vein length than with its typicality of the genus, as far as can be determined in this sample collection (cf. *Pouteria, Ocotea*).

5. *E. Form of the ramifications,*
   F. Thickness of the ramifications,
   G. Frequency of the ramifications – vein length/cm²

The criteria describing the ramification type fail to reveal any visible relation to the vein length.

As shown by the relation between the characteristic of meshes and vein length/cm^2, a high vein length of the species concerned correlates with the regularity in form and size of the vein islets. The islets are comparatively small and reticulate. It follows, therefore, that this kind of arrangement guarantees the most advantageous exploitation of the lamina for the development of vascular bundles, leading to an optimal supply to the leaf of water and nutritive salts for assimilation.

This result of the investigation is confirmed by the quotient of vein length/cm^2 : vein islets/cm^2 (Q_{ID} cf. Synopsis of the ecological significance of the taxonomical investigations). For the material investigated in *Pouteria aff. anibaefolia* and *P. venosa* this result is very low, and it is even lower for *Piranhea longepedunculata*.

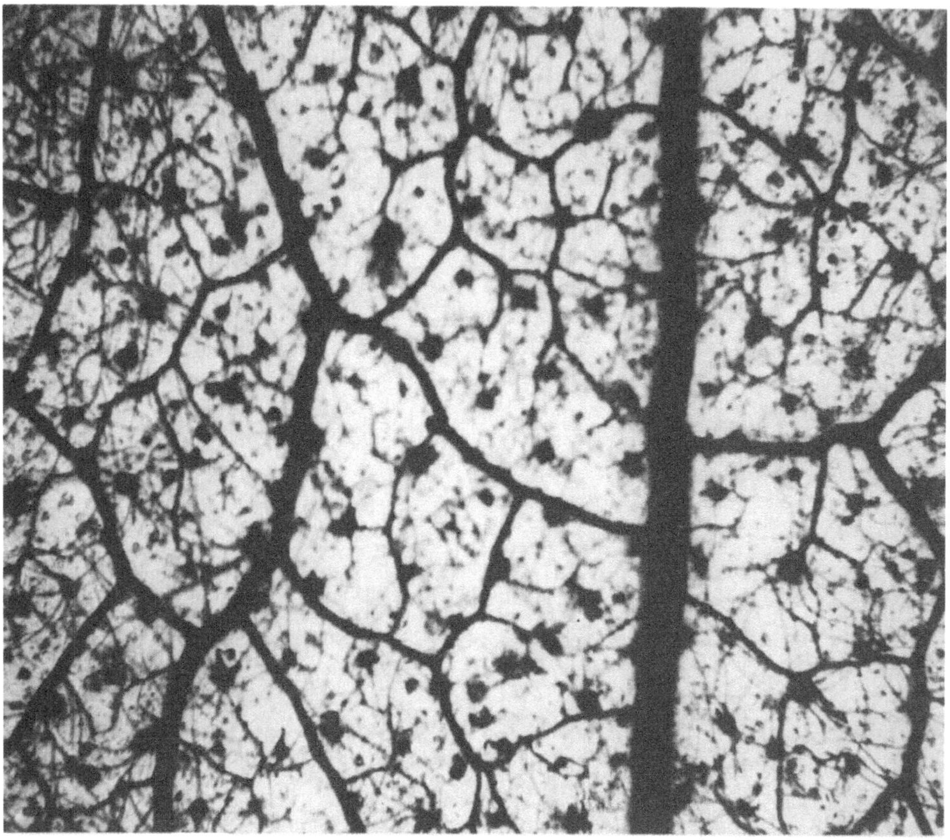

1 mm

Fig. 50. Croton matourensis (leaf of a young tree) — venation pattern
6, 3 × 4

Taxonomical validity of venation measurement (leaves of adult trees)

A (measurable) characteristic of the venation is considered the more typical of a species, the less the average mean of all investigated leaves of a single species deviates from the standard.

Whether an investigation criterion is taxonomically valid is only ascertainable for those characteristics which I could study on the basis of a larger number of leaves from one species.

The number of secondary veins of the lamina has proved to be typical of the species, because here the lowest deviations occur: mean deviation from the average are 9.33% (Lauraceae), 10.59% (Sapotaceae), 11.49% (Euphorbiaceae).

Within the values for the average distances of the secondaries, also measured in all leaves of adult trees of the species, the mean deviation is nearly 15% in the three families (Sapotaceae: 14.83%, Lauraceae: 15.31%, Euphorbiaceae: 15.47%). This characteristic of venation is therefore, with some restrictions, less variable within a species.

According to Levin's results (1929), the number of vein islets per unit area is a venation characteristic which is typical of the species:

> For any given species the vein-islet number varies within narrow limits, the variation being such as one might reasonably expect when dealing with a biological subject. (Levin, 1929, p. 42)

Levin had determined the islet number/mm^2 in different parts of the lamina.

The counting of vein islets/cm^2 in my research was carried out on 2—3 preparations from the middle of the leaf. Each time I selected a middle-sized leaf of an adult tree. The average of all mean deviations of the leaves of one family, with regard to the islet number, is also roughly 15%: 14.14% (Euphorbiaceae), 15.25% (Lauraceae), 16.66% (Sapotaceae). These results confirm Levin's observations.

The comparatively high deviations of the ramifications per islet prove that this venation characteristic is not very typical of the species. But the high mean deviations are the result of the small amount of data upon which the calculation of the average rate of ramification was based. The real range of the ramification number is much smaller, as is shown by the averages of the given species. Considering the range of variation in living organisms quoted by Levin (1929, p. 42), I too regard this venation characteristic as typical of the species.

In spite of the high values for the veinlet termination number/cm^2, the mean deviations are generally rather low. The Sapotaceae in particular, classified as very homogeneous, show a low deviation of 13.37% in comparison to the other two families. Thus, the investigated leaves of adult Sapotaceous plants have a fairly constant rate of termination within the species. In contrast, the mean deviation of the Euphorbiaceae of 18.38% reveals greater differences among the species. These differences become even more apparent in the Lauraceae, where the deviation amounts to 26.98% on average.

On the whole, I regard the number of secondaries per lamina as the most characteristic feature for the given species, followed by the criterion of "average distance of the secondaries" and the vein islet number/cm^2.

Synopsis of the ecological significance of the anatomical investigations

Synopsis of the values measured for the leaves from adult trees

The average of all mean values for the species from one family serves as the standard of comparison (cf. Table 41).

The average leaf size of all species belonging to the same family differs only slightly, on comparing the three families. The Euphorbiaceous leaves have a mean size about 60 cm^2, those of the Lauraceae are roughly 55 cm^2, the foliage leaves of the Sapotaceae finally show nearly 47 cm^2 in size.

While the number of secondary veins per lamina amounts to 20—30 in the Euphorbiaceae and Lauraceae, the Sapotaceae have about 42 of these ramifications on average.

The lowest area index of the three families (quotient of area of the lamina : number of secondaries) is found in the Sapotaceae (1.38 cm^2) compared with the much higher values for the Euphorbiaceae (2.25 cm^2) and for the Lauraceae (2.48 cm^2).

The Sapotaceae display the smallest mean distances for secondaries (0.67 cm). The Euphorbiaceae and Lauraceae follow with 0.91 and 1.25 cm.

For the criterion of the vein islets per cm^2 of the lamina, the samples from the Lauraceae with 345 meshes/cm^2 differ greatly from the data of the Sapotaceae (603) and Euphorbiaceae (535).

The mean maximum of ramifications per vein islet is found in the Euphorbiaceae (2.22 ramifications). The Sapotaceae and Lauraceae follow with 1.99 and 1.15 ramifications. (In order to provide a better basis for comparison the exact mathematical averages were given.)

The numbers of veinlet terminations/cm^2 range from 1001 (Sapotaceae) to 806 (Lauraceae). Here the Euphorbiaceae (999) have an intermediate position.

While the average vein length per cm^2 is almost equal in the Sapotaceae (58.90 cm/cm^2) and Euphorbiaceae (57.99 cm/cm^2), the value for the Lauraceae (46.29 cm/cm^2) is slightly lower.

A general survey of the three families with regard to the number of secondaries and the area index reveal data within the Sapotaceae which provide a clear contrast to those of the two other families. The vein islet number of the Lauraceae also differs from that of the Sapotaceae and Euphorbiaceae. The remaining criteria fail to

Table 41. Measurements of the venation: comparison of the averages for Sapotaceae, Lauraceae and Euphorbiaceae.

Family[1]	Leaf area (cm^2)	Number of secondaries/ leaf	Area[1] index (cm^2)	Mean distance of secondaries (cm)	Vein[2] islets/cm^2	Ramifications/ islet	Veinlet termina- tions/cm^2	Vein length/cm^2 (cm)
Sapotaceae leaves$_{AT}$)	17.24—47.02 −76.8	22.14—41.82 −61.5	0.27—1.38 −2.49	0.3—0.67 −1.04	222.57—602.93 −983.29	1.42—1.95 −2.48	711.28—1000.83 −1290.38	38.95—58.9 −78.85
	(±63.33%)	(±47.06%)	(±80.43%)	(±55.22%)	(±63.09%)	(±27.18%)	(±28.93%)	(±33.87%)
(leaves$_{YT}$)	62.01—100.26 −138.51	34.06—52.69 −71.32	0.93—2.21 −3.49	0.43—0.73 −1.03	124.65—665.75 −1206.85	on average ca. 1	130—150 −170 (2 species)	34.11—52.12 −70.13
	(±38.15%)	(±35.36%)	(±57.92%)	(±41.1%)	(±81.28%)		(±13.33%)	(±34.55%)
Lauraceae (leaves$_{AT}$)	15.78—55.38 −94.98	13.47—23.08 −32.69	1.36—2.48 −3.6	0.78—1.25 −1.72	106.61—345 −583.39	0.7—1.15 −1.6	340.03—806 −1271.97	33.73—46.29 −58.85
	(±71.51%)	(±41.64%)	(±45.16%)	(±37.6%)	(±69.1%)	(±39.13%)	(±57.81%)	(±27.13%)
(leaves$_{YT}$)	44.41—140.97 −237.53	17.62—24.07 −30.52	2.38—6.48 −10.58	0.92—1.78 −2.64	154.42—237 −319.58	1.18—1.43 −1.68	764.35—883.33 −1002.31	38.79—49.5 −60.21
	(±68.5%)	(±26.8%)	(±63.27%)	(±48.31%)	(±34.84%)	(±17.48%)	(±13.47%)	(±21.64%)
Euphorbiaceae (leaves$_{AT}$)	5.72—59.73 −113.74	19.61—28.9 −38.19	0.49—2.25 −4.01	0.49—0.91 −1.33	−389.49—534.82 −1459.13	0.62—2.22 −3.82	503.06—998.75 −1494.44	35.68—57.99 −80.3
	(±90.42%)	(±32.15%)	(±78.22%)	(±46.15%)	(±172.83%)	(±72.07%)	(±49.63%)	(±38.47%)
(leaves$_{YT}$)	34.18—87.41 −140.64	22.53—34.55 −46.57	0.86—2.29 −3.72	0.46—1.05 −1.64	70.84—473.2 −1017.24	0.79—1.92 −3.05	226.38—658 −1089.62	35.54—49.71 −63.88
	(±60.9%)	(±34.79%)	(±62.45%)	(±56.19%)	(±114.97%)	(±58.85%)	(±65.6%)	(±28.51%)

[1] Abbreviations: leaves$_{AT}$ = leaves of adult trees
leaves$_{YT}$ = leaves of young trees

show any extremely high or low result in any of the three families compared.

Although the gross conductive volumes per cm^2 of the lamina (Tables 9, 22, 35) only yield approximate data due to the amount of the non-supporting tissues included, I should draw the reader's attention to the averages of the gross volumes for each family: considered on average, the Lauraceae possess the greatest gross conductive volume (0.0014 cm^3/cm^2 of the lamina — cf. Table 22) and also the largest range of single measurements, borne out by a nearly 60% deviation. The Sapotaceae in the second position (0.0012 cm^3/cm^2 — cf. Table 9) have in contrast only a mean deviation of about 33%, while the Euphorbiaceae, with the comparatively smallest gross conductive volume of 0.0010 cm^3/cm^2 (cf. Table 35), show an astonishingly low deviation of only 20%.

An examination of the investigation methods proposed in Section 3.1 will show how far these results for the net conductive volume can be corroborated.

In comparison, the mean data of the gross conductive volumes of the Sapotaceae (0.0012 cm^3/cm^2), Lauraceae (0.0014 cm^3/cm^2) and Euphorbiaceae (0.0010 cm^3/cm^2) differ only slightly.

The averages for the net conductive volumes, taken from random samples, support the trend described above, the differences themselves are much lower: Sapotaceae and Euphorbiaceae 0.0002 cm^3/cm^2 of the lamina, Lauraceae 0.0001 cm^3.

The results of the quantitative comparison of the water conductive systems of different species under the same ecological conditions, published by Huber (1924) indicate the same direction:

Versuchen wir schließlich auch für verschiedene Arten, das Leitungssystem quantitativ zu vergleichen, so fällt (. . .) eine überraschend große Gleichförmigkeit innerhalb ökologisch einheitlicher Gruppen auf. (Huber, 1924, p. 96)

Clear differences emerge, if we use as the basis for comparison the percentage proportion of the net conductive volume of the gross conductive volume: in the samples of the Lauraceae the net data make up only ca. 4% to ca. 17% of the gross volumes; in the Sapotaceae it lies between ca. 11 and about 31%, in the Euphorbiaceae it is nearly 11 to about 38%.

The almost similar ranges of the net volumes in the Sapotaceae and Euphorbiaceae must be considered in conjunction with the equal averages of the net volumes (see above) and the mean data of the vein length/cm^2 (cf. Table 41).

Relation between measured data and tree height (leaves of adult trees)

This section refers only to those species whose stratum could categorically be identified as "A", "a" or "aa".

The classification of high or low values for all the leaves investigated is based on the measured data as standard.

1. *Number of secondaries — tree height*

Although the number of secondaries in the height category "A" (22 samples) and "a" (7 samples) differs considerably, the 7 highest numbers appear in the Sapotaceous specimens from "A" (*Manilkara sp., M. bidentata*/"Purguo"/"Purguo morado"/ "Purguo blanco", "Pendarito", "Purguillo" and the Euphorbiaceae *Sapium sp.*. The single sample from "aa" (*Pausandra flagellorhachis*) with 35 secondary veins does not allow a comparable assessment.

2. *Average distance of the secondaries — tree height*

In "A" comparatively more species with larger distances are found than in "a". This could be due to the fact that from the height category "A" more samples were available.

3. *Vein islets/cm² — tree height*

This comparison defied any attempt at classification.

4. *Ramifications/islet — tree height*

This comparison also defied any conclusions as to correlation.

5. *Veinlet terminations/cm² — tree height*

Whereas the veinlet terminations of the species from "a" move in the range from 1460 (*Conceveiba guyanensis*) to 590 (*Beilschmiedia curviramea*), the leaves of adult trees from "A" show extremely low data (130 — *Aniba riparia*, 340 — *Piranhea longepedunculata*, 470 — *Drypetes variabilis*) as well as very high values (1910 — *Ocotea martiana*, 1700 — *Hieronyma laxifolia*).

6. *Vein length/cm² — tree height*

The comparison of vein length and height of tree serves as verification of the Zalenski law (Zalenski, 1904),[1] according to which the vein length per unit area increases with the increasing height of the leaves on the plant.

In the 22 species, whose leaves of adult trees belong to the height category "A", vein lengths are found between 102.82 cm/cm² (*Piranhea longepedunculata*) and 21.38 cm/cm² ("Pendarito"), including four measured data which would have been more appropriate to "aa": "Pendarito", *Aniba riparia, A. excelsa* and *Chaetocarpus schomburgkianus*. Five species posses a vein length which could, unknowingly, have been classified as "a": *Manilkara bidentata*/"Purguo"/"Purguo blanco" and *Drypetes variabilis*. This last group might include trees whose height just reaches the lower limit of "A".

The vein lengths of the leaves of the adult trees from "a" cover an intermediate range: from 55.16 cm/cm² (*Conceveiba guyanensis*) to 37.59 cm/cm² (*Ocotea duotincta*). Only the extremely high value of 96.25 cm/cm² (*Pouteria aff. anibaefolia*) is out of place in the series of the height category "a". *P. aff. anibaefolia* could belong possibly to the upper limit of "a".

[1] Quoted according to Roth (1980, p. 511), since the original text could not be procured, in spite of every possible effort.

The only example of the category "aa" (*Pausandra flagellorhachis*) has a vein length (17.52 cm/cm²) which is so low, in comparison to all other measurement results, that it corresponds to the smallest value anticipated for the lowest stratum.

7. *Conductive capacity/cm² — tree height*

Although in this sample collection there seems to be no correlation between gross conductive volume and height category, it is evident that within the Sapotaceae the gross conductive volumes/cm² of the representatives of "A" have much higher values than within the Euphorbiaceae. The gross volumes of the five "A" species of the Lauraceae comprise the maximum and minimum data of all investigated leaves of adult trees.

The species, whose net conductive volume was taken from random samples, belong mostly to "A", some species to "a". The specimens from "a" only reveal two different data, which are also found in the middle range of the "A" values. Maximum and minimum numbers emerge in the species of the height category "A". Apart from "Purguillo" (Sapotaceae), which has the highest net conductive volume of all "A" species, there are no evident differences between the Sapotaceae and Euphorbiaceae: the net volumes are closer to each other than the gross volumes. The net conductive volumes of the Lauraceous species are the lowest found in the middle of the range of the investigated leaves.

Different anatomical adaptations must be taken into consideration in order to offset the atypically short vein lengths of the height category "A" (cf. the nine species, mentioned above, whose adaptations Roth (1984) established during her studies of the leaf structures).

Thus, "Pendarito" for example possesses a "drip-tip" (according to Roth, 1980, a long extended tip of the leaf, which is separated clearly from the lamina) and a shiny leaf surface (Roth, 1984). From the present point of view, these anatomical characteristics guarantee a faster drying of the surface, which accelerates the beginning of photosynthesis and transpiration, in spite of relatively high atmospheric humidity.

Gessner (1956a) proved by experiment that especially for plants in the tropical rain forest, drip-tips provide a decisive advantage in selection, as they ensure faster drying of the leaves, and transpiration as well as assimilation of CO_2 are guaranteed.

The experiments of Dean and Smith (1978) also confirm these advantageous effects which the drip-tip has on the plant metabolism.

Vareschi (1980) has come out against this allegedly clear and sole function of drip-tips. According to his investigations, even in evergreen tropical rain forests those forms of adaptation are missing. On the other hand they appear in plants where they have no apparent value in their environment (*Ficus religiosa* L. — Moraceae, Vareschi (1980, p. 85)). Vareschi found a high percentage of drip-tips with the function, as described by Gessner (1956a) and Dean and Smith (1978), in rain forests with seasonal periods of rain and drought. There the tips accelerate the quick runoff of water from xeromorphic leaves. Thus, Vareschi regards a drip-tip only as the morphological formation of a leaf without discriminating between mere anatomical feature and functional characteristic with a selective value (cf. Vareschi, 1980, pp. 85 ff.).

According to Roth (1980), the development of a drip-tip could be explained by retarded growth in width of the lamina in contrast to a favoured longitudinal and apical growth of the lamina.

Aniba riparia and *A. excelsa* compensate for their relatively slight supporting system of vascular bundles with a correspondingly larger lamina. Both species have an upper one-layered epidermis with water storing cells. Finally, *A. excelsa* has a high number of stomata close to the vascular bundles (Roth, 1984), also important for the water supply. This high number of stomata near the veins (according to Roth, 1984, p. 303, up to $400/mm^2$) ensures a fast, intensive water supply by transpiration and a full exchange of gas. In relation to the surface of the lamina this species is abundantly equipped with stomata.

Chaetocarpus schomburgkianus possesses water storing tracheids, which surround the veins like a sheath in order to support the water household (Roth, 1984).

The large lamina (in *Manilkara bidentata*/ "Purguo") as well as the mesophyll cells, interpreted by Roth as water storing, can serve *M. bidentata*/"Purguo blanco" and *Chrysophyllum auratum* as compensation for the vein length/cm^2 which is slight, considering the tree height category.

In *Ocotea nicaraguensis* the large lamina together with water storing cells of the upper epidermis counteract the low innervation which is atypical of "A" (Roth, 1984).

Roth also considers the upper epidermis of *Drypetes variabilis* to be a water storing tissue.

General comparison of the measurements on leaves of young and adult trees (Table 41)

The average of all mean values of the adult leaves of young and adult trees of the family serves as the standard of comparison.

On the whole, the average size of the leaf is greater in the leaves of young plants than in the adult ones. The leaves of the young Lauraceous trees (ca. 141 cm^2) are noticeably large.

The number of secondaries in the leaves of young plants also exceeds that in the adult trees in the three families. Here, attention must be drawn to the ca. 53 secondary veins of the Sapotaceous leaves of young plants contrasting with the much lower number of ca. 35 and ca. 24 in the Euphorbiaceae and Lauraceae, respectively.

Although the area indexes of the leaves of young trees are higher than those of the leaves of their adult trees, the differences of the indexes between leaves of young and adult trees show clear contrasts: while a secondary vein of a young Lauraceous plant supplies 4 cm^2 of the lamina more than that of the adult tree, these difference only amounts to 0.83 cm^2 (Sapotaceae) and to 0.04 cm^2 (Euphorbiaceae) in favour of the leaves of the young plants. The relatively high area index of 6.48 cm^2 (Lauraceae) stands out against the remaining groups of leaves of young trees.

The average distances of the secondaries are only a little greater in the leaves of young trees than in the adult ones: the maximum is +0.53 cm (Lauraceous leaves of young plants). In this criterion too, the data for the leaves of young trees of the Lauraceae represent the maximum value as compared to the two other groups.

Apart from the Sapotaceous leaves of young trees, where a higher density of vein islets is found than in the leaves of the adult trees, the vein islet number is lower in the leaves of the young plants. They possess approx. 100 meshes/cm^2 of the lamina (Lauraceae), and about 60 islets (Euphorbiaceae) fewer than their adult trees.

If an approximate value is considered for the ramifications of a vein islet, then the islets of the Lauraceous and Euphorbiaceous leaves of young trees contain as many ramifications as those of adult plants. In contrast to these, there are only half as many ramifications in the Sapotaceous leaves of young plants.

Only on the Lauraceous leaves of young trees did I count more veinlet terminations per cm^2 as compared with the adult trees (+77); this is also the highest result within the group of leaves of young trees. The leaves of young plants of the Sapotaceae and Euphorbiaceae only have ca. 15% and 66%, respectively, of the terminations in relation to the foliage leaves of the adult trees.

Only in the Lauraceae does an insignificantly higher vein length per cm^2 of the lamina (+3.21 cm/cm^2) appear. The leaves of young trees of the Sapotaceae and Euphorbiaceae are more weakly veined than those of the adult plants (Sapotaceae: −6.78 cm/cm^2, Euphorbiaceae: −8.28 cm/cm^2).

The following differences between the leaves of young trees (YT) and adult trees (AT) occur when comparing the total mean data (cf. Table 41) and especially in the direct comparison of the pairs of leaves of young and adult plants, belonging to one of the three families.

Sapotaceae (Tables 14, 41):
size of the lamina$_{YT}$ > size of the lamina$_{AT}$
number of secondaries$_{YT}$ > number of secondaries$_{AT}$
area index$_{YT}$ > area index$_{AT}$
ramifications/vein islet$_{YT}$ < ramification/vein islet$_{AT}$

Lauraceae (Tables 27, 41):
size of the lamina$_{YT}$ > size of the lamina$_{AT}$
area index$_{YT}$ > area index$_{AT}$
distance of the secondaries$_{YT}$ > distance of the secondaries$_{AT}$
vein islet number$_{YT}$ < vein islet number$_{AT}$

Euphorbiaceae (Tables 40, 41):
area index$_{YT}$ > area index$_{AT}$
distance of the secondaries$_{YT}$ > distance of the secondaries$_{AT}$
vein islet number$_{YT}$ < vein islet number$_{AT}$
veinlet termination number$_{YT}$ < veinlet termination number$_{AT}$
vein length$_{YT}$ < vein length$_{AT}$

A difference between the leaves of young and adult trees found in all three families, is the constantly higher area index of the leaves$_{YT}$ compared with their adult plants, which means that a greater part of the lamina is supplied by a single secondary vein.

In the Lauraceae and Euphorbiaceae the average distance of the secondaries is larger in the group of the leaves of young trees.

The readings for the leaves$_{YT}$ in each of the three families — either fewer ramifications/vein islet (Sapotaceae), fewer meshes/cm^2 (Lauraceae, Euphorbiaceae) or fewer veinlet terminations/cm^2 (Euphorbiaceae) and a weaker supply with veins/cm^2 (Euphorbiaceae) — prove that generally their venation network has not yet reached the degree of discrimination and fineness which characterizes the mature leaf of the adult tree.

Correlations between the groups of measuring data and the height category of the young trees, as shown in part within the leaves$_{AT}$, were not observed.

The negative correlation between the leaf size and the vein islet and veinlet termination number per unit area (increase of these data with decreasing size of the leaf), proved by Gupta (1961a) also applies to some extent to the leaves of young trees in comparison with their adult plants.

Comparison of the vein density of leaves of adult and young trees

Vein length/cm^2 of the lamina

An examination of the assumption that the leaves

of adult trees in general have a higher vein density than the leaves of young trees only revealed a corresponding value of 100% for the vein length per cm^2 within the Euphorbiaceae (three pairs of leaves of adult and young plants).[1] The three pairs of the Sapotaceae show a higher vein length for two leaves of the adult trees (66.67%). The supposition is valid for only 33.3% of the Lauraceous three pair samples. Seen as a whole, there is a 66.67% verification of the hypothesis for the nine pairs of the three families.

Conductive volume/cm² of the lamina

The comparison of the gross conductive volume/cm^2 (without deducting the percentage of non-conducting tissues — cf. Section 3.1) of the leaves of adult and young trees also reveals the tendency for the conductive system of the leaves of adult plants to display a greater capacity than that of the corresponding $leaf_{YT}$: while the three Lauraceous groups of leaves of adult trees display a higher gross conductive volume than the leaves of the young plants, this can only be confirmed for 2/3 of the pairs of the remaining families examined. For the nine pairs of $leaves_{AT/YT}$ together, 77.78% of the leaves of the adult plants show the higher gross volume per cm^2 of the lamina in relation to that of the young tree.

A comparison of the net conductive volume of leaves of adult and young trees was not carried out, since the net volume, as well as the gross conductive volume, are only based on approximate data, and a comparison would therefore produce only very rough approximate values. An exact comparison of the conductive volume and the conductive capacity (cf. Section 3.1) of the leaves of young and adult plants would be an interesting investigation in the scope of a subsequent study, the results of which would serve as a valuable complement to my findings, from an ecological as well as a physiological point of view.

Quotient vein length/cm² : vein islets/cm² (Q_{ID})

The quotient of the vein length/cm^2 and the vein islets/cm^2 relation leads to a clearer result than the isolated consideration of the vein length. The Q_{ID} defines the islet density.

While within the Sapotaceae only two of the three leaves of adult trees exhibit lower quotients than the leaves of young plants, this value is constantly lower for the leaves of adult trees of the other two families. The lower (higher) the quotient, the more (fewer) vein islets are present in relation to the vein length per cm^2 of the lamina. In 89.89% of the available pairs of leaves, the leaves of adult plants possess more islets in relation to the vein length per cm^2 than the corresponding leaves of young trees.

Table 42. Leaves of adult trees of the Sapotaceae.

Species	Quotient vein length/cm² : vein islets/cm²
Pouteria egregia	0.078
Pouteria venosa	0.068
Manilkara sp.	0.230
Manilkara bidentata/ "Purguo"	0.082
Manilkara bidentata/ "Purguo morado"	0.070
Manilkara bidentata/ "Purguo blanco"	0.085
Ecclinusa guianensis	0.226
Chrysophyllum auratum	0.045
"Pendarito"	0.314
"Capurillo negro"	0.398
"Purguillo"	0.205
Pouteria aff. anibaefolia	0.084
Pouteria cf. trilocularis	0.111
Pouteria sp.	0.111
	0.048—0.151—0.254 (±68.21%)

Table 43. Leaves of young trees of the Sapotaceae.

Species	Quotient vein length/cm² : vein islets/cm²
Chrysophyllum sp.	0.050
Chrysophyllum auratum	0.069
Manilkara bidentata/ "Purguo"	0.365
Ecclinusa guianensis	0.185
	0.042—0.167—0.292 (±74.85%)

[1] "Pair" means that leaves of a young and an adult tree were available.

Table 44. Leaves of adult trees of the Lauraceae.

Species	Quotient vein length/cm^2 : vein islets/cm^2
Ocotea martiana	0.273
Ocotea nicaraguensis	0.126
Aniba riparia	0.143
Aniba excelsa	0.210
Nectandra grandis	0.217
"Laurel canelo"	0.067
Ocotea duotincta	0.090
Ocotea aff. subalveolata	0.089
Beilschmiedia curviramea	0.303
"Laurel rastrojero"	0.127
Endlicheria cocuirey	0.301
	0.093—0.177—0.261 (±47.46%)

Table 45. Leaves of young trees of the Lauraceae.

Species	Quotient vein length/cm^2 : vein islets/cm^2
Ocotea nicaraguensis	0.166
Beilschmiedia curviramea	0.441
Ocotea duotincta	0.142
	0.114—0.250—0.386 (±54.4%)

Table 46. Leaves of adult trees of the Euphorbiaceae.

Species	Quotient vein length/cm^2 : vein islets/cm^2
Sapium sp.	0.210
Pogonophora sagotii	0.184
Pera schomburgkiana	0.527
Drypetes variabilis	0.236
Hieronyma laxifolia	0.228
Piranhea longepedunculata	0.030
Chaetocarpus schomburgkianus	0.421
Conceveiba guyanensis	0.148
Pausandra flagellorhachis	1.251
Mabea taquiri	0.168
"Kerosén blanco"	0.173
	0.005—0.325—0.645 (±98.46%)

Table 47. Leaves of young trees of the Euphorbiaceae.

Species	Quotient vein length/cm^2 : vein islets/cm^2
Mabea piriri	0.195
Drypetes variabilis	0.304
Piranhea longepedunculata	0.043
Hieronyma laxifolia	0.252
Croton matourensis	0.204
	0.113—0.200—0.287 (±43.5%)

Thus the quotient of vein length and vein islet number is a standard for the greater or smaller density of the reticulated venation of a foliage leaf in relation to the innervation of the lamina per unit area.

Among the Euphorbiaceae especially, the leaves of adult trees are seen to include some with very large vein meshes, which is also reflected in the relatively high mean quotient of 0.325. Within this family the quotients differ considerably (cf. mean deviation).

With regard to the tree height of the leaves of adult plants, it is noticeable that the higher and lower extremes of the quotient are to be found in category "A". The highest value for the whole sample collection was measured for *Pausandra flagellorhachis* ("aa"); but this is no basis for a comparison. Within the group of the leaves of young trees, there is no apparent tendency suggesting a relation between quotient and height of tree.

Quotient vein length/cm^2: vein islets + veinlet termination/cm^2 (Q_{VD_1})

The relationship between vein length and the vein islet number + veinlet terminations (values per cm^2 of the lamina) constitutes the third decisive parameter of the vein density. The Q_{VD_1} is therefore a standard for the vein density of a species.

Due to the indistinct venation pattern the veinlet termination number of some species could not be counted; thus, the corresponding quotients are not available.

The single pair of leaves from the Sapotaceae

Table 48. Leaves of adult trees of the Sapotaceae.

Species	Quotient vein length/cm^2 : vein islets + veinlet terminations/cm^2
Pouteria egregia	0.037
Pouteria venosa	0.036
Manilkara sp.	0.075
Manilkara bidentata/ "Purguo"	0.025
Manilkara bidentata/ "Purguo morado"	0.041
Manilkara bidentata/ "Purguo blanco"	0.028
Ecclinusa guianensis	0.062
Chrysophyllum auratum	0.027
"Pendarito"	
"Capurillo negro"	
"Purguillo"	0.042
Pouteria aff. anibaefolia	0.043
Pouteria cf. trilocularis	0.034
Pouteria sp.	0.032
	0.026—0.040—0.054 (±35.0%)

Table 49. Leaves of young trees of the Sapotaceae.

Species	Quotient vein length/cm^2 : vein islets + veinlet terminations/cm^2
Chrysophyllum sp.	0.045
Chrysophyllum auratum	0.062
Manilkara bidentata/ "Purguo"	
Ecclinusa guianensis	

shows a lower quotient in the leaf of the adult tree which displays a higher vein density compared to the leaf of the young plant. Within the Lauraceous and Euphorbiaceous pairs 66.67% of the leaves of the adult trees (two out of three pairs) reveal the lower quotient. Taking the seven pairs from the three families together, 71.43% of the leaves of adult trees have the higher vein density per unit area.

A comparison between these results and the quotient calculated above (vein length/cm^2 : vein islets/cm^2) shows that among these seven pairs, too, the higher quotient is to be found in the leaves

Table 50. Leaves of adult trees of the Lauraceae.

Species	Quotient of vein length/cm^2 : vein islets + veinlet terminations/cm^2
Ocotea martiana	0.030
Ocotea nicaraguensis	0.040
Aniba riparia	0.087
Aniba excelsa	
Nectandra grandis	0.044
"Laurel canelo"	0.037
Ocotea duotincta	0.033
Ocotea aff. subalveolata	0.030
Beilschmiedia curviramea	0.061
"Laurel rastrojero"	0.067
Endlicheria cocuirey	0.055
	0.030—0.048—0.066 (±37.5%)

Table 51. Leaves of young trees of the Lauraceae.

Species	Quotient vein length/cm^2 : vein islets + veinlet terminations/cm^2
Ocotea nicaraguensis	0.050
Beilschmiedia curviramea	0.062
Ocotea duotincta	0.027
	0.031—0.046—0.061 (±32.61%)

Table 52. Leaves of adult trees of the Euphorbiaceae.

Species	Quotient vein length/cm^2 : vein islets + veinlet terminations/cm^2
Sapium sp.	
Pogonophora sagotii	0.055
Pera schomburgkiana	0.042
Drypetes variabilis	0.069
Hieronyma laxifolia	0.039
Piranhea longepedunculata	0.027
Chaetocarpus schomburgkianus	
Conceveiba guyanensis	0.030
Pausandra flagellorhachis	
Mabea taquiri	0.041
"Kerosén blanco"	0.062
	0.032—0.046—0.060 (±30.43%)

Table 53. Leaves of young trees of the Euphorbiaceae.

Species	Quotient of vein length/cm^2 : vein islets + veinlet terminations/cm^2
Mabea piriri	0.043
Drypetes variabilis	0.067
Piranhea longepedunculata	0.037
Hieronyma laxifolia	0.089
Croton matourensis	0.029
	0.031–0.053–0.075 (±41.51%)

of young plants. It follows that these leaves present a less dense vein network than the leaves of adult trees.

When taking account of the height of the trees to which the leaves of the adult trees belong, results show that regarding the quotient of vein length and vein islet number per unit area, the upper and lower extremes are to be found in “A”. The leaves of the young plants fail to reveal any trend.

A comparison of the genera of which I was able to study several samples of leaves of adult trees shows that the quotients of *Pouteria* and *Ocotea* have a much smaller range than *Manilkara* (this pertains to the sample collection only).

Quotient vein length/cm^2 : vein islets + ramifications + veinlet terminations/cm^2 (Q_{VD_2})

Since the veinlet termination number cannot be equated with the ramifications of a vein islet, which means that the ramifications are a further factor influencing the vein density, the criterion “ramifications per islet” must also be considered in a quotient form.

Multiplying the vein islets/cm^2 with the number of ramifications/islet gives the ramification number/cm^2.

Leaves of young trees of the Sapotaceae.

Due to the lack of measurement data the quotient could not be calculated.

Results are not available for the pairs of leaves of adult and young plants of the Sapotaceae due to the lack of data in the group of the leaves of young trees.

Table 54. Leaves of adult trees of the Sapotaceae.

Species	Quotient vein length/cm^2 : vein islets + ramifications + veinlet terminations/cm^2
Pouteria egregia	0.019
Pouteria venosa	0.018
Manilkara sp.	0.049
Manilkara bidentata/ “Purguo”	0.017
Manilkara bidentata/ “Purguo morado”	0.023
Manilkara bidentata/ “Purguo blanco”	0.020
Ecclinusa guianensis	0.039
Chrysophyllum auratum	
“Pendarito”	
“Capurillo negro”	
“Purguillo”	0.028
Pouteria aff. anibaefolia	0.019
Pouteria cf. trilocularis	0.017
Pouteria sp.	0.020
	0.014–0.024–0.034 (±41.67%)

Table 55. Leaves of adult trees of the Lauraceae.

Species	Quotient vein length/cm^2 : vein islets + ramifications + veinlet terminations/cm^2
Ocotea martiana	0.025
Ocotea nicaraguensis	0.031
Aniba riparia	0.061
Aniba excelsa	
Nectandra grandis	0.033
“Laurel canelo”	0.025
Ocotea duotincta	0.027
Ocotea aff. subalveolata	0.020
Beilschmiedia curviramea	0.045
“Laurel rastrojero”	0.051
Endlicheria cocuirey	0.045
	0.023–0.036–0.049 (±36.11%)

Analogously to the evaluation of the quotient “vein length : vein islets + veinlet terminations”, the same result occurs after taking the ramifications of the meshes into consideration, too: two of the three pairs available for comparison

Table 56. Leaves of young trees of the Lauraceae.

Species	Quotient vein length/cm^2 : vein islets + ramifications + veinlet terminations/cm^2
Ocotea nicaraguensis	0.033
Beilschmiedia curviramea	0.051
Ocotea duotincta	0.022
	0.023—0.035—0.047 (±34.29%)

Table 57. Leaves of adult trees of the Euphorbiaceae.

Species	Quotient vein length/cm^2 : vein islets + ramifications + veinlet terminations/cm^2
Sapium sp.	
Pogonophora sagotii	0.036
Pera schomburgkiana	0.029
Drypetes variabilis	0.049
Hieronyma laxifolia	0.030
Piranhea longepedunculata	0.019
Chaetocarpus schomburgkianus	
Conceveiba guyanensis	0.016
Pausandra flagellorhachis	
Mabea taquiri	0.024
"Kerosén blanco"	0.044
	0.020—0.031—0.042 (± 35.48%)

Table 58. Leaves of young trees of the Euphorbiaceae.

Species	Quotient vein length/cm^2 : vein islets + ramifications + veinlet terminations/cm^2
Mabea piriri	0.029
Drypetes variabilis	0.036
Piranhea longepedunculata	0.026
Hieronyma laxifolia	0.059
Croton matourensis	0.024
	0.022—0.035—0.048 (±37.14%)

(Lauraceae and Euphorbiaceae), produce the lower quotient within the leaf of the adult tree (= 66.67%). This pertains to the same species whose leaves from the adult plants also show the lower value without the ramifications (*Ocotea nicaraguensis, Beilschmiedia curviramea* — Lauraceae, *Piranhea longepedunculata, Hieronyma laxifolia* — Euphorbiaceae).

This result confirms a generally higher vein density in the leaves of the adult trees measured in this sample collection.

The average quotient of the leaves of adult plants, once again, is the highest within the family of the Sapotaceae. The specimens from the Lauraceae have a lower vein density than the Euphorbiaceous and Sapotaceous species of the sample collection. The clearly lowest average was calculated for the Sapotaceae. On the whole, the differences of the mean values, which include the ramifications, are higher among the three families in comparison to the quotients without ramifications.

The comparison between tree height and the Q_{VD_2} reveals no relation between either the leaves of adult plants or those of the young trees.

A comparative consideration of the three quotients and the vein length/cm^2 as the sole defining criterion for the vein density, indicates that in relation to the vein length, the vein islet number and the veinlet terminations per cm^2 must in any case be included in the calculation. The mere vein length is not sufficient in order to define the density, nor is the calculation of vein length and vein islet number (Q_{ID}). The quotient vein length/cm^2 : islets + *ramifications* + veinlet terminations/cm^2 (Q_{VD_2}) serves to confirm the result attained by the quotient vein length/cm^2 : vein islet number + veinlet terminations/cm^2 (Q_{VD_1}). In the Q_{VD_2} all the elements making up the vein density are allowed for.

4. SUMMARY OF THE RESULTS

4.1. Taxonomical utilization of the characteritics in comparison with the leaf anatomy

Within the investigated characteristics of the

venation, the pattern and arrangement of the vein islets has proved to be the criterion with the greatest discrimination significance for the evaluation of the species.

The Sapotaceous species can be differentiated reliably one from the other by the typical cross section of the medianus of the species.

The manner of the ending of the secondary veins at the leaf margin is only suitable for classification, if the course of the secondaries at the margin is considered too. Thus, within the different families, characteristic forms of brochidodromous and camptodromous venation occur which may also be typical of the genera of Venezuelan Guayana (*Manilkara, Aniba*).

Although a comparison of the thickness of the secondaries is generally no reliable means of differentiating exactly, the typical alternating of stronger and finer secondary veins is significant in *Manilkara* and two other Sapotaceous species.

Unique in the whole sample collection is the relation between the formation of the leaf margin and the venation type in two Euphorbiaceous species, whose leaf margin is coarsely dentate (*Conceveiba guyanensis, Pausandra flagellorhachis*). The leaf margin has a mixture of closed and open venation.

Pera schomburgkiana is the only example of leaves with open vein endings at the margin.

Conceveiba guyanensis alone possesses the craspedodromous venation type.

A further characteristic, which only occurs occasionally, is the branch-like refined ending of the medianus at the leaf tip ("Capurillo negro", in part also *Drypetes variabilis* and "Kerosén blanco").

As well as venation characteristics, some species have other anatomical peculiarities which distinguish them from other species: the Euphorbiaceae *Chaetocarpus schomburgkianus* and *Pausandra flagellorhachis* show sclereids which are arranged reticulately or parallel to the leaf-surface, while *Croton matourensis* (leaf of a young tree) displays stellate trichomes on the lower surface of the leaf.

Within the sample collection it may be concluded that a high vein length/cm^2 presupposes an arrangement of the vein islets that is regular in form and size. Here, relatively small meshes occur. However it is not possible to infer a high vein length from these venation characteristics.

The leaves of young trees of the Sapotaceae have a finer structure of islets with coarser meshes than the venation of the leaf of the adult tree. Nevertheless, the pattern of the adult plant as regards form and arrangement, is already visible.

Within the Lauraceous samples too, the finer patterns of the juvenile venation are basically identical to those of the leaf of the adult tree.

In contrast, the pairs of leaves of young and adult trees of the Euphorbiaceae are noticeable for their remarkable differences in the venation patterns.

The number of secondaries per lamina has proved to be the most typical (measurable) venation characteristic of the species. A somewhat less important degree of specification can be applied to the criterion "average distance of the secondaries" and the vein islet number/cm^2.

While the xylems of the Sapotaceous species, whose cross sections of the vascular bundles were used for the calculation of the conductive volume, possess a kidney-shaped outline with transversal elongation, the xylems of the Lauraceae are shaped like a semicircle or a half oval, the cross sections of the leaves of adult trees of the Euphorbiaceae fail to show any common characteristics.

4.2. Ecological significance of the anatomical investigations

On average, the Euphorbiaceous leaves of adult plants are the largest, whereas the Lauraceae have the largest leaves of young trees.

The greatest number of secondary veins per lamina by far was counted for the Sapotaceae (leaves of adult and young plants).

The Lauraceae possess the greatest mean distances between the secondaries (leaves of adult and young trees).

The densest vein network/cm^2 of the leaf blade is found in the Sapotaceae (leaves of adult and young plants).

On average the highest rate of ramification/vein islet is exhibited by the Euphorbiaceae (leaves of adult and young trees).

The Sapotaceous leaves of adult plants as well as those of the young trees of the Lauraceae have the greatest number of veinlet terminations/cm^2 on average.

Considered alone the comparison of the area indexes (quotient of leaf area : number of secondaries) of leaves of adult and young trees leads to the same result in all three families: the data are higher in the groups of the leaves of young plants. That means that a greater area of the blade is supplied by one secondary vein than in the leaves of the adult plants. The reason is to be found in the larger leaf area of the $leaf_{YT}$ (Sapotaceae) or the wider distances between the secondaries (Lauraceae, Euphorbiaceae). This result of the comparison is the first demonstration of the generally more intense, denser supply of veins in the leaves of the adult trees.

A comparative consideration of the tree height category to which the leaves of the adult plants belong, and the results of the venation measurements fail to reveal any correlations, regarding the height category in relation to the vein islets/cm^2 and the ramifications per islet. The remaining data indicate tendencies which could be proved with the help of a larger sample collection. The following tendencies are discernible: maxima of secondaries/lamina in the height category "A"; in "A" more species with wider distances between the secondaries; mean values of veinlet terminations/cm^2 in "a", as well as an average range, lower and higher extremes in "A"; highest vein lengths/cm^2 in "A"; a classification of gross conductive volume and tree height not applicable to all three families, but higher values within the "A" species of the Sapotaceae in comparison to the Euphorbiaceae; gross volumes of the Lauraceae from "A" include lower and higher extremes; extreme values for the net conductive volume for the three families in "A", a difference between the Sapotaceae and Euphorbiaceae with regard to the gross volume, hardly visible for the net volume.

The atypically small vein lengths/cm^2 for nine species from "A" may be compensated for by different anatomical adaptations (cf. Roth, 1984).

The consideration of the gross volumes of the specimens from each of the families reveals both for the leaves of the adult trees as well as for those of the young plants, the highest mean values within the Lauraceae, followed by the Sapotaceae and Euphorbiaceae.

While the net conductive volumes of the Lauraceae comprise only ca. 4% to ca. 17% of the gross volume, this lies between ca. 11% and ca. 31% within the Sapotaceae and ca. 11% to ca. 38% for the Euphorbiaceae.

Comparison of the vein density of leaves of adult and young trees

66.67% of the leaves of adult trees from the nine pairs compared, have a higher vein length/cm^2 than the leaves of young plants.

A comparison of the gross conductive volumes/cm^2 also reveals higher data in the group of the adult trees: among the nine pairs the gross volume is higher in 77.78% within the leaves of adult trees.

The quotient vein length/cm^2 and vein islets/cm^2 (Q_{ID}) yields lower data for 89.89% of the available pairs of leaves in the group of adult trees. It follows that there is a greater vein islet number per cm^2 in relation to the vein length among the leaves of adult plants as opposed to the young trees.

A consideration of the height category to which the leaves of adult trees belong, underlines the fact that the upper and lower extreme values of the quotient appear mainly in "A".

The calculation of the quotient vein length/cm^2 : vein islets + veinlet terminations/cm^2 (Q_{VD_1}) leads to the result that within the pairs of leaves studied, 71.43% of the leaves of adult plants possess a higher vein density in comparison to the corresponding leaf of the young tree.

The highest and smallest quotients were also ascertained for the leaves of adult trees from "A".

The result of the higher vein density of the leaves of adult trees is borne out by another quotient (vein length/cm^2 : vein islets + ramifications + veinlet termination/cm^2, Q_{VD_2}).

This quotient failed to reveal any correlation with the height of the tree.

Within the group of the leaves of young plants no correlation between the height of trees and the quotients could be found.

The quotients vein length/cm^2 : vein islets (+ ramifications) + veinlet terminations/cm^2 ($Q_{VD_{1/2}}$) have proved to be the most suitable for the determination of the vein density and its comparison with the leaves of adult and young trees.

5. CORRELATION BETWEEN VEIN DENSITY AND STRATIFICATION OF THE TREES – FURTHER STUDY

In order to answer the question raised in the first study, whether there is a relation between the origin of the leaves from different strata of the rain forest and the vein density, this aspect was studied more deeply within a much larger sample collection of the same region. The determination of the vein density was carried out as a blind study, also on permanent preparations of the middle of the leaf.[1] As the taxonomical origin of the species is not important for the problem mentioned, this point of view is not taken into consideration.

This sample collection comprises 84 tropical representatives from 79 species (including eight species only known by their vernacular name) which belong to at least 43 genera and 15 families. Four families yield 11 (adult) leaves of young trees.

Instead of the vein length per cm^2 of the lamina as one of the parameters determining the vein density I have chosen, for this analysis, the number of points of intersection which the vascular bundles and the borders of a square millimeter have in common. The other two parameters, which are necessary for the calculation of the quotient of the vein density are once again the vein islet and veinlet termination number per mm^2 of the blade. A verification on typical examples of the previous study proved that this modified method, too, allows valid statements about the vein density – provided that one is aiming only at a comparative study.

The three parameters of the vein density were assessed by putting a piece of transparent foil of millimetre paper over the preparations. The numbers of vein islets, terminations and points of intersection were counted for 1 or 4 mm^2. After performing the division "points of intersection : vein islets + veinlet terminations" I had 84 quotients of the vein density. These were divided into groups according to the 10 decimal places. The number of the decimal places was equally distributed to the height categories "A", "a" and "aa". Thus, the smallest quotients (smallest meshed and densest venation) were presumed as belonging to the trees of the "A" stratum. Mean data must thus characterize "a" trees, while the highest quotients (fewer vein islets and terminations, less dense venation) fit the height category "aa". Blind study means that the values of the vein density were classified without taking a prior look at the strata of the sample leaves. The classification was introduced on the assumption that higher vein densities occur in the upper strata (cf. discussion on page 179). The assessed heights of the trees (Roth, 1984) were then compared with the expected ones and the percentage of correct classifications was determined. The height categories expected are listed in Table 59. This table also contains the results of the validation by means of the real tree strata.

The verification was made from three points of view:

(a) absolute agreement of expected and assessed height category
(b) quotients which can be regarded as transitions between two strata
(c) quotients leading to a tree height, which differs completely from Roth's (1984) data [cf. Table 59: (!)]

[1] I thank Prof. I. Roth for providing the permanent slides.

Table 59. Assignment of the quotients of the vein density (Q_{VD}) to the stratification of the forest.

Family	Species	Q_{VD}[1)]	Expected tree height[2)]	Assessed stratum[3)]
Apocynaceae	Himatanthus articulata	0.784	A	A
	Aspidosperma album	0.697	A	A
	"Leche de burra"	1.155	a	A
Burseraceae	Protium decandrum? ("Caraño blanco")	1.948	aa	a
	Protium decandrum ("Azucarito")	1.033	a (transition A → a)	A
	Protium neglectum ("Caraño")	0.827	A	a
	Protium neglectum ("Azucarito blanco")	0.598	a	A
	Protium sp.	0.916	A	A
	Tetragastris panamensis? ("Caraño negro")	0.945	A (transition a → A)	a
	Tetragastris or Dacryodes ("Aracho blanco")	0.817	A	a
	"Maro"	0.678	A	a
Caesalpiniaceae	Peltogyne porphyrocardia	1.022	a (transition A → a)	A
	Peltogyne pubescens	0.578	A	A ?
	Peltogyne sp.	1.386	a	A
	Crudia glaberrima	0.704	A	A
	Crudia oblonga	0.771	A	A
	Sclerolobium paniculatum	0.699	A	A
	Sclerolobium sp.	0.965	A	A
	Mora excelsa	1.364	a	A
	Hymenaea courbaril	1.089	a (transition A → a)	A
	Dialium guianense	0.843	A	A

Table 59 (Continued)

Family	Species	Q_{VD}[1)]	Expected tree height[2)]	Assessed stratum[3)]
	Brownea latifolia	0.9	A (transition a → A)	a
Lecythidaceae	Eschweilera chartacea	0.631	A	A
	Eschweilera grata	1.052	a (transition A → a)	A
	Eschweilera odora	0.668	A	A
	Eschweilera corrugata	0.961	A	A
	Eschweilera subglandulosa	0.943	A	A
	Eschweilera cf. trinitensis	1.03	a (transition A → a)	A
	Eschweilera sp.	0.944	A	A
	Lecythis davisii	0.793	A	A
	Couratari multiflora	1.073	a (transition A → a)	A
	Couratari pulchra (young tree)	0.745	A (!)	aa
Melastomaceae	Mouriria sideroxylon	1.465	a	A
	Mouriria huberi	0.841	A	A
	"Saquiyak"	0.763	A (!)	aa
Meliaceae	Trichilia propingua	1.82	aa	a
	Trichilia smithii	1.51	aa	a
	Trichilia schomburgkii	1.016	a	a
	Guarea schomburgkii (young tree)	1.646	aa	aa
Mimosaceae	Inga capitata	0.642	A	A
	Inga rubiginosa	0.672	A (!)	aa
	Inga scabriuscula	0.513	A	a
	Inga alba	0.629	A	A

Table 59 (Continued)

Family	Species	Q_{VD}[1)]	Expected tree height[2)]	Assessed stratum[3)]
	Inga sp. ("Guamo blanco")	0.752	A	A
	Inga sp. ("Guamo caraoto")	0.564	A	A
	Piptadenia psilostachya	0.874	A	A
	Pithecellobium cf. claviflorum	0.598	A	A
	Parkia pendula	2.459	aa (!)	A
	Parkia oppositifolia	0.762	A	A
	Pentaclethra macroloba	0.927	A (transition a → A)	a
Moraceae	Clarisia racemosa	1.168	a	A
	Brosimum ? ("Charo peludo")	0.727	A	A ?
	Helicostylis tomentosa (young tree)	1.184	a	aa
Myrtaceae	Eugenia compta	0.887	A	A
	Calycorectes sp.	1.308	a	a
	"Guayabito piedrero"	0.803	A	a
	"Guayabito piedrero chiquito"	0.736	A	a ?
	"Guayabito negro"	1.328	a	a
	"Guayabito zaba"	0.826	A	a
Ochnaceae	Ouratea sagotii	1.109	a	a
Rosaceae	Licania hypoleuca	0.511	A	A
	Licania sp.	1.354	a	A
	Licania alba	0.757	A	A
	Licania densiflora	0.454	A	A

Table 59 (Continued)

Family	Species	Q_{VD}[1]	Expected tree height[2]	Assessed stratum[3]
	(adult tree)			
	(young tree)	1.518	aa	aa
	Prunus	1.077	a	A
	shaerocarpa (young tree)		(transition A → a)	
	Couepia glandulosa	1.138	a	A
	Hirtella americana	0.582	A	A
	Hirtella davisii	0.884	A	A
	Parinari rodolphii (adult tree)	0.978	A	A
	(young tree)	1.28	a	a
	Parinari excelsa (adult tree)	0.668	A	A
	(young tree)	0.937	A (!)	aa
Rubiaceae	Genipa americana	1.55	aa	aa
	Duroia sp.	0.578	A (!)	aa
	Amaioua guianensis (young tree)	0.92	A (transition a → A)	a
	Faramea torquata (young tree)	1.525	aa (transition a → aa)	a
Sterculiaceae	Sterculia pruriens	0.633	A	A
Tiliaceae	Apeiba echinata (adult tree)	0.627	A	A
	(young tree)	0.925	A (!)	aa
Vochysiaceae	Erisma uncinatum	0.794	A	A
	Vochysia lehmannii	1.186	a	A
	Qualea dinizii (adult tree)	0.563	A	A
	(young tree)	0.61	A	A

1) Points of intersection /mm^2: number of vein islets + veinlet terminations/mm^2.
2) Classification of the transitional values as correct assignments in consideration of a certain variability within the vein density.
3) According to Roth (1984).

Results

The consideration of the 84 quotients of the vein density (Q_{VD}) led to the classification as shown in Figure 51.

Quotients of correct assignment of the height categories expected, cf. Figure 52.

Whole sample collection

For 45 leaves of the 84 tropical representatives studied the assignment was correct (= 53.6%).

This percentage is distributed as follows to the three strata:

54 A-species: 37 correct assignments (= 68.5%)
20 a-species: 5 correct assignments (= 25%)
10 aa-species: 3 correct assignments (= 30%)

Including the 12 transitional quotients (= 14.3%) between the height categories "A" → "a" (7/8.3%), "a" → "A" (4/4.8%) and "a" → "aa" (1/1.2%) the number of, in principle, correct classifications rises to 57 (= 67.9%):

54 A-species: 7 transitions from "A" → "a" (= 13%)
20 a-species: 4 transitions from "a" → "A" (= 20%)
1 transition from "a" → "aa" (= 5%)

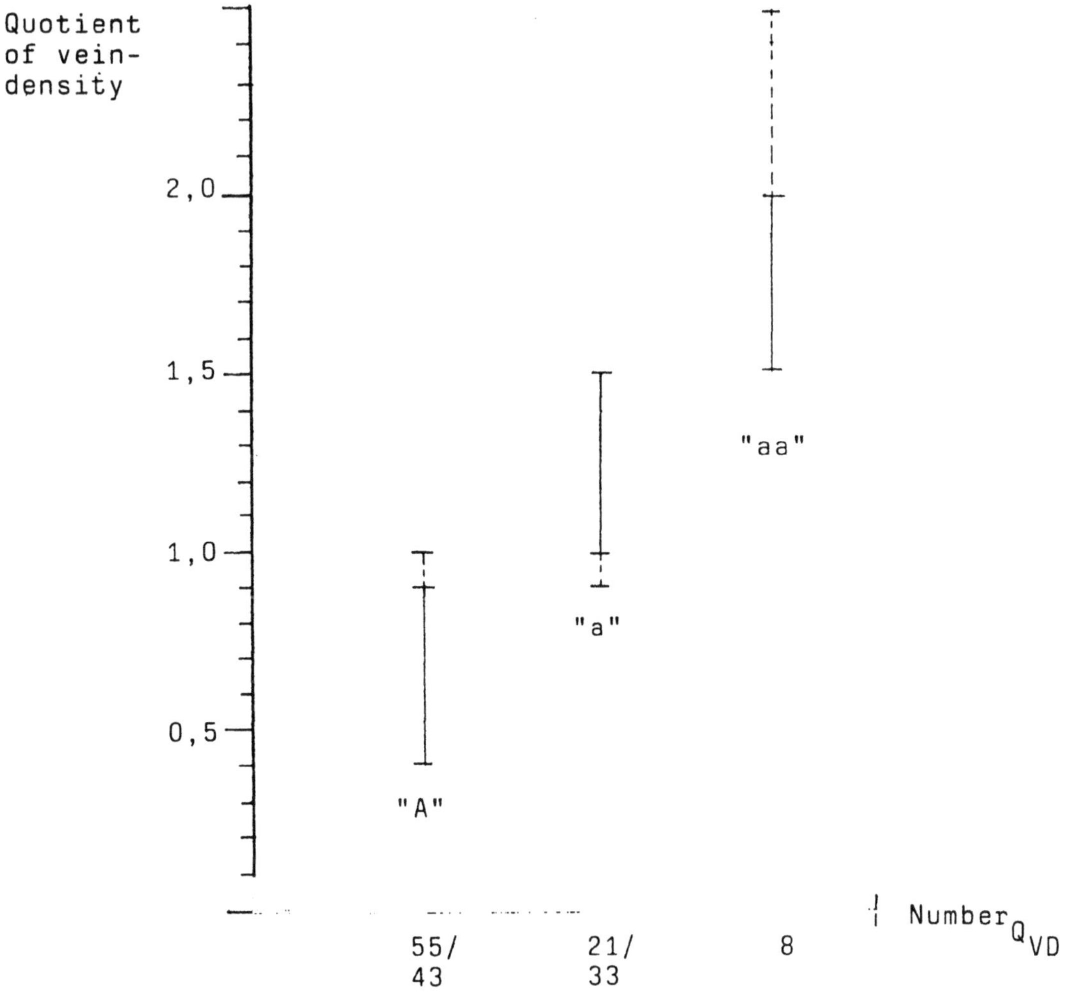

Fig. 51.

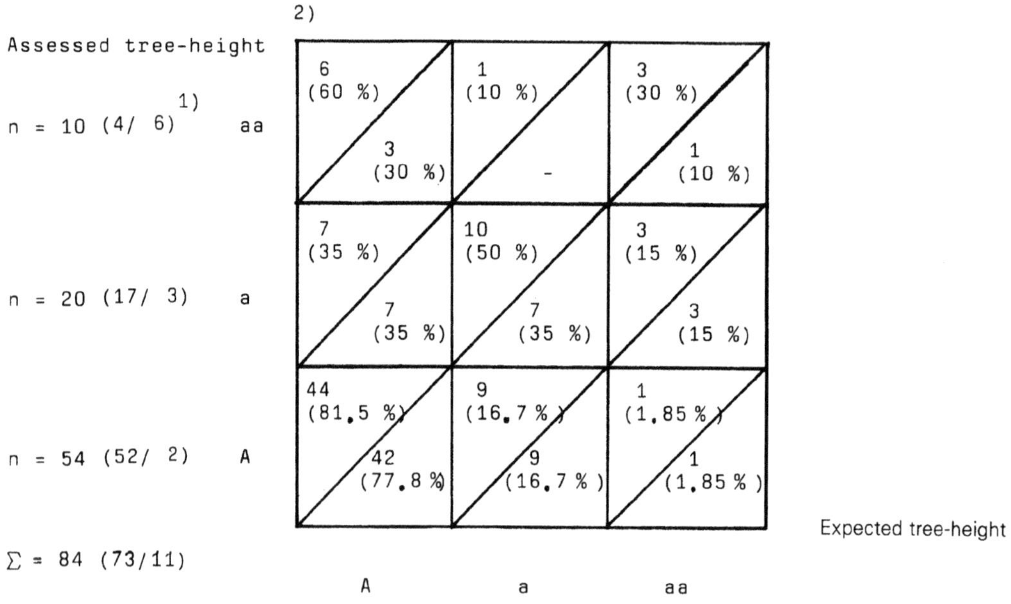

Fig. 52. Distribution of the assignments of assessed and expected tree height (transitions between two strata included). Note: (1) Number of samples of the height category (leaves of adult trees/leaves of young trees). (2) On the left: total of samples (of the height cateory), on the right: number of leaves of adult trees (of the height category).

10 aa-species: —

20 leaves (= 23.8%) were classified wrong, differing between two neighbouring strata:

54 A-species: 9 wrong assignments between neighbouring strata (= 16.7%)

20 a-species: 10 wrong assignments between two neighbouring strata (= 50%)

10 aa-species: 1 wrong assignment between two neighbouring strata (= 10%)

For 7 leaves (= 8.3%) the classification expected differed so much that the adjacent height was omitted:

54 A-species: 1 omitted height category (= 1.85%)

20 a-species: —

10 aa-species: 6 omitted height categories (= 60%).

Leaf material of the adult trees

For 41 leaves of 73 representatives (= 56.2%) the assignment was correct:

52 A-species: 36 correct classifications (= 69.2%)

17 a-species: 4 correct classifications (= 23.5%)

4 aa-species: 1 correct classification (= 25%)

Including the 9 transitional quotients (= 12.3%) between "A" → "a" (6/8.2%) and "a" → "A" (3/4.1%), the number of principally correct classifications rises to 50 leaves (= 68.5%):

52 A-species: 6 transitions from "A" → "a" (= 11.5%)

17 a-species: 3 transitions from "a" → "A" (= 17.6%)

4 aa-species:-

19 samples were assigned wrongly (= 26%), differing between two neighbouring strata:

52 A-species: 9 wrong classifications (= 17.3%)

17 a-species: 10 wrong classifications (= 58.8%)

4 aa-species:-

For 4 leaves (= 5.5%) the classification expected differed so much that the adjacent height was omitted:

52 A-species: 1 omitted height category (= 1.92%)

17 a-species:-
4 aa-species: 3 omitted height categories (= 75%)

Discussion

The assignment of the quotients of the vein density to the three horizontal strata shows a clear tendency of a relation between the vein density and the height of the trees. A verification with the assessed height categories of the leaves studied (cf. Roth, 1984) led to this result.

The two quotes of more than 65% allow us to reason — within this sample collection — that the vein density per unit area of the lamina correlates with the height of the tree. The tendency is improved by considering only the leaves of the adult plants.

From the positive correlation between the Q_{VD} and the stratification of the forest the not unexpected conclusion can be drawn that the vein density increases with increasing height of the tree.

As it follows from the distribution of the wrong assignments (Figure 52), the transition of the vein density/mm^2 seems to be rather fluid between "A" and "a", because here the maximum of wrong classifications is visible.

A possible reason might be that the individuals concerned are somewhat higher than 30 m (already "A") or slightly below this height (still "a").

Figure 52 reveals that the wrong classifications from "aa" to "A" belong mostly to the group of the leaves of young trees — a fact which requires further study.

The observation in my previous study that the extreme data of the Q_{VD} occur in the stratum "A" does not hold true for this larger collection of leaf material. Thus, it is demonstrated once again that extreme values of the quotient in "A" are exceptions which have nothing to do with the character of the leaf. Those Q_{VD} decrease in number according to the amount of the available sample leaves.

ACKNOWLEDGEMENTS

I would like to express my gratitude to Prof. Dr I. Roth for having placed the sample collection and the subject to my disposal. In particular I thank her and also Prof. Dr H. Lieth with his team as well as the members of the section "Zoophysiologie" for their advice and cooperation as to functional questions.

Finally, I am very grateful to Mrs. B. Teschke who revised this English translation critically.

REFERENCES

Avita, S., Rao, N. V. and Inamdar, J. A. (1981): Studies on the leaf architecture of the Ranunculaceae. In: Flora, 171 (3), pp. 280—298.

Banerjee, G. (1980): Mature venation pattern in certain members of Amaranthaceae. In: Flora, 169 (1), pp. 104—110.

Biebl, R. and Germ, H. (1967): Praktikum der Pflanzenanatomie. 2. ed., Wien, New York, Springer.

Boyd, G. A. (1955): Autoradiography in biology and medicine. New York, Academic Press Inc.

Braune, W., Leman, A. and Taubert, H. (1983): Pflanzenanatomisches Praktikum I. 4. ed., Stuttgart, G. Fischer.

Buckley, R. C., Cortlett, R. T. and Grubb, P. J. (1980): Are the xeromorphic trees of tropical upper montane rain forests drought-resistent? In: Biotropica, 12, pp. 124—136.

Busse, A. (1914): Vergleichende Untersuchungen der Blumen-, Kelch- und Laubblätter der Ranunculaceen. Thesis, Kiel.

Charlier, A. (1906): Contribution à l'étude anatomique des plantes à guttapercha et d'autres Sapotacées. In: Journ. de Botanique, 20, pp. 22—77.

Chonan, N., Kaneko, M., Kawahara, H. and Matsuda, T. (1981): Ultrastructure of the large vascular bundles in the leaves of rice plant. In: Jap. J. Crop Sci., 5 (3), pp. 323—331.

Cooper, A. (1980): An ecological and phytogeographic study on the embankment flora of the Linden highway. Guyana. In: Carib. J. Sci. 15, pp. 139—147.

Cronquist, A. (1945a): Studies in the Sapotaceae — I. The North American species of *Chrysophyllum*. In: Bull. Torrey Bot. Club, 72, No. 2, pp. 192—205.

Cronquist, A. (1945b): Studies in the Sapotaceae — IV. The North American species of *Manilkara*. In: Bull. Torrey Bot. Club, 72, No. 6, pp. 550—562.

Cronquist, A. (1946a): Studies in the Sapotaceae — V. The South American species of *Chrysophyllum*. In: Bull. Torrey Bot. Club, 73, No. 3, pp. 286—311.

Cronquist, A. (1946b): Studies in the Sapotaceae — VI. Miscellaneous notes. In: Bull. Torrey Bot. Club, 73, No. 5, pp. 465—471.

Czihak, G., Langer, H. and Ziegler, H. (Eds.) (1981): Biologie. Ein Lehrbuch für Studenten der Biologie. 3. ed., Berlin, Heidelberg, New York, Springer.

Datta, P. C. and Biswas, Ch. (1968): Pharmacognostic study of the leaf and bark of *Barleria prionitis* Linn. In: Quart. J. Crude Drug Research, 8, pp. 1161 ff.

Datta, P. C. and Maiti, R. K. (1968): Pharmacognostic study on *Ecbolium linneanum* Kurz var. *dentata* Clarke. In: Quart. J. Crude Drug Research, 8, pp. 1189—1194.

Dean, J. M. and Smith, A. P. (1978): Behavioral and morphological adaptations of a tropical plant to high rainfall. In: Biotropica, Vol. 10, No. 2, pp. 152—154.

Delevoryas, T. (1967): Prinzipien der Pflanzenphylogenie. München, Basel, Wien: Bayerischer Landwirtschaftsverl. (Moderne Biologie).

Dilcher, D. L. (1974): Approaches to the identification of angiosperm leaf remains. In: Bot. Rev., Vol. 40, No. 1, pp. 1—157.

Encke, F., Buchheim, G. and Seybold, S. (1984): Zander. Handwörterbuch der Pflanzennamen. 13. ed., Stuttgart, Eugen Ulmer.

Engler, A. (1890): Beiträge zur Kenntnis der Sapotaceae. In: Botanische Jahrbücher für Systematik, Pflanzengeschichte und Pflanzengeographie, 12, pp. 496—525.

Engler, A. (1964): Syllabus der Pflanzenfamilien. II. Bd. Angiospermen, Übersicht über die Florengebiete der Erde. 12. ed., Berlin-Nikolassee, Gebr. Borntraeger.

Ensign, M. R. (1921): Area of vein-islets in leaves of certain plants as an age determinant. In: Am. J. Bot., 8, pp. 433—441.

Esau, K. (1969): Pflanzenanatomie. 2. ed., Stuttgart, G. Fischer.

Esau, K. (1977): Anatomy of seed plants. 2. Auflage. New York, Santa Barbara, London, Sydney, Toronto, J. Wiley and Sons.

Eschrich, W. (1976): Strasburger's kleines Botanisches Praktikum für Anfänger. 17., ed. Stuttgart, New York, G. Fischer.

Ettinghausen, C. Ritter von (1854): Über die Nervation der Blätter und blattartigen Organe bei den Euphorbiaceen, mit besonderer Rücksicht auf die vorweltlichen Formen. In: Sitzungsber. der kaiserl. Akademie der Wiss. mathemat.-naturwiss. Cl., 12 (1), pp. 138—154.

Ettinghausen, C. (1861): Die Blatt-Skelette der Dikotyledonen. Mit besonderer Rücksicht auf die Untersuchung und Bestimmung der fossilen Pflanzenreste. Wien, Druck und Verl. d. Kaish.- Kön. Hof- u. Staatsdruckerei.

Fisher, D. B. (1967): An unusual layer of cells in the mesophyll of the soybean leaf. In: Bot. Gaz., 128, pp. 215—218.

Fisher, D. G. (1985): Morphology and anatomy of the leaf of *Coleus blumei* (Lamiaceae). In: Am. J. Bot. 72 (3), pp. 392—406.

Fitting, H. (1926): Die ökologische Morphologie der Pflanzen im Lichte neuerer physiologischer und pflanzengeographischer Forschungen. Jena, Fischer.

Fitting, H. (Eds.) (1934): Jahrbücher für wissenschaftliche Botanik. Vol. 79, Leipzig, Gebr. Borntraeger.

Foster, A. S. (1949): Practical plant anatomy. 2. ed. Toronto, New York, London: D. van Nostrand Company, Inc.

Foster, A. S. (1950b): Venation and histology of the leaflets in *Touroulia guianensis* Aubl. and *Froesia tricarpa* Pires. In: Am. J. Bot., 37, No. 10, pp. 848—862.

Foster, A. S. (1950c): Techniques for the study of venation patterns in the leaves of angiosperms. In: Proceedings of the International Biological Congress, 7, pp. 586f., Stockholm.

Foster, A. S. (1952): Foliar venation in angiosperms from an ontogenetic standpoint. In: Am. J. Bot., 39, pp. 752—766.

Foster, A. S. and Gifford, E. M. Jr. (1974): Comparative morphology of vascular plants. 2. ed., San Francisco, Freeman and Company.

Fremdwörterbuch, naturwissenschaftlicher und mathematischer Begriffe. Vol. 1 and 2, 4. ed., Köln, Aulis, Deubner & Co. KG, 1982.

Geiger, D. R. and Cataldo, D. A. (1969): Leaf structure and translocation in sugar beet. In: Plant Physiol., 44, pp. 45 ff.

Geographie Heute, Juni (1983): Die Grenzen der Tropen. No. 17, Seelze, Friedrich.

Gerlach, D. (1984): Botanische Mikrotechnik. Eine Einführung. 3. ed., Stuttgart, Thieme.

Gessner, F. (1956a): Die Wasseraufnahme durch Blätter und Samen. In: Ruhland, W., Ed., Handb. der Pflanzenphysiol., Vol. 3, Berlin, Göttingen, Heidelberg, Springer, pp. 215—246.

Gessner, F. (1960): Die Assimilationsbedingungen im tropischen Regenwald. In: Ruhland, W., Ed., Handb. der Pflanzenphysiol., Vol. 5 (2), Berlin, Göttingen, Heidelberg, Springer, pp. 492—505.

Glück., H. (1919): Blatt- und blütenmorphologische Studien. Jena, G. Fischer.

Gupta, B. (1961a): Correlation of tissues in leaves. I. Absolute vein islet numbers and absolute veinlet termination numbers. In: Ann. Bot., 25, pp. 65 ff.

Gupta, M. (1978): Cotyledon architecture in Trifolieae. In: Acta Bot. Indica, 6, pp. 171—176.

Gupta, B. and Kundu, B. C. (1965): Determination of average vein islet, veinlet termination and stomatal numbers of a leaf. In: Planta medica, 13, pp. 247—256.

Haberlandt, G. (1909): Physiologische Pflanzenanatomie. 4. ed., Leipzig, Engelmann.

Hall, J. P. and Melville, C. (1951): Veinlet termination number. A new character for the differentiation of leaves. In: J. Pharm. Pharmacol., 3, pp. 934—941.

Hall, J. P. and Melville, C. (1954): Veinlet termination number some further observations. In: J. Pharm-Pharmacol., 6, pp. 129—133.

Herbst, D. (1971): Disjunct foliar veins in Hawaiian Euphorbias. In: Science, 171, pp. 1247—1248.

Hickey, L. J. (1979): A revised classification of the architecture of dicotyledonous leaves. In: Metcalfe, C. R., Chalk, L., Anatomy of the dicotyledons, Vol. 1, 2. ed., Oxford, Clarendon Press, pp. 25—39.

Hickey, L. J. and Doyle, J. A. (1973): Fossil evidence on evolution of angiosperm leaf venation. (Abstract), In: Am. J. Bot., 59, p. 661.

Hifny Saber, A., Mahran, G. H. and El-Alfy, T. S. (1968): The macro- and micromorphology of *Ranunculus scleratus* L. growing in Egypt. Bull. Fac. Pharm. Cairo Univ., 7, pp. 119—134.

Hifny Saber, A., Mahran, G. H. and Salah Ahmed, M. (1969): *Thevetia nereifolia* Juss. growing in Egypt. Part III Macro- and micromorphology of the stems and leaves. Bull. Fac. Pharm. Cairo Univ., 8, pp. 145—159.

Hilgemann, W., Kettermann, G. and Hergt, M. (1975): dtv-Perthes-Weltatlas. Großräume in Vergangenheit und Gegenwart. Vol. 3 Südamerika., München, Darmstadt, Deutscher Taschenbuchverl. &. J. Perthes.

Hill, R. S. (1980): A numerical taxonomic approach to the study of angiosperm leaves. Bot. Gaz., 141 (2), pp. 213—229.

Högermann, C. (1990): Untersuchungen zur Korrelation zwischen Leitbündeldichte und Waldschichtung an tropischen Vertretern von 79 Species aus dem Regenwald von Venezolanisch Guayana. In: Flora, 184, 1990, in press.

Höster, H.-R. (1962): Das Adernetz der Blätter. Methoden zur Darstellung des Adernetzes. Mikrokosmos, 51, pp. 6—8.

Holle, G. (1892): Über den anatomischen Bau des Blattes in der Familie der Sapotaceen und dessen Bedeutung für die Systematik. Inaugural-Dissertation, Erlangen.

Huber, B. (1924): Die Beurteilung des Wasserhaushaltes der Pflanze. Ein Beitrag zur vergleichenden Physiologie. Jahrb. f. wiss. Bot., 1, pp. 1—120.

Huber, B. (1961): Grundzüge der Pflanzenanatomie. Berlin, Göttingen, Heidelberg, Springer.

Hueck, K. (1966): Die Wälder Südamerikas. Stuttgart, G. Fischer.

Index Kewensis (1977ff): An enumeration of the genera and species of flowering plants. Koenigstein, Otto Koeltz Science Publishers.

Jurasky, K. A. (1943): Kutikular-Analyse. Grundlegendes zur folgerichtigen Auswertung einer Methode. Biologia generalis, 10, pp. 383—402.

Kapoor, S. L. and Mitra, R. (1979): Epidermal and venation studies in Apocynaceae — VI. Bull. Bot. Surv. India, 21 (1—4), pp. 68—80.

Keller, B. A. and Keller, E. F. (1927): Materialen zur ökologischen Anatomie der Gattung *Betula* und *Pinguicula vulgaris* L. Veröffentlichungen des Geobotanischen Institutes Rübel in Zürich, 4, pp. 96 ff.

Killian, Ch. (1942): *Bromus rubens* L. Contribution à l'étude des plantes annuelles xérophytiques du désert. Ber. Schweiz. Bot. Ges., 52, pp. 215 ff.

Kim, K. S. and Kim, M. H. (1984): Systematic studies on some Korean woody plants — venation patterns of Lauraceae. Korean J. Bot. 27 (1), pp. 15—24.

Lam, H. J. (1939): On the system of the Sapotaceae, with some remarks on taxonomical methods. Rec. Trav. Bot. Néerl., 36, pp. 509—525.

Larcher, W. (1983): Physiological plant ecology. 2. ed., Berlin, Springer.

Larcher, W. (1984): Ökologie der Pflanzen. 4. ed., Stuttgart, UTB 232 Ulmer.

Lawrence, G. H. M. (1965): Taxonomy of vascular plants. New York, Macmillan Company.

Lems, K. (1964): Evolutionary studies in the Ericaceae. II. Leaf anatomy as a phylogenetic index in the Andromedeae. Bot. Gaz., 125 (3), pp. 178 ff.

Levin, F. A. (1929): The taxonomic value of vein islet areas. Quart. J. Pharm. Pharmacol., 2, pp. 17—43.

Metcalfe, C. R. and Chalk, L. (1965): Anatomy of the dicotyledons. Vol. I and II, Oxford, Clarendon Press and 2. ed. from 1972.

Manze, U. (1967): Die Nervaturdichte der Blätter als Hilfsmittel der Paläoklimatologie. Thesis, Köln.

Melville, R. (1976): The terminology of leaf architecture. Taxon, 25 (5/6), pp. 549—561.

Ministro de Agricultura y Cria (1961): Direction de Recursos naturales renovables V.A. Giménez Landinez, Atlas forestal de Venezuela. Caracas, Ministro de Agricultura y Cria.

Mohan, J. S. S. and Inamdar, J. A. (1983): Studies on leaf architecture of the Oleaceae with a note on the systematic position of the genus *Nyctanthes*. Feddes Repert., 94 (3, 4), pp. 201—211.

Mouton, J.-A. (1966a): Sur la systématique foliaire en paléobotanique. Bull. Soc. bot. Fr., 113 (9), pp. 492 ff.

Mouton, J.-A. (1966b): Les types biologiques foliaires de Raunkiaer. Etat actuel de la question. Bull. Soc. bot. Fr., Mémoires, pp. 28 ff.

Mouton, J.-A. (1967): Architecture de la nervation foliaire. 92e Congrés national des sociétés savantes, Strasbourg et Colmar, pp. 165 ff.

Mouton, J.-A. (1972): Une nouvelle méthode d'isolement de la nervation des feuilles d'arbres. Bull. Soc. bot. Fr., 119, pp. 581 ff.

Mouton, J.-A. (1976): La biométrie du limbe: mise au point de nos connaissances. Bull. Soc. bot. Fr., 123 (3/4), pp. 145 ff.

Müller, E. (1944): Die Nervatur der Nieder- und Hochblätter. Botanisches Archiv. Zeitschrift für die gesamte Botanik und ihre Grenzgebiete, Vol. 45, edited by A. Seybold and W. Troll, pp. 1—92.

Napp-Zinn, K. (1966): Anatomie des Blattes. I. Blattanatomie der Gymnospermen. In: Handbuch der Pflanzenanatomie Vol. VIII, Part 1, 2. ed. Berlin-Nikolassee, Gebr. Borntraeger.

Napp-Zinn, K. (1973): Anatomie des Blattes. II. Blattanatomie der Angiospermen. In: Handbuch der Pflanzenanatomie, Vol. VIII, Part 2A, 1, 2. ed., Berlin, Stuttgart, Gebr. Borntraeger.

Napp-Zinn, K. (1974): Anatomie des Blattes. II. Blattanatomie der Angiospermen. In: Handbuch der Pflanzenanatomie, Vol. VIII, Part 2A, 2, 2. ed., Berlin, Stuttgart, Gebr. Borntraeger.

Payne, R. W. and Preece, D. A. (1977): Incorporating checks against observer error into identification keys. New Phytologist, 79, pp. 203—209.

Perry, D. R. (1985): Die Kronenregion des tropischen Regenwaldes. Spektrum der Wissenschaft, 1, pp. 76—85.

Perthes-Transparente (undated): Südamerika I und II. Darmstadt, J. Perthes-Geographische Verlagsanstalt.

Pisek, A. and Cartellieri, E. (1931): Zur Kenntnis des Wasserhaushaltes der Pflanzen. I. Sonnenpflanzen. Jahrb. f. wiss. Bot., 75 (2), pp. 195—251.

Pisek, A. and Cartellieri, E. (1933): Zur Kenntnis des Wasserhaushaltes der Pflanzen. III. Alpine Zwergsträucher. Jahrb. f. wiss. Bot., 79 (1), pp. 131—190.

Plymale, E. L. and Wylie, R. B. (1944): The major veins of mesomorphic leaves. Am. J. Bot. 31, pp. 99—106.

Prantl, K. (1883): Studien über Wachstum, Verzweigung und Nervatur der Laubblätter, insbesondere der Dicotylen. Ber. Dtsch. Bot. Ges., 1, pp. 280—288.

Pray, T. R. (1954): Foliar venation of angiosperms. I. Mature venation of *Liriodendron*. Am. J. Bot., 41, pp. 663—670.

Pray, T. R. (1955a): Foliar venation of angiosperms. II. Histogenesis of the venation of *Liriodendron*. Am. J. Bot., 42, pp. 18—27.

Pray, T. R. (1955b): Foliar venation of angiosperms. III. Pattern and histogenesis of the venation of *Hosta*. Am. J. Bot., 42, pp. 611—618.

Pray, T. R. (1955c): Foliar venation of angiosperms. IV. Histogenesis of the venation of *Hosta*. Am. J. Bot., 42, pp. 689—706.

Pray, T. R. (1963): Origin of vein endings in angiosperm leaves. Phytomorphology, 13, pp. 60 ff.

Precht, M. (1977): Bio-Statistik. Eine Einführung für Studierende der biologischen Wissenschaften. München, Oldenbourg.

Pyykkö, M. (1979): Morphology and anatomy of leaves from some woody plants in a humid tropical forest of Venezuelan Guayana. Acta Bot. Fennica, 112, pp. 1—41.

Ravindranath, K. and Inamdar, J. A. (1982): Leaf architectural studies in the Asteraceae. Pak. J. Bot., 14 (2), pp. 143—154.

Record, S. J. (1938): The American woods of the family Euphorbiaceae. Tropical Woods, 54, pp. 7—40.

Record, S. J. (1939): American woods of the family Sapotaceae. Tropical Woods, 59, pp. 21—51.

Richards, P. W. (1952): The tropical rain forest. An ecological study. Cambridge, University Press (1952 and 2. ed., 1976).

Riedl-de Haen (undated): Mikroskopie-Chemikalien: Farbstoffe, Farbstofflösungen, Hilfsmittel. Seelze.

Riedel, W. (1983): Die Tropen heute. Geographie heute, 17, 4. year's set, pp. 2—14.

Rollet, B. (1969a): Etudes quantitatives d'une forêt dense humide sempervirente de plaine de la Guayane Vénézuélienne. Thèse Doctorat, Toulouse.

Rollet, B. (1969b): Etudes quantitatives d'une forêt dense humide sempervirente de plaine en Guayane Vénézuélienne. Travaux du Laboratoire Forestier de Toulouse, T1, Vol. 8, Art. 1, pp. 1—36.

Rollet, B. (1969c): La régénération naturelle en forêt dense humide sempervirente de plaine de la Guayane Vénézuélienne. Revue Bois et Forêts des Tropiques, No. 124, pp. 19—38.

Rollet, B. (to appear): Stratification of tropical forests as seen in leaf structure. Part 2, Morphology.

Roth, I. (1964): Microtécnica vegetal. Caracas, Imprenta universitaria.

Roth, I. (1977a): Anatomia y textura foliar de plantas de la Guayana Venezolana. Acta Bot. Venez., 12, pp. 79—146.

Roth, I. (1977b): Fruits of angiosperms. Handb. der Pflanzenanatomie, Vol. X, Part 1, Berlin, Stuttgart, Gebr. Borntraeger.

Roth, I. (1980): Blattstruktur von Pflanzen aus feuchten Tropenwäldern. Bot. Jahrb. Syst., 101 (4), pp. 489—525.

Roth, I. (1981): Structural patterns of tropical barks. Handb. der Pflanzenanatomie, Vol. XVI. Berlin, Stuttgart, Gebr. Borntraeger.

Roth, I. (1984): Stratification of tropical forests as seen in leaf structure. Den Haag, Dr. W. Junk Publishers (Tasks for vegetation science; 6).

Roth, I. and Mérida de Bifano, T. (1971): Morphological and anatomical studies of leaves of the plants of a Venezuelian cloud forest. I. Shape and size of the leaves. Acta Biol. Venez., 7 (2), pp. 127—155.

Roth, I. and Mérida de Bifano, T. (1979): Morphological and anatomical studies of leaves of the plants of a Venezuelan cloud forest. II. Stomata density and stomatal patterns. Acta Biol. Venez., 10 (1), pp. 69—107.

Roth, I. and Mérida de Bifano, T. (in press): Morphological and anatomical studies of leaves of the plants of a Venezuelan cloud forest. III. Leaf anatomy.

Russin, W. A. and Evert, R. F. (1985): Studies of the leaf of *Populus deltoides* (Salicaceae): quantitative aspects, and solute concentrations of the sieve-tube members. Am. J. Bot., 72 (4), pp. 487—500.

Sachs, J. (1875): Geschichte der Botanik vom 16. Jahrhundert bis 1860. München, Oldenbourg Verl., (Geschichte der Wissenschaften in Deutschland. Neuere Zeit. 15 Vol.).

Schuster, W. (1908): Blattaderung des Dicotylenblattes und ihre Abhängigkeit von äußeren Einflüssen. Ber. Dtsch. Bot. Ges., Vol. XXVI, pp. 194—237.

Seybold, A. (1957): Träufelspitzen? Beitr. Biol. Pflanz., 33, pp. 237—264.

Shields, L. M. (1950): Leaf xeromorphy as related to physiological and structural influences. Bot. Rev., 16 (8), pp. 399—447.

Smith, W. (1909): The anatomy of some Sapotaceous seedlings. Transactions of the Linn. Soc. of London, 7, pp. 189—200.

Steinegger, E. and Kritikos, P. G. (1949): Die Beziehung der Blattinselzahl zur Größe des Blattes und ihre Bedeutung für die Unterscheidung verschiedener Spezies und Varietäten. Festschrift Paul Casparis, Zürich, pp. 211—220.

Stocker, O. (1923): Die Transpiration und Wasserökologie nordwestdeutscher Heide- und Moorpflanzen am Standort. Zeitschr. f. Bot., Vol. 15, pp. 1—41.

Stocker, O. (1926): Über transversale Kompaßpflanzen. Flora, 120, pp. 371—376.

Stocker, O. (1937): Über die Beziehungen zwischen Wasser- und Assimilations haushalt. Ber. Dtsch. Bot. Ges., 55, pp. 370—376.

Stocker, O. (1954): Der Wasser- und Assimilationshaushalt südalgerischer Wüstenpflanzen. Ber. Dtsch. Bot. Ges., 67, pp. 288—298.

Stocker, O. (1957): Grundlagen, Methoden und Probleme der Ökologie. Ber. Dtsch. Bot. Ges., 70, pp. 411—423.

Stocker, O., Rehm, S. and Schmidt, H. (1943): Der Wasser- und Assimilationshaushalt dürreresistenter und dürreempfindlicher Sorten landwirtschaftlicher Kulturpflanzen. I. Hafer, Gerste und Weizen. II. Zuckerrüben. Jahrb. f. wiss. Bot., No. 1, Vol. 91, pp. 1—53, and No. 2, Vol. 91, pp. 278—330.

Strasburger, E. (1891): Über den Bau und die Verrichtungen der Leitungsbahnen in den Pflanzen. Jena, G. Fischer.

Strasburger, E., Noll, F., Schenck, H. and Schimper, A. F. W. (1983): Lehrbuch der Botanik für Hochschulen. 32. ed., Stuttgart, New York, G. Fischer.

Sturm, M. (1971): Die eozäne Flora von Messel bei

Darmstadt. I. Lauraceae. Palaeontographica, Vol. 134, Part B Paläophytologie, Parts 1–3, pp. 1–60.

Thompson, D. W. (1961): Growth and form. Vol. I and II., Cambridge, University Press, Vol. 1, Vol. 2, pp. 465–1116.

Troll, W. (1938/39): Vergleichende Morphologie der höheren Pflanzen, Vol. 1, Part 2, 1. part, pp. 957–1148, 3 part, pp. 1424–1724. Berlin, Gebr. Borntraeger.

Troll, W. (1939): Vergleichende Morphologie der höheren Pflanzen. Vol. 1: Vegetationsorgane. Part 2. Berlin, Gebr. Borntraeger. Authorized reprint in Koenigstein-Taunus, Koeltz Verl. (1967).

Unruh, M. (1941): Blattnervatur und Karpellnervatur. Beitr. Biol. Pflanz., 27, pp. 232—241.

Unterricht Biologie (April 1985): Regenwald. Part 103, 9. year's set, Seelze, Friedrich.

Vareschi, V. (1980): Vegetationsökologie der Tropen. Stuttgart, Verl. Eugen Ulmer.

Walter, H. (1931): Die Hydratur der Pflanze und ihre physiologisch-ökologische Bedeutung (Untersuchungen über den osmotischen Wert). Jena, G. Fischer.

Walter, H. (1973): Die Vegetation der Erde in öko-physiologischer Betrachtung. Vol. I: Die tropischen und subtropischen Zonen. 3. ed., Stuttgart, G. Fischer.

Walter, H. (1979): Vegetation of the earth and ecological systems of the geo-biosphere. 2. ed., New York, Heidelberg, Berlin, Springer.

Walter, H. and Lieth, H. (1960ff): Klimadiagramm — Weltatlas. Jena, VEB — G. Fischer.

Wanner, H. and Soerohaldoko, S. (1979): Transpiration types in montane rain forest. Ber. Schweiz. Bot. Ges., 89 (3/4), pp. 193—210.

Westermann (1982/83): Diercke Weltatlas. 185. ed., Braunschweig, Westermann Verl.

Whittenberger, R. T. and Kelner, A. (1945): Rubber in *Cryptostegia* leaf chlorenchyma. Am. J. Bot., 32, pp. 619—627.

Williams, M. A. (1977): Quantitative methods in biology. (Practical methods in electron microscopy. Volume 6). Amsterdam, New York, Oxford, North-Holland Publishing Company.

Wylie, R. B. (1938): Concerning the conductive capacity of the minor venation of foliage leaves. Am. J. Bot., 25, pp. 567—572.

Wylie, R. B. (1939): Relations between tissue organization and vein distribution in dicotyledon leaves. Am. J. Bot., 26, pp. 219—225.

Wylie, R. B. (1943a): The leaf organization of *Hedera helix*. Proc. Iowa Acad. Sci., 50, pp. 199—207.

Wylie, R. B. (1943b): The role of the epidermis in foliar organization and its relations to the minor venation. Am. J. Bot., 30, pp. 273—280.

Wylie, R. B. (1946a): Conduction in dicotyledon leaves. Proc. Iowa Acad. Sci., 53, pp. 195—202.

Wylie, R. B. (1946b): Relation between tissue organization and vascularization in leaves of certain tropical and subtropical dicotyledons. Am. J. Bot., 33, pp. 721—726.

Wylie, R. B. (1950): Foliar organization and vascularization of *Tolmiea menziesii*. Proc. Iowa. Acad. Sci., 57, pp. 149—155.

Wylie, R. B. (1952): The bundle sheath extension in leaves of dicotyledons. Am. J. Bot., 39, pp. 645—651.

Zalenski, V. (1902): Über die Ausbildung der Nervation bei verschiedenen Pflanzen. Ber. Dtsch. Bot. Ges., 20, pp. 433—440.

Zalenski, V. (1904): Materials for the study of the quantitative anatomy of different leaves on the same plant. Mem. Polytech. Inst. Kiev, 4, pp. 1—203. (quoted according to Roth, 1984).

Zimmermann, W. (1959): Die Phylogenie der Pflanzen. 2. ed., Stuttgart, Fischer.

INGRID ROTH

PECULIAR SURFACE STRUCTURES OF TROPICAL LEAVES FOR GAS EXCHANGE, GUTTATION, AND LIGHT CAPTURE

(Fissures in the epidermis, lenticels, giant stomata, glands, ocelli, papillas, and surface sculpturing)

CONTENTS

1. INTRODUCTION

The material studied in this contribution comes from Venezuelan Guiana and was collected by forest engineer Dr. Bernnard Rollet.

In the volume "Tropical forests as seen in leaf structure" (Roth 1984), the 232 species collected are anatomically described, taking into account about 50 different leaf characteristics. From these studies, it became obvious that a certain stratification of the tropical forest becomes apparent in the leaf structure, influenced by the microclimate. While, in the undergrowth, where shade, humidity and a still air prevail, the hygromorphic shade leaf is more common, the xeromorphic sun leaf becomes more characteristic of the crown region where — at least at times — insolation, air movement and drought may be of importance.

Certain characteristics, such as fissures in the epidermis, lenticels, giant stomata, ocelli, papillas, and surface sculpturing have been mentioned briefly in the volume cited above, but they will be studied more closely in the present contribution in order to discuss their occurrence, structure, development and possible function.

2. FISSURES IN THE EPIDERMIS

A phenomenon that is relatively frequently observed in the epidermis of the plant species studied in Venezuelan Guiana are fissures of varying shape and size. The origin of such epidermal fissures may be twofold. A fissure may arise through schizogeneous separation of epidermis cells from one another by tearing forces. Fissures of this kind may develop as long as growth in the surface is still in progress. The other manner of fissure formation is by chemical changes in the wall material at certain spots of the epidermis. During these processes decomposition of cell walls begins with a swelling of the wall material, the appearance of pectic substances or mucilage and a subsequent dissolution of the walls, in a more or less lysigeneous manner. The phenomenon somewhat recalls the processes of "gummosis" in peach fruits. Before the cells separate from one another, a thick, less compact wall mass becomes

B. Rollet et al.: Stratification of Tropical Forests as seen in Leaf Structure, Part 2, pp. 185—238.

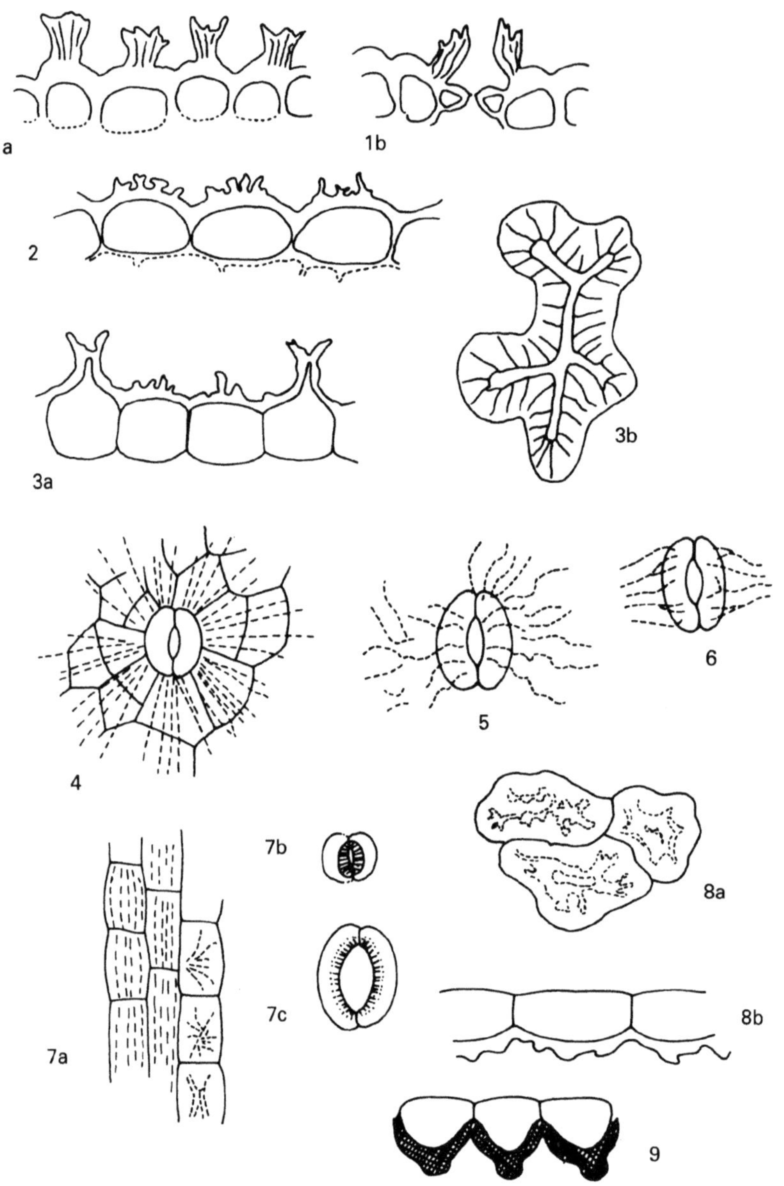

Fig. 1. **1.** *Aspidosperma excelsum*, Apocynaceae, transection of lower epidermis: (a) showing protuberances of outer cell walls, (b) stoma protected by protuberances which form a "calm chamber" above the stoma. **2.** *Jacaranda obtusifolia*, Bignoniaceae, lower epidermis in transverse section demonstrating protuberances on the outer wall surface. **3.** *Capparis guaguaensis* (a) lower epidermis in transection with scales rising above the epidermis level, (b) "peltate" scale in surface view showing crests in the center and radiating striation. **4.** *Qualea dinizii*, Vochysiaceae, lower epidermis in surface view with large stoma and characteristic arrangement of surrounding cells originating from periclinal cell divisions. Note the cuticular striation in the form of stripes radiating from the stoma. **5.** *Mabea taquiri*, Euphorbiaceae, lower epidermis in surface view with stoma and cutinized wall foldings in the form of undulated stripes. **6.** *Rheedia* aff. *spruceana*, Guttiferae, stoma of lower epidermis in surface view with strong cuticular folds. **7.** *Eschweilera subglandulosa*, Lecythidaceae, (a) Surface view of lower epidermis showing striation of cells above nerves (parallel stripes) and above normal epidermis cells (irregular ribbing). (b) Normal stoma with strengthened cuticular ledges, (c) Large stoma (same enlargement scale) with weaker developed ledges and wide open aperture. **8.** *Beilschmiedia curviramea*, Lauraceae, (a) Surface view of lower epidermis cells with strong surface sculpturing. (b) Lower epidermis cells with crests on outer walls, as seen in transverse section. **9.** "Rabo pelado", Olacaceae, papillas 6—10 μm long with very thick outer walls, situated at the lower epidermis.

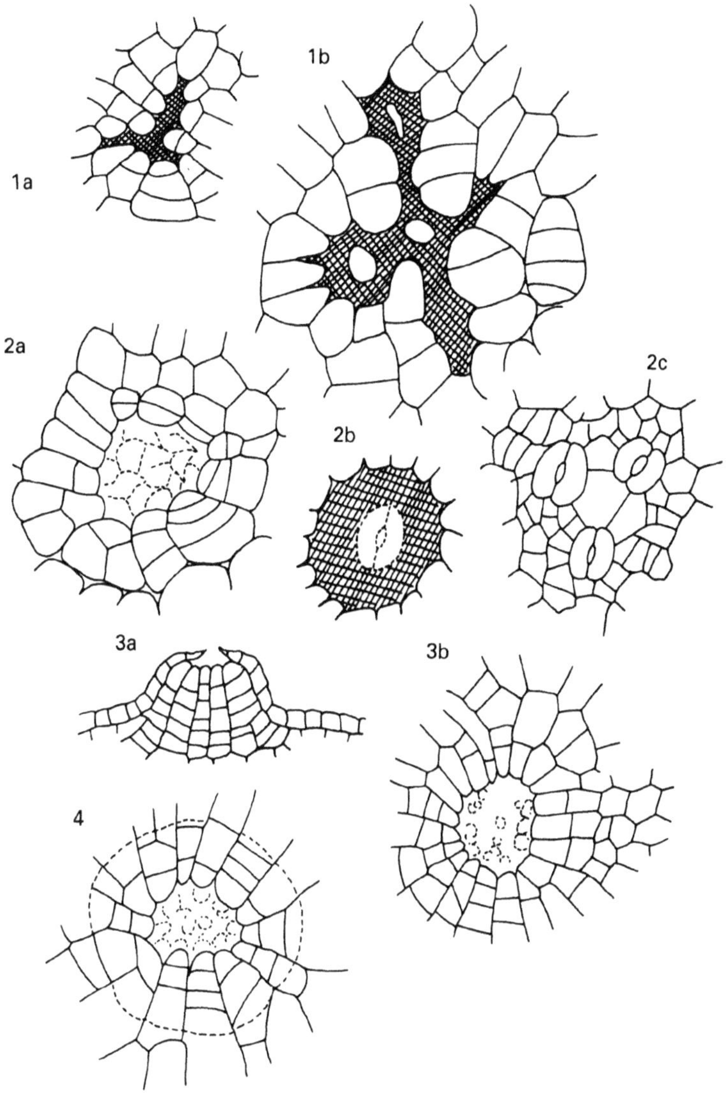

Fig. 2. **1** (a and b), *Eschweilera chartaceae*, Lecythidaceae, formation of fissures in the lower epidermis, as seen in surface view. In (a) Swollen cell walls stain dark-blue with toluidine blue (center), while the surrounding cells begin to divide periclinally. No stoma is visible as a starting point in the center for lenticel formation. (b) Later stage of lenticel development. The channels in the center extending between the cells ramified in different directions and still consist in dissolved wall material; some cell remainders are visible within this decomposing material, but cell division around them is progressing. **2**. *Eschweilera* cf. *trinitensis*, Lecythidaceae, (a) Lower epidermis with young lenticel. Note obliterated cells in the center around which living cells are dividing periclinally to form some kind of a "wreath". (b) Formation of a hole of 0.1 mm in diameter in which the remainder of a stoma is still visible. Around the hole the lenticel will begin to form. (c) Group of stomata already partly obliterating around which cell division starts to initiate a lenticel. **3**. *Eschweilera chartacea*, Lecythidaceae, (a) Lenticel of lower epidermis in transverse section: Note anticlinal cell rows arisen from the subepidermal layer and covering epidermis which is ruptured in the center above the lenticel. (b) Lenticel in surface view with obliterated cells in the central hole and surrounding radial cell rows. Diameter of the hole about 0.04 mm, entire lenticel 0.17 mm in diameter. **4**. *Unonopsis glaucopetala*, Annonaceae, lenticel in surface view: radiating cell rows which originated from periclinal divisions surround a hole in which remnants of cell walls are visible. The cells of the radial rows are distinguished by staining dark-violet with toludine blue. The entire cell group is surrounded by a darkly stained margin contrasting with the almost colorless normal epidermis cells.

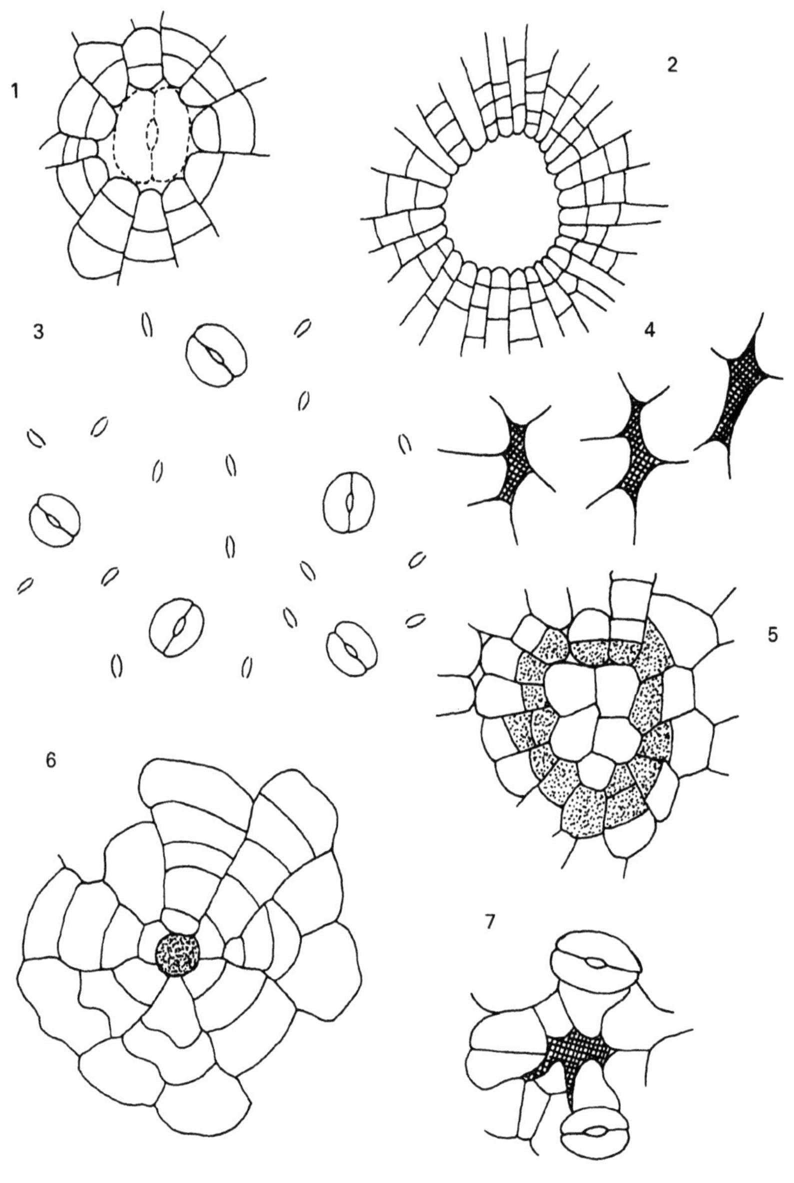

Fig. 3. **1**. *Peltogyne porphyrocardia*, Caesalpiniaceae, giant stoma in a decomposing state surrounded by periclinally dividing cells, indicating the origin of lenticel formation. **2**. *Simarouba amara*, Simaroubaceae, lenticel of the lower epidermis in surface view with a large hole in the center, surrounded by radiating cell rows. **3**. *Cordia exaltata*, Boraginaceae, surface view of lower epidermis indicating well stained large (giant) stomata and stained ledges of normal stomata. **4**. *Zinowiewia australis*, Celastraceae, showing fissures in the epidermis, as seen in surface view. **5**. *Eschweilera chartacea*, Lecythidaceae, Lenticel formation, as seen in surface view. Darkly stained cells (dotted) surround a group of 4 thin-walled stainless epidermis cells which later will decompose. **6**. *Iryanthera lancifolia*, Myristicaceae, Surface view of a beginning lenticel with cell rows radiating from a center which could be interpreted as an obliterating hair base. **7**. *Hymenaea courbaril*, Caesalpiniaceae, Surface view of lower epidermis with two stomata, a darkly stained spot in the center probably composed of pectic wall material and epidermis cells of irregular shape which partly divide. The dark spot in the center later transforms into a fissure.

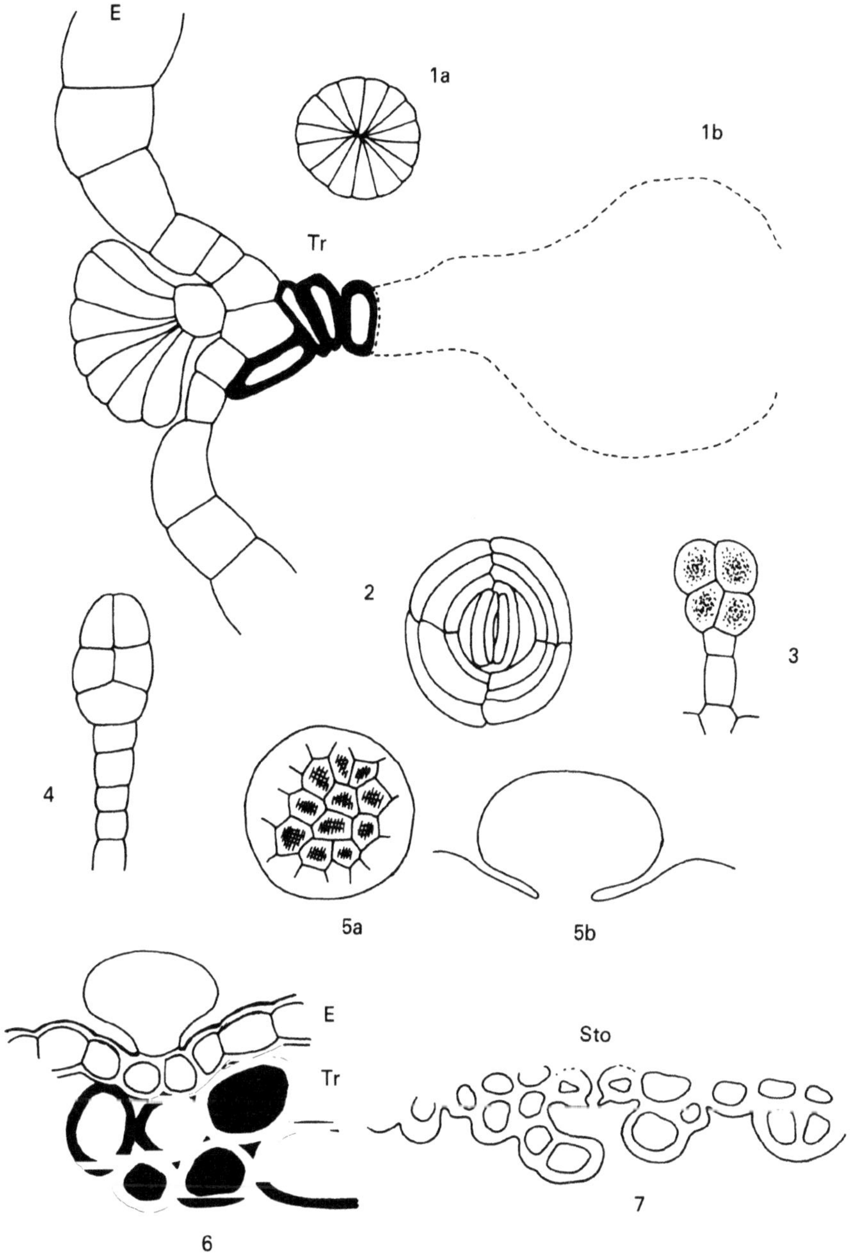

Fig. 4. **1**. Coccoloba sp. ("Uvero"), Polygonaceae, glandular hydathode: a) seen from above, b) in longitudinal section, showing epidermis (E), tracheidal cells (Tr) below the foot cells of the gland, and connection with the vascular bundle (Vb). **2**. Coccoloba sp. ("Arahuceque"), stomatal hydathode in surface view, surrounded by a wreath of calotte-like cells. **3**. Brosimum sp., Moraceae, stalked gland of the lower leaf side. **4**. Inga capitata, Mimosaceae, club-shaped headed gland, found mainly above veins. **5**. Centrolobium paraense, Papilionaceae, giant gland of the lower leaf surface with a diameter of 0.18 mm or 180 μm (ca. 3 per mm^2) with a yellowish content (dying orange-red with Sudan III). (a) In surface view, (b) in transverse section. **6**. Vitex stahelii, Verbenaceae, headed gland with a diameter of 56 μm in connection with vascular bundle (Vb) through tracheidal "water cells" (Tr), (approx. one per mm^2). E = Lower epidermis, Vb = Vascular bundle. **7**. Virola surinamensis, Myristicaceae, lower epidermis with protrusions which partly protect the stomata (Sto).

obvious between them. Both types of fissure formation are related to one another insofar as, in the first case, dissolution of the middle lamella also possibly precedes wall separation.

2.1. Separation of cell walls in a schizogeneous way

Formation of fissures was observed in the epidermis of *Zinowiewia australis*, Celastraceae (Figure 3, 4). A group of 5—6 cells is torn apart, probably due to arising tensions during surface growth. The clefts are extended in one direction, but remain relatively small. Holes or clefts were also observed in the upper epidermis of *Trichilia propingua*, Meliaceae. In the middle of a group of 5—6 cells a fissure appears by separation of the cells. But, later on, periclinal divisions may occur around the hole so that some kind of a lenticel is initiated. In *Licania* sp. (Pilón nazareno), Rosaceae, separation of cells and formation of a hole likewise occur in the upper epidermis. However, later on, periclinal divisions arise around the hole and lenticel formation is prepared. Formation of a hole in the center of a cell group could also be found in the upper epidermis of *Lecointea amazonica*, Caesalpinaceae. The predestined cell groups are distinguished from normal epidermis cells by their stronger affinity to stains. A wreath of dark-blue stained cells thus surrounds the hole when toluidine blue is applied.

2.2. Separation of cell walls through lysigeneous processes

Separation of cells preceded by swelling of walls likewise occurs in *Lecointea amazonica*, so that both types of cell separation may be observed in one and the same species. In the middle of a cell group cell walls begin to swell and become slimy. In this stage of development the wall clot has a great affinity for stains. Later on, however, the slimy wall mass may degenerate and a hole or cleft arises. Large fissures may develop in the lower epidermis of *Eschweilera subglandulosa*, Lecythidaceae, by wall swelling through chemical changes, probably of the pectic material. The swollen cell wall material stains intensely with toluidine blue. Elongated fissures originate by complete wall degeneration. In *Eschweilera chartacea*, Lecythidaceae, thickening and swelling of cell walls may lead to the formation of ramified "channels" in the epidermis which extend over larger cell groups. The wall material, which may swell so much as to reach the width of an entire epidermis cell, stains dark blue with toluidine blue. Later, cell divisions may originate around the ramified wall clots. Fissures more frequently occur in the lower than in the upper leaf epidermis of *Hymenaea courbaril*, Caesalpiniaceae (Figures 3, 7). Holes likewise develop in the lower epidermis of *Tovomita brevistamina*, Guttiferae, by dissolution of cells. Cells surrounding the hole are distinguished from ordinary epidermis cells by their affinity for stains. They start with periclinal divisions around the hole so that, later on, radial cell rows spread from the fissure in all directions. Cell lysis is furthermore observed in the epidermis of *Unonopsis glaucopetala* before lenticel development starts, equally as in many other examples of lenticel formation (see the section on lenticels or cork warts below).

Both of these fissure types, the schizogeneous in the same way as the lysigeneous, are independent of the presence of stomata. They may occur on the upper as well as on the lower leaf surface. Even transitions between both fissure types may be observed. The middle lamella may be dissolved in the schizogeneous type and surface tensions may play their part in the lysigeneous type, besides chemical changes and transformation of wall material.

Fissures in the epidermis were also repeatedly observed in other leaf samples of species not growing in Venezuelan Guiana and therefore not mentioned here. They arise with greater frequency in spots where stronger surface tensions act, especially so in the lower epidermis, where the formation and the development of stomata as well as that of hairs may lead to surface tensions.

The fissures and clefts in the leaf epidermis may function in the same way as stomata and lenticels, i.e. they may serve for gaseous exchange without, however, any capacity for regulation by opening

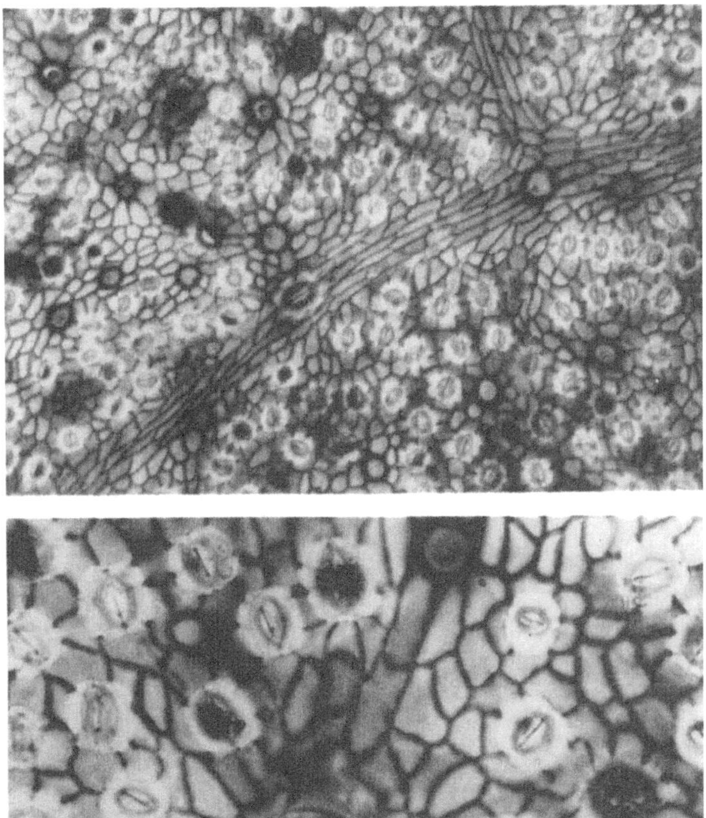

Fig. 5. Above: "Maro", Burseraceae, lower leaf epidermis with normal stomata and a large stoma of giant size above the nerve (note difference in staining affinity of small and large stomata (3.2 × 2.5 × 6.3) Middle: "Maro", lower leaf epidermis with small stomata and a giant stoma with much staining affinity (16 × 2.5 × 3.2). Below: Rollinia multiflora, Annonaceae, t. s. of leaf, showing papillas and a stoma in the lower epidermis (2.5 × 25 × 3.2).

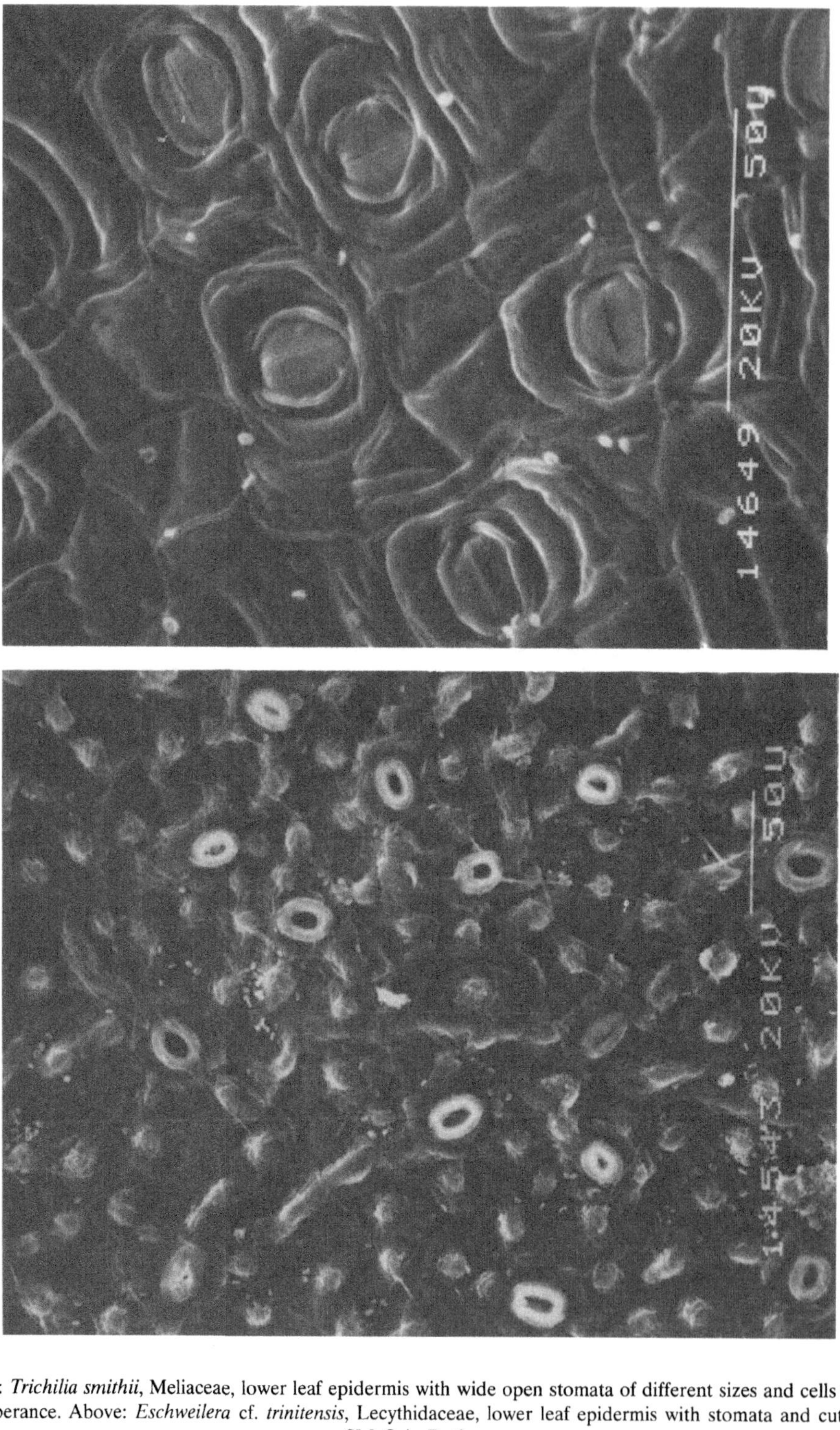

Fig. 6. Below: *Trichilia smithii*, Meliaceae, lower leaf epidermis with wide open stomata of different sizes and cells with a small central protuberance. Above: *Eschweilera* cf. *trinitensis*, Lecythidaceae, lower leaf epidermis with stomata and cuticular folds. SM. Orig. Roth.

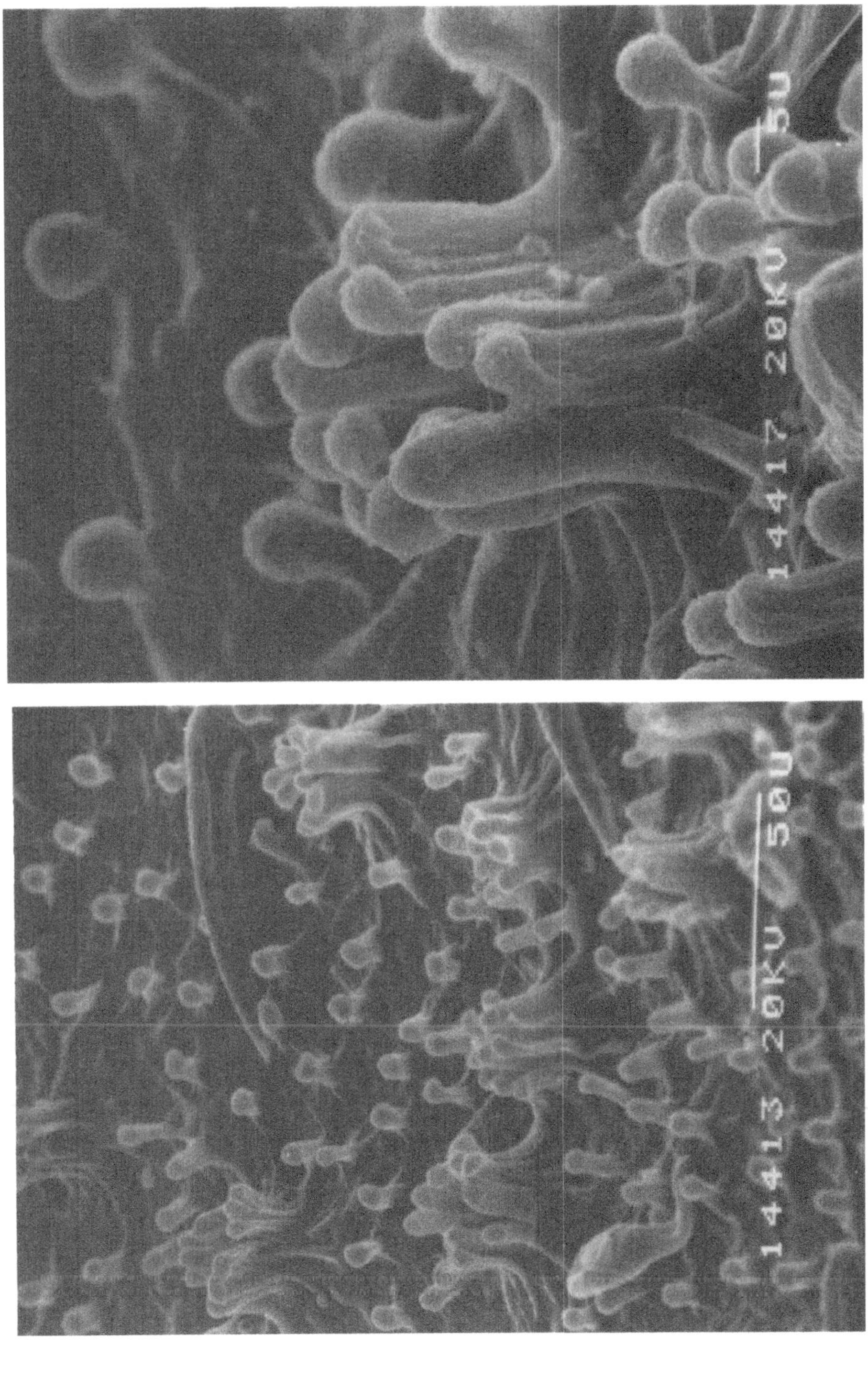

Fig. 7. Helicostylis tomentosa, Moraceae, lower leaf epidermis with papillas and stomata surrounded by elongated papillas. SM. Orig. Roth.

and closing. It is furthermore possible that water in the form of droplets is secreted through the holes when the atmosphere is vapour saturated.

3. LENTICELS OR CORK WARTS

In general, lenticels may occur on the lower as well as on the upper leaf surface. However, more frequently they are found on the lower side, usually developing around a stoma. Lenticel formation is often initiated by radial elongation of the surrounding cells and subsequent periclinal divisions. The stoma or stomata in the center progressively obliterate and in this way a hole arises which is surrounded by cell rows radiating from the hole in all directions. Each cell row may correspond to an enlarged epidermis mother cell repeatedly divided periclinally or parallel to the hole in the center. But cell division not only occurs in one plane, parallel to the surface, likewise in vertical direction the same processes occur. The subepidermal cells below the respiratory cavity of the stoma or stomata become extended in an anticlinal direction to the epidermis surface and later on repeatedly divide periclinally. Not infrequently, remnants of the obliterated cells in the center of the lenticel are visible.

3.1. Occurrence of lenticels on leaves

Lenticel formation has been observed in the following species studied from Venezuelan Guiana, belonging to the following families:

Eschweilera c.f. *trinitensis* *Eschweilera chartacea* *Lecythis davisii* *Eschweilera grata*	Lecythidaceae
Peltogyne porphyrocardia *Lecointea amazonica* *Hymenaea courbaril*	Caesalpiniaceae
Tovomita brevistamina *Rheedia* sp. (Cozoiba picuda)	Guttiferae

Iryanthera lancifolia, Myristicaceae
Tapirira guianensis, Anacardiaceae
Simarouba amara, Simaroubaceae
Licania sp. (Pilón nazareno), Rosaceae
Unonopsis glaucopetala, Annonaceae
Trichilia propingua, Meliaceae

In total, 15 species belonging to 9 different families were observed sporadically to have lenticels on leaves. Considering the total number of the species studied in Venezuelan Guiana only 6.5% develop lenticels, which is a very low percentage.

Occurrence of lenticels on the lower leaf side

Lenticels were observed on the lower leaf side only in the following species: *Eschweilera* c.f. *trinitensis, Eschweilera chartacea, Eschweilera grata, Lecythis davisii, Peltogyne porphyrocardia, Lecointea amazonica, Hymenaea courbaril, Tovomita brevistamina, Rheedia* sp. (Cozoiba picuda), *Tapirira guianensis, Simarouba amara, Unonopsis glaucopetala.*

In 12 species altogether lenticels were found on the lower leaf surface only.

Occurrence of lenticels on the upper leaf side

Lenticels were observed on the upper leaf side only in *Iryanthera lancifolia.*

Occurrence of lenticels on the upper as well as on the lower leaf side

Cork warts were found on the upper as well as on the lower leaf surface in *Trichilia propingua* and *Licania* sp. (Pilón nazareno).

Lenticels, therefore, occur more frequently on the lower than on the upper leaf surface. When arising on the upper leaf surface, they form independently of stomata, as usually no stomata occur on the upper leaf side, and often appear in a reduced form. This reduction becomes evident in a reduced number of lenticels as compared with the lower surface where they are more frequent, in a reduced radial extension of surrounding cells, and in a reduced number of periclinal divisions in the radially extended cells.

As lenticels were found in very different plant families it is suggested here that lenticel formation is more dependent on the habitat in which the plants grow than on pariental relations.

3.2. Surface tensions and epidermal cell division

In the developing epidermis, cells different in shape and consistency from the ordinary epidermis cells may act as "obstacles", such as stone cells, hairs, glands, idioblasts, and even stomata. Surface tensions may arise around them so that neighbouring cells are extended radially, and cell divisions may follow as a consequence of the tensions. Frequently, these cell divisions proceed in a centripetal direction so that radial cell rows originate around the "obstacle". This is a very frequently observed phenomenon around stone cells or hair bases, producing a cell arrangement in a "spider-like" form. Apparently, meristematic activity is initiated here by surface tensions. Surface tensions which lead to a fissure or cleft in the epidermis may likewise evoke cell division activity around the hole. Lenticel-like formations may thus develop around hair bases (Figure 3, 6: *Iryanthera lancifolia*), around an injury in the epidermis through an insect sting, around epidermal fissures (Figure 3, 4: *Zinowiewia australis*), or, finally, around stomata (e.g. in *Aniba riparia*) or distinct epidermal cell groups.

3.3. Development of true cork warts

Regular development on the lower leaf surface

In the most simple case, cells surrounding a stoma become extended in a radial direction — with the stoma as a center — so that a wreath of elongated cells radiates from the stoma. The elongated cells may then divide periclinally to the center and form walls parallel to the surface of the stoma, describing in this way the arc of a circle. Each cell may give rise to 2—3 or even more such walls in a centripetal direction. The outermost cells may later become extended tangentially to the center. In this way radial cell rows arise around the stoma (Figures 2 and 3). However, cell division does not only progress in one plane, parallel to the epidermis surface. Likewise the subepidermal cells bordering the respiratory cavity become extended in an anticlinal direction to the leaf surface and then repeatedly divide periclinally to form anticlinal cell rows below the stoma. The subepidermal rows push the epidermis upwards so that a wart develops above the surface. Simultaneously, the substomatal cavity becomes occluded. A meristematic complex has thus developed around the stoma becoming apparent by the very regular arrangement of its cells in the form of radial rows. Between the subepidermal cell rows, however, intercellular spaces arise so that the gaseous exchange is ensured. The walls of the radial cell rows may finally become impregnated with suberine taking on a brownish colour. At the same time necrotic processes take place within the stoma until it is completely obliterated. Sometimes, remnants of the walls of the stoma are still visible in the center of the lenticel (Figure 3, 1), but at the end of the process the stoma completely disappears leaving a hole in the center of the lenticel.

The meristematic processes not always develop their activity around a single stoma, but an entire cell complex including several stomata as well as neighbouring epidermis cells may become surrounded by dividing cells. The further development is, however, the same as in the case of a single stoma, only the resulting hole in the center will be larger. However, the formation of radiating cell rows may not be as regular as in the case of a single stoma, but several stories of periclinally divided cells may finally surround the cell complex in the center. This means that cells further remote from the central cell complex are also included in the meristematic activity (Figure 2, 2c).

The abovementioned type of development which is considered the regular one, mainly refers to the lower leaf surface where stomata are present.

Development on the upper leaf surface

Development of lenticels on the upper leaf surface

is different from that on the lower surface, since stomata are usually absent on the upper leaf side. In this case, lenticel development may start with necrotic processes in the walls of certain epidermis cells, as observed, for example, in *Eschweilera chartacea* (Figure 2, 1). The walls of some contiguous epidermis cells become more and more thickened, probably undergoing chemical changes. Possibly pectic substances play an important part during wall decomposition, the middle lamellae becoming dissolved, while the swelling walls lose their compactness. Finally, a thick gelatinous mass of wall material becomes obvious around which the neighbouring cells begin to divide in different directions. Owing to surface tensions which are due to the physical and chemical changes of the wall material in the center, the surrounding cells become extended and start dividing in different directions. Later on, the processes of cell extension and division may become more ordered so that cell rows radiate from the center. The cells in the center finally separate from one another and degenerate leaving a hole in the center.

In *Eschweilera chartacea*, another additional way of lenticel formation could be observed too. A complex of 4—5 ordinary epidermis cells may be distinguished from the surrounding cells (Figure 3, 5). At the beginning, the cells become conspicuous by their lower affinity for stains. While they remain colourless (when treated with toluidine blue), the surrounding cells stain with toluidine blue. Then the surrounding cells begin to divide periclinally to the central cell group, they may also elongate in a radial direction and other adjoining cells may become meristematic as well and divide periclinally. The cell group in the center finally disintegrates (Figure 2, 3b).

In other cases, meristematic activity may start around an intercellular space in the epidermis itself (e.g. in *Trichilia propingua*, see also the paragraph dealing with "fissure formation"). The hole or cleft in the epidermis may develop by surface tensions, through an injury or insect sting or even through an abscised hair. The cells surrounding the hole undergo the same radial extension and the same periclinal divisions as mentioned before. In *Trichilia propingua*, the cells surrounding a hair base may start to divide periclinally due to surface tensions in a radial direction around the hair base. However, similar processes also occur on the upper leaf surface independently of the existence of hairs.

We thus may distinguish three different processes which lead to the formation of cork warts: (a) Periclinal divisions around a stoma or a group of stomata, which later degenerate, (b) periclinal divisions around a group of normal epidermis cells which likewise decompose later on, and (c) formation of a hole or fissure in the epidermis and following meristematic activity around the hole. The hole may arise through dissolution of cell walls and consequent cell division around the necrotic spot in which living cells may be included too. Cell separation and formation of a hole may also occur schizogeneously without swelling of cell walls, but a strict distinction between schizogeneous and lysigeneous processes is always difficult, as dissolution of the middle lamella is involved in schizogeneous processes too. When the radial cell rows are formed and the meristematic activities are settled, cell wall impregnation with suberine may follow, especially in old cork warts.

3.4. Size of lenticels

As lenticels commonly develop around a single stoma or even a group of stomata it is evident that they form relatively large spots on the leaves.

In *Eschweilera* c.f. *trinitensis* the lenticels reach a diameter of 0.15—0.18 mm. The hole in the center may have a diameter of up to 0.1 mm.

In *Eschweilera chartacea* the diameter of the entire lenticel may measure 0.17 mm, the diameter of the central hole 0.04 mm, and the height of the lenticel above the leaf surface may attain 0.05 mm. *Lecythis davisii* has lenticels with an entire diameter of 0.1—0.5 mm and a diameter of the hole of 0.02—0.08 mm.

The lenticel diameter of *Simarouba amara* oscillates between 0.24 and 0.65 mm, while the diameter of the hole reaches 0.026—0.05 mm. In *Lecointea amazonica* a hole diameter of 0.04 mm has been measured.

3.5. Lenticel density per mm^2

In some of the species studied lenticel density could be measured, as follows:

Eschweilera c.f. *trinitensis*	0.8	lenticels per mm^2
Lecythis davisii	3— 9	" " "
Eschweilera grata	3— 9	" " "
Eschweilera chartacea	13	" " "
Peltogyne porphyrocardia	13—27	" " "

Lenticels are usually scarce on the leaf surface and may be found accidentally here and there. However, in five species (see above) lenticel occurrence is relatively high, oscillating between less than one up to as many as 27 per mm^2. This is a lenticel density that almost reaches a low stomata density.

3.6. Leaf types in which lenticels have been found

The leaves which bear cork warts are of different types. However, most leaves are of the sun type and have some xeromorphic characters. In total, 10 species belong to this category. Two of them are actually very pronounced sun leaves, and one of these two is xero- and scleromorphic at the same time. Of the five remaining species, one is of the sun type but mesomorphic. Two are located between the sun and the medium type, and one of them is xeromorphic, while the other one is mesomorphic. Another one is of the medium type and mesomorphic, while the last one which is xeromorphic may be placed between the medium and the shade type. The xeromorphic sun leaf type thus prevails. It is striking that a number of species in which lenticels have been observed develop a water-storing tissue in the leaves, such as *Eschweilera* c.f. *trinitensis, Peltogyne porphyrocardia, Hymenaea courbaril, Tovomita brevistamina, Tapirira guianensis, Simarouba amara, Licania* sp., *Unonopsis glaucopetala*, and *Trichilia propingua*. Nine species out of 15, or 60%, thus possess a water storing tissue in the leaves. Seven of these species develop a water storing upper epidermis, while two of them have an upper hypodermis which serves as a water reservoir. Some of these species have large stomata as well, which may possibly be interpreted as hydathodes.

Of the 15 species with lenticels, 12 are high trees belonging to height category A, while three are of height category a. Lenticels were not found in leaves of small trees or shrubs.

3.7. Function of lenticels

Although the usual function of lenticels is thought to be aeration and gaseous exchange, cork warts with other functions have been described previously by Areschoug (1902). Water secretion was suggested as a further function. Areschoug draws attention to the so-called "lenticel-hydathodes", especially in mangrove plants. He suggested that chlorides are excreted in this way. He mentions lenticel-hydathodes in the following species: *Carapa obovata, Conocarpus erecta*, several species of *Rhizophora*, several species of *Bruguiera, Kandelia rheedii*, two species of *Sonneratia, Lumnitzera racemosa, Avicennia nitida, Scyphiphora caryophyllacea, Acanthus ilicifolius, Excoecaria agalocha*, and *Aegiceras majus*. Areschoug observed destruction of cell groups in the epidermis and following hole formation. The leaves may become completely perforated in this way. Around the necrotic cell group a cork layer is developed which separates the water secreting tissue from the remaining tissue functioning as a physiological sheath. The lenticel-hydathodes are, however, not in direct connection with the vascular bundles.

In the present material studied lenticels were quite frequently observed above veins. It is therefore possible that these lenticels have an excretory function as well, replacing the stomatal hydathodes in their function. This assumption is supported by the fact that several lenticel-bearing species additionally possess large stomata which may have the same water secreting function.

4. HYDATHODES

On principle, hydathodes may appear in the form of two distinct types: either as "glandular trichomes" or as more complex structures composed of a secreting tissue called epithema combined with special openings in the epidermis, such as modified stomata, fissures or holes (Roth and Clausnitzer 1972). Glandular trichomes may occur scattered all over the leaf surface or may be concentrated at certain spots, while epithema hydathodes are well-known of leaf tips (e.g. in *Bambusa*) or of leaf margins. Quite often the epithema — often a chlorophyll-free parenchyma — is penetrated by prominent intercellular spaces. Release of water through hydathodes may either occur passively under the control of the root pressure or, when an epithema is present, this may actively secrete water.

The process of water secretion is known as "guttation". Experiments suggest that guttation mainly serves for mineral nutrition of the plant when transpiration is suppressed, especially in a warm and humid climate (Klepper and Kaufmann 1966). Consequently, hydathodes should be most frequently found in humid tropical forests, such as the rain forest and the montane cloud forest, where vapour-saturated air may impede transpiration, particularly during the night.

In the material studied of Venezuelan Guiana, both types of hydathodes seem to be present. A third type is represented by so-called "giant stomata" (Roth 1984) which are suspected to secrete water passively.

4.1. Stomatal hydathodes

In checking the surface slides of the leaves collected in Venezuelan Guiana I was surprised by the size irregularities of stomata. In many species (see list below) I noted a great size variation, larger or very large stomata being interspersed between smaller ones (Figure 1). In plants from temperate regions, we do not observe such a great difference between small and large stomata in one and the same species, and less so in one and the same leaf; on the contrary, we are used to finding little variation in stomatal size when we observe the surface of a given leaf.

In the following I present a list of plant species in which the stomatal size of leaves varies a great deal. All the species come from the same humid forest in Venezuelan Guiana (see Roth 1984).

Of the 232 species studied in Venezuelan Guiana 76 have stomata of a very variable size, representing 32.75%. Of these, 49 species or 64.47% belong to height category A, 24 species or 31.57% are of height category a, and 3 species or 3.94% belong to height category aa. About 67% of the species with variable stomatal size have leaves of the sun type with xeromorphic characteristics, 14.4% have sun type leaves with mesomorphic characters, and 9.2% have leaves of the medium type with xeromorphic characters. All other combinations are rare.

In total, 33 plant families are involved. Some families supply a larger number of species with stomata of variable size, such as the Burseraceae (with six species), the Caesalpiniaceae (with five), the Lecythidaceae (with seven), the Myrtaceae (with five), the Rosaceae (with four), Guttiferae (with five), Lauraceae (with three), Vochysiaceae (with three), and finally the Euphorbiaceae (with three species).

It is surprising that about one third of the species studied have stomata with a very variable size (Figures 3 and 5). Included in this list are only those examples in which a difference between the smallest and the largest stomata observed on the leaves of a given species varied at least by 6 μm. Most species, however, show much greater differences. The greatest difference measured occurred in *Sloanea guianensis* with 54 μm, an enormous deviation considering the width of a thin hand slide of only 20 μm. Stomata in crypts, on the other hand, for example in Rosaceae and Melastomaceae, which are usually small, do not vary much in size. When speaking of a variable size of stomata, one expects a certain number of stomata with different sizes on 1 mm^2. Only seldom, the larger stomata are very rare. Furthermore, variation in size may often be more or less continuous without an abrupt distinction. In other cases, the normal stomata may be small, and giant stomata may be large without any gradual transitions.

We may ask what the reason for these size differences of stomata could be in the species studied. Is this simply a primitive characteristic, as is often found in tropical humid forest plants, or is some difference in function connected with the

Species with a variable stomatal size (length) in μm

No.	Species	Type	Length
1.	*Homalium racemosum*, Flacourtiaceae	(a)	13—30
2.	*Banara nitida*, 〃	(A)	24—36
3.	"Maspara", 〃	(a)	22—34
4.	*Capparis amplissima*, Capparidaceae	(A)	13—39
5.	*Vismia macrophylla*, Guttiferae	(a)	18—29
6.	*Tovomita brevistamina*, 〃	(a)	20—45
7.	*Rheedia* sp. (Cozoiba picuda), 〃	(A)	21—60
8.	*Rheedia* sp. (Cozoiba rebalsera), 〃	(A)	39—48
9.	*Calophyllum brasiliense*, 〃	(A)	36—60
10.	*Apeiba echinata*, Tiliaceae	(A)	18—24
11.	*Byrsonima aerugo*, Malpighiaceae	(A)	29—38
12.	"Manteco Adol", 〃	(A)	27—34
13.	"Erizo", Rutaceae	(A)	34—47
14.	*Simaba multiflora*, Simaroubaceae	(A)	23—34
15.	*Simarouba amara*, 〃	(A)	23—65
16.	*Picramnia macrostachya*, 〃	(a)	13—23
17.	*Protium neglectum*, Burseraceae (Caraño)	(a)	13—29
18.	*Protium decandrum*, 〃 (Azucarito)	(A)	13—36
19.	*Protium* sp. (Sipuede), 〃	(A)	15—47
20.	"Maro", 〃	(a)	18—48
21.	*Protium decandrum* ?, 〃 (Caraño blanco)	(a)	26—36
22.	*Tetragastris* or *Dacryodes*, 〃 (Aracho blanco)	(a)	27—50
23.	*Trichilia smithii*, Meliaceae	(a)	22—45
24.	*Trichilia propingua*, 〃	(a)	21—45
25.	*Goupia glabra*, Celastraceae	(A)	24—34
26.	*Toulicia guianensis*, Sapindaceae	(a)	13—44
27.	*Meliosma herbertii*, Sabiaceae	(a)	45—56
28.	*Tapirira guianensis*, Anacardiaceae	(A)	17—57
29.	*Pentaclethra macroloba*, Mimosaceae	(a)	15—24
30.	*Inga rubiginosa*, 〃	(aa)	22—34
31.	*Hymenaea courbaril*, Caesalpiniaceae	(A)	29—40
32.	*Sclerolobium paniculatum*, 〃	(A)	17—35
33.	*Mora excelsa*, 〃	(A)	30—57
34.	*Peltogyne* sp. 〃 (Zapatero negro)	(A)	26—44
35.	*Peltogyne porphyrocardia*, 〃	(A)	26—48
36.	*Lonchocarpus sericeus*, Papilionaceae	(A)	18—25
37.	*Licania* sp. (Pilón nazareno), Rosaceae	(A)	36—45
38.	*Hirtella davisii*, 〃	(A)	11—35
39.	*Couepia glandulosa*, 〃	(A)	13—27
40.	*Licania densiflora*, 〃	(A)	16—23
41.	*Terminalia amazonica*, Combretaceae	(A)	18—35
42.	*Buchenavia capitata*, 〃	(A)	20—36
43.	*Calycorectes* sp. (Terciopelo), Myrtaceae	(a)	17—44
44.	*Eugenia anastomosans*, 〃	(a)	11—23
45.	*Eugenia compta*, 〃	(A)	11—23
46.	"Guayabito piedrero chiquito, 〃	(a)	22—32
47.	"Guayabillo negro", 〃	(a)	13—23
48.	*Eschweilera odora*, Lecythidaceae	(A)	18—35
49.	*Eschweilera* c.f. *trinitensis* 〃	(A)	26—35
50.	*Eschweilera* sp. (Majaguillo erizado) 〃	(A)	23—51
51.	*Eschweilera subglandulosa*, 〃	(A)	27—44
52.	*Eschweilera chartacea*, 〃	(A)	19—26
53.	*Couratari multiflora*, 〃	(A)	17—44
54.	*Lecythis davisii*, 〃	(A)	21—48
55.	*Genipa americana*, Rubiaceae	(aa)	24—34
56.	*Pouteria trilocularis* (Rosado), Sapotaceae	(a)	24—38
57.	*Manilkara bidentata* (Purguo morado) 〃	(A)	22—30
58.	*Diospyros lissocarpioides*, Ebenaceae	(aa)	27—36
59.	*Cordia alliodora*, Boraginaceae	(A)	22—39
60.	*Cordia exaltata*, 〃	(A)	29—42
61.	*Tabebuia* sp. (Araguaney puig), Bignoniaceae	(A)	23—34
62.	*Coccoloba* sp. (Arahueque), Polygonaceae	(a)	18—27
63.	*Iryanthera lancifolia*, Myristicaceae	(A)	18—29
64.	*Ocotea nicaraguensis*, Lauraceae	(A)	22—32
65.	*Aniba riparia*, 〃	(A)	17—30
66.	*Beilschmiedia curviramea*, 〃	(a)	23—39
67.	*Sandwithia guayanensis*, Euphorbiaceae	(a)	21—48
68.	*Drypetes variabilis*, 〃	(A)	45—52
69.	*Chaetocarpus schomburgkianus*, 〃	(A)	35—52
70.	*Erisma uncinatum*, Vochysiaceae	(A)	18—35
71.	*Vochysia lehmannii*, 〃	(A)	23—44
72.	*Qualea dinizii*, 〃	(A)	21—52
73.	*Sloanea guianensis*, Elaeocarpaceae	(A)	11—65
74.	*Agonandra brasiliensis*, Opiliaceae	(A)	26—35
75.	*Unonopsis glaucopetala*, Annonaceae	(a)	20—48
76.	*Touroulia guianensis*, Quiinaceae	(a)	17—32

size differences? In certain plant species, size differences are enormously great, and the very large, so-called "giant stomata" are distinguished from the other "normal" stomata by the following characteristics:

(a) by their larger size
(b) by their scarcer occurrence
(c) by their possible affinity to stains or their coloration different from that of normal stomata
(d) by their preferred position above veins or in the close vicinity of veins, and
(e) possibly by a wide open porus.

Considering the species with a variable stomatal size, not all of them develop "giant stomata" in the abovementioned sense. In the following, the species with giant stomata are listed.

As may be seen from the data below, the size (length) of giant stomata oscillates between 29 and 65 μm. Some giant stomata reach a respectable width of up to 87 μm. It must, however, be emphasized that the size of the giant stomata has to be seen in relation with the normal stomata. A

Size (length) of giant stomata in μm

1.	*Tovomita brevistamina*, Guttiferae	(a)	44—45	19. *Eschweilera odora*, Lecythidaceae	(A)	30—35
2.	*Rheedia* sp. (Cozoiba picuda), "	(A)	60	20. *Eschweilera* sp. " (Majaguillo erizado)	(A)	44—51
3.	*Calophyllum brasiliense*, "	(A)	60	21. *Eschweilera subglandulosa*, "	(A)	44
4.	"Erizo", Rutaceae	(A)	54	22. *Couratari multiflora*, "	(A)	44
5.	*Simarouba amara*, Simaroubaceae	(A)	65	23. *Lecythis davisii*, "	(A)	44—48
6.	*Protium neglectum* (Caraño), Burseraceae	(a)	29	24. *Cordia alliodora*, Boraginaceae	(A)	35—39
7.	*Protium* sp. (Sipuede) "	(A)	44—47	25. *Aniba riparia*, Lauraceae	(A)	30
8.	"Maro", "	(a)	44—48	26. *Beilschmiedia curviramea*, "	(a)	35—39
9.	*Protium decandrum*? (Caraño blanco), "	(a)	36	27. *Sandwithia guayanensis*, Euphorbiaceae	(a)	35—48
10.	*Tetragastris* or *Dacryodes*, " (Aracho blanco)	(a)	50	28. *Drypetes variabilis*, "	(A)	52
11.	*Trichilia smithii*, Meliaceae	(a)	38—45	29. *Chaetocarpus schomburgkianus*, "	(A)	50
12.	*Trichilia propingua*, "	(a)	44—45	30. *Erisma uncinatum*, Vochysiaceae	(A)	35
13.	*Toulicia guianensis*, Sapindaceae	(a)	30—44	31. *Vochysia lehmannii*, "	(A)	44
14.	*Tapirira guianensis*, Anacardiaceae	(A)	44—57	32. *Qualea dinizii*, "	(A)	48—52
15.	*Sclerolobium paniculatum*, Caesalpiniaceae	(A)	35	33. *Sloanea guianensis*, Elaeocarpaceae	(A)	44—65
16.	*Mora excelsa*, "	(A)	57	34. *Agonandra brasiliensis*, Opiliaceae	(A)	35
17.	*Peltogyne porphyrocardia*, "	(A)	48	35. *Unonopsis glaucopetala*, Annonaceae	(a)	48
18.	*Hirtella davisii*, Rosaceae	(A)	35			

stoma of 22 μm in length may appear as enormously much larger than a stoma of 11 μm in length. There is no absolute size which would define a giant stoma, only its relative size in comparison with the normal stomata is of importance. This comparison implies that smaller and larger stomata are present at the same time, the smaller stomata usually prevailing in number. Sometimes the difference between large and small stomata may be very great, amounting to as much as 54 μm. In other cases, however, size differences are not too large, but other criteria add more weight. Very common is the greater affinity of giant stomata for stains. Slides coloured with toluidine blue or methylene blue show differences in the staining affinity of normal and giant stomata. Giant stomata usually show a much deeper blue or even a colour different from that of the normal stomata, e.g. a staining towards a more violet or purple colour. In certain instances the porus of the giant stomata is wide open, in contrast to the normal stomata where it is smaller. The stomata may even be extended in a lateral direction so that a "ring-porus" with a wide open mouth results. Not infrequently the position of the giant stomata in relation to the veins is very conspicuous. Especially large giant stomata are often found in the closest vicinity of the veins or even above them. They are thus near the vascular bundles and possibly more closely connected with them by intercellular spaces. Their function as water secreting organelles seems not to be too far fetched when considering this relation.

Within the 232 species studied of Venezuelan Guiana, about 35 or 15% were found with giant stomata of the kind described above. The 35 species belong to very different plant families (altogether 17), whereby some families prevail, such as the Burseraceae, Guttiferae, Lecythidaceae, Caesalpiniaceae, which are considered more primitive, while more advanced families are in the minority (e.g. the Euphorbiaceae). It is perhaps not misleading to assume that giant stomata are a more primitive character occurring more frequently in more primitive families.

Considering the height categories of the species with giant stomata, 24 are of height category A, while 11 are of height category a.

Of the 35 species with giant stomata, 23 have leaves of the sun type with xeromorphic characteristics, corresponding to 65.7%, five or 14.28% have sun leaves with mesomorphic characters, three or 8.57% have leaves of the medium type with xeromorphic characteristics, two or 5.7% have shade leaves with some xeromorphic characters, only one species (2.85%) has leaves of the medium type with mesomorphic characteristics, and another species (2.85%) has shade leaves with mesomorphic characteristics. Leaves of the sun type with some xeromorphic characters thus prevail. In total, 80% possess sun leaves, and 20% have leaves of the medium or shade type. We may thus suppose that mainly high trees are furnished with giant stomata.

A special shape of the giant stomata occurs in the form of the so-called "ring-pores", which were also observed in the material studied from Venezuelan Guiana. In several cases it is very obvious that the stoma, possibly together with its companion cells, is laterally extended by surface tensions. An example is supplied by *Beilschmiedia curviramea*, Lauraceae. Giant stomata with a length of 35—44 μm and a width of the entire stomatal apparatus (including the companion cells) of 65—87 μm and even 130 μm were measured above the veins. It is surprising that the largest stomata usually occur above the veins, while smaller giant stomata may be found more remote from them. Generally, the stomata are situated with their porus parallel to the longitudinal extension of the veins so that the lateral expansion of the stomata always occurs at right angles to the vascular course. Ring-pores were likewise observed in *Aniba excelsa*, Lauraceae, where only few stomata are extended in width about to the double of the normal size. In this case, the porus remains wide open. In *Eschweilera subglandulosa*, Lecythidaceae, I found ring-pores with a wide open porus, about 9 per mm^2. The giant stomata reached a size of 44 μm and were distinguished from the other stomata by their lesser staining affinities being of lighter colour than the normal stomata. A weak affinity for stains may be connected with processes of dedifferentiation. However, this lighter staining was observed only rarely in giant stomata. In *Trichilia smithii*, Nyctaginaceae, I observed very large stomata, 39—44 μm long, deeply staining and with a wide open porus.

Staining differences between giant and ordinary stomata have been observed in most cases. In *Cordia alliodora*, Boraginaceae, the normal stomata with a length of 17—26 μm regularly alternate with larger deeply staining stomata of violet colour, 35 up to 39 μm long. The occlusive cells of the normal stomata remain colourless and only their inner wall thickenings in the form of ledges stain with toluidine blue. A regular pattern of large deeply stained stomata alternating with small colourless stomata with coloured inner wall thickenings results (Figure 3, 3). The large stomata reach a relatively high density of 22 up to 54 per mm^2, while the complete stomata density (including the large stomata) amounts to 400 per mm^2. The amount of giant stomata interspersed between the normal ones profoundly changes from species to species. While in the "Erizo", Rutaceae, the number of giant stomata is relatively high, large stomata (65 μm long and 57 μm in width) are scarce in *Simarouba amara*, Simaroubaceae. In *Erisma uncinatum*, Vochysiaceae, only few larger stomata, up to 35 μm long, are interspersed within the normal ones, which reach 21—26 μm in length. In *Vochysia lehmannii*, the larger stomata (44 μm long) not only stain darker than the normal ones (21—26 μm long), but also differ from them in their colour. While the normal stomata stain blue with toluidine blue, the large stomata adopt a violet colour. Frequently, the larger stomata are surrounded by a wreath of cells with less staining affinity or of almost colourless epidermis cells. Surprisingly, radial cell extension of neighbouring cells which surround the stomata may precede formation of giant stomata and even periclinal divisions in the radially extended cells may occur. These activities very much recall the processes during lenticel formation. Lighter spots regularly distributed over the lower leaf surface may thus be distinguished. However, lighter spots sometimes also become visible on the upper leaf side when stained, but without any relation to stomata or cell divisions.

Stomatal size is very variable in *Capparis amplissima* oscillating between 13 and 39 μm. The larger stomata show much affinity to stains. Large dark spots are visible here and there on the lower leaf surface when stained, representing a stoma surrounded by cells which underwent periclinal divisions. Expansion of cells as an inducing agent here precedes cell division. Particularly large stomata with much staining affinity are conspicuous in *Couratari multiflora*, Lecythidaceae, and often occur above the nerves. While stomata density oscillates between 360 and 372 per mm^2, the giant stomata reach only a density of 1.15 per mm^2. A difference in the staining affinity of normal and giant stomata could also be observed in *Eschweilera odora*, Lecythidaceae. While the occlusive cells of the normal stomata (17—26 μm in length) stain blue with toluidine blue, those of the giant stomata (about 35 μm long) adopt a violet or purple colour. Their density reaches only 1 per mm^2, as compared with the normal stomata of a density of 800—1060 per mm^2. In *Toulicia guianensis*, Sapindaceae, stomata size oscillates between 13 and 44 μm. The giant stomata are distinguished from the normal stomata by their strong affinity to stains. In this example, not only the occlusive cells dye deeply with toluidine blue, but also the companion cells of the giant stomata. The leathery leaves of the Burseracea "Maro" shows a stomata density of 320—600 per mm^2, while the giant stomata reach a density of 5.7 per mm^2. The normal stomata have a length of 18—21 mm, whereas the giant stomata reach up to 48 μm in length. Relatively frequently they occur above the veins.

Sanwithia guayanensis, Euphorbiaceae, of height category a, has a more or less hygromorphic shade leaf with normal stomata of a length of 21—26 μm and giant stomata which reach 44—48 μm in length. The giant stomata, about double the size of the normal stomata, have a strong affinity to stains. The xeromorphic sun leaf of *Qualea dinizii*, Vochysiaceae, on the other hand, has normal stomata of a length of 21—30 μm and giant stomata of 48—52 μm in length. The large stomata with open porus mostly occur above or in the vicinity of veins. The large stomata are furthermore distinguished from the ordinary stomata by the arrangement of the neighbouring cells in the form of a wreath. The neighbouring cells extending radially from the stoma as a center possibly form periclinal walls around the stoma. A certain meristematic activity is apparently involved in the formation of the cell group which immediately surrounds the giant stomata. The cuticular ribbing of the neighbouring cells likewise radiates from the giant stomata in all directions indicating certain surface tensions. The last two mentioned examples prove that hygromorphic shade leaves (*Sanwithia guayanensis* of height category a) as well as xeromorphic sun leaves (*Qualea dinizii* of height category A) may develop giant stomata.

In *Unonopsis glaucopetala*, Annonaceae, the normal stomata of about 26—35 μm in length almost remain colourless, while the giant stomata, up to 48 μm long, stain violet with toluidine blue. A deeply violet or purple stained ring of cells occurs around the giant stomata due to deeper staining of the surrounding cells. The giant stomata of *Peltogyne porphyrocardia*, Caesalpiniaceae, about 48 μm long, are distinguished from the normal stomata, about 21—39 μm in length, by their strong affinity to stains and often are found near or above the veins. Some of them are surrounded by a wreath of almost colourless cells. The density of the giant stomata is relatively high in this case, amounting up to 13—27 per mm^2. The giant stomata of *Tapirira guianensis*, Anacardiaceae, are surrounded by cells which underwent periclinal divisions in a similar way as observed during lenticel formation (see the paragraph dealing with lenticels). It is therefore not improbable that certain types of giant stomata later transform into lenticels, particularly when cell divisions around them become obvious (e.g. in some Lecythidaceae). As a consequence, lenticels of this origin may likewise function as hydathodes.

4.2. Glands functioning as hydathodes (Figures 21 and 22)

As has been mentioned already, hydathodes may occur in the form of glandular hairs, usually with a stalk and a uni- or pluricellular head. This type is described in the following.

1. *Brosimum* sp. ("Charo negro"), Moraceae, develops stalked glands with a pluricellular head (Figure 4.3) which may correspond to water-secreting hydathodes. They only occur on the lower surface.

2. *Coccoloba* sp. ("Uvero"), Polygonaceae, has headed glands with a head of 54—60 μm in diameter, mostly situated above the veins or in

closest vicinity of them. They are found on the upper as well as on the lower leaf surface, being more frequent on the lower leaf side where a density of 12 per mm^2 may be reached. The glands are always in connection with the vascular bundles by pitted extensions which correspond to tracheidal cells. The multicellular heads of the glands (Figure 4.1) resemble those of the digesting glands of *Pinguicula*. On leaves of young trees the diameter of the heads is increased in size up to 74 μm, but the density is the same. It is very probable that these glands function as hydathodes (see also Roth 1984).

3. *Coccoloba* sp. ("Arahueque"), Polygonaceae. Very few glandular hydathodes occur on the upper leaf side. On the lower leaf surface very few large peltate hairs, 113 μm in diameter, situated on top of the nerves may have developed from hydathodes. Of special interest are stomatal hydathodes surrounded by a wreath of calotte-like cells which occur here and there on the lower leaf surface (Figure 4.2).

4. *Helicostylis tomentosa*, Moraceae. Few stalked glands similar in their structure to those of the "Charo negro" with a pluricellular head occur on the lower leaf surface. They may correspond to hydathodes.

5. *Apeiba echinata*, Tiliaceae. Headed glands are found on the lower leaf surface, occurring infrequently. It is not known whether they secrete only water.

6. *Cusparia trifoliata*, Rutaceae. Stalked glands with a pluricellular head occur on the lower leaf surface of this shrub. The leaf is of the hygromorphic shade type.

7. *Protium neglectum* ("Caraño"), Burseraceae. Pluricellular headed glandular hairs infrequently occur on the lower leaf surface.

8. "Maro", Burseraceae. Few headed glands are found on the lower leaf side.

9. *Trichilia propingua*, Meliaceae. Glandular hairs occur on the upper as well as on the lower leaf surface. They are, however, more frequent on the lower leaf side. The glands are already dried out on the grown-up leaf. The leaf is of the hygromorphic shade type.

10. *Inga capitata*, Mimosaceae. Club-shaped glands occur on the lower leaf surface (Figure 4.4). Their density amounts up to 36 per mm^2. They are mainly found above the veins.

11. *Inga* sp. ("Guamo blanco"), Mimosaceae. Stalked glandular hairs occur at a density of 30—40 per mm^2 on the lower leaf surface. On the upper leaf side they are rare.

12. *Inga* sp. ("Guamo caraota"), Mimosaceae. Club-shaped glandular hairs, mainly occurring near or above the veins, reach a density of up to 9 per mm^2 on the lower leaf surface.

13. *Centrolobium paraense*, Papilionaceae. Giant glands with a head diameter of 180 μm reach a density of 3 per mm^2 on the lower leaf surface (Figure 4.5). The heads are pluricellular and contain a yellow secretion which stains pink-orange with Sudan III.

14. "Saquiyak", Melastomaceae. Very few remnants of glandular hairs are found on the lower leaf surface.

15. *Tabebuia* sp., ("Araguaney puig"), Bignoniaceae. Headed glands of a density of 30 per mm^2 occur on the upper surface of leaves of adult trees, while on the lower leaf side the density reaches up to 50—60 per mm^2. On leaves of young trees, on the other hand, the glands are scarcer with a density of only 15 per mm^2 on the lower leaf side.

16. *Tabebuia serratifolia*, Bignoniaceae. Headed glands reach a density of 6 per mm^2 on the upper leaf surface, and of 40 per mm^2 on the lower leaf side.

17. *Tabebuia stenocalyx*, Bignoniaceae. Headed glands with a density of 21 per mm^2 occur on the upper leaf surface, but reach a density of 45 per mm^2 on the lower leaf side. It is noteworthy that glands occur in the form of two different kinds: the larger ones with thicker cell walls are less frequent, while the smaller ones have very thin walls and contain crystals of the rhomboid type in their head cells. The crystals may consist in Ca carbonate, which is not infrequently found in Bignoniaceae. The leaf is of the mesomorphic shade type.

18. *Vitex stahelii*, Verbenaceae. Large glands with a head of up to 56 μm in diameter occur on

the lower leaf surface (Figure 4.6). Their density is about 1 per mm^2 or even less. Glands on top or in the vicinity of veins are in connection with the vascular bundles by tracheidal "water cells" and may function as hydathodes. The so-called water cells are larger thick-walled cells with abundant pits. The glands show uni- as well as pluricellular heads. They are more frequent in leaves of young trees.

19. *Ficus* sp. ("Higuerón"), Moraceae. Headed glands are found on the upper as well as on the lower leaf surface.

20. *Ficus* sp. ("Matapalo"), Moraceae. Very few two-celled club-shaped hairs are found on the lower leaf surface. They may function as glands.

Glandular hairs, often with a stalk and a uni- or pluricellular head, are only rarely found in the material studied (about 8%). They generally occur with a higher frequency on the lower than on the upper leaf surface, or reach a higher density on the lower leaf side, when present on both surfaces. It is very probable that most of these glands function as hydathodes. The original function of glandular hairs was certainly only water secretion. But later during evolution glands may have become specialized in secreting other substances, organic or inorganic. Nectaries may have developed in this way. That the fluid secreted by hydathodes is not necessarily pure water, but a solution containing salts, minerals, sugars and so on, is well known. However, density of glandular hairs is usually low. The highest frequency observed is 50—60 per mm^2 in the "Araguaney", Bignoniaceae. Giant glands with a head diameter of up to 180 μm were found in *Centrolobium paraense*, Papilionaceae. Glands may possibly be more frequent in the juvenile leaves, i.e. in leaves of young trees, than in the leaves of adult grown-up trees, for example, in *Vitex stahelii*, Verbenaceae (see also Roth 1984).

In the literature, almost nothing is known of the function of the glands in the species studied. Glandular hairs of *Jacaranda*, Bignoniaceae, serve as hydathodes, according to Metcalfe and Chalk (1950), those of *Tabebuia* sometimes secrete Ca carbonate together with water. Secretion of Ca carbonate may also prove right for *Tabebuia stenocalyx* of the material studied in Venezuelan Guiana, as crystals, probably of Ca carbonate, were found in the head cells of the smaller glands (Figures 21 and 22). As has been reported already, the glandular hairs of certain species of the genus *Inga* seem to secrete an oily substance of a fishy smell (Roth 1984). A yellowish probably oily content is also found in the glands of *Centrolobium paraense*.

From the above given lists it becomes obvious that most of the species with glandular hairs on the leaves are not furnished with giant stomata — with the exception of the Burseraceae. Although the function of the glands as well as that of the giant stomata has not been proved experimentally as yet, it is possible that both structures replace one another in their function. The fact that the glands more frequently occur on leaves of young trees and that they are functional as long as the leaves are young could be explained by the consideration that young growing organs need more minerals and that water turnover, therefore, must be more vigorous in them.

From the anatomical point of view, different kinds of hydathodes may thus be distinguished. On the one hand, there are glandular hairs or trichomes, mostly of epidermal origin, composed of a foot, a stalk, and of head cells, which secrete the water through the cuticle — either by rupture or by perforations of the cuticle. On the other hand, certain necrotic processes may play their part in the formation of "water ways". Examples are found in the lenticel development, when stomata degenerate to give way to the formation of larger openings in the epidermis. Furthermore, so-called "apical openings" occur on the leaf tips of some monocotyledonous water plants. Stomata degenerate or the entire epidermis cells are destroyed in this region so that a hole develops where the tracheidal endings of the vascular bundles terminate. In this hole water collects, even in the absence of an epithema. A further type of hydathodes has a more complicated structure with a secreting layer or epithema, which is usually covered by the epidermis so that the water may be released through the stomata. Endings of tracheids are usually in close contact with the epithema. Johnson (1937) observed hydathodes on the adaxial leaf side of species of *Equisetum*. They often had a reduced, poorly developed epithema and were "studded with pores resembling stomata in general appearance but differing from them in their position above a vein". The "water stomata" did not show chloroplasts and had very thin walls, as compared with the normal stomata. Although related in their position with the foliar vein, they are not associated with the free tips of vascular elements. However, Johnson could prove water secretion through the "water stomata". Already in 1924, Haberlandt drew attention to the fact that stomata of this last type of hydathodes have often a shape different from that of ordinary stomata and that they may be distinguished by their enormous size. Zimmermann and Bachmann-Schwegler (1953) found stomata of different sizes in leaves of species of *Anemone*. The larger stomata leave larger distances between them producing a "gross-pattern", while the smaller stomata forming a "small-pattern" are more densely packed. The same was observed by Maercker in *Ilex aquifolium* leaves (1965). Developmentally, the larger stomata are formed first. Their substomatal cavity is connected with a large intercellular space and their porus is usually wide open. Stomata which are distinguished by their larger size, their special shape ("ring-pores"), and by their wide-open apertures were repeatedly described in fruits (Arzt 1962, Sauer 1933, Fischer 1921 and the following papers. See also Roth 1977 and the literature cited there). Arzt (1962) suggested that larger stomata with large apertures give the impression of hydathodes. Large stomata distinguished from the ordinary stomata by their apertures, their size or their divergent shape were repeatedly suspected to be real hydathodes, even in the absence of an epithema. Eames and MacDaniels (1947), for example, consider certain types of enlarged stomata which possibly serve for water secretion "water stomata". These authors emphasize that hydathodes may resemble stomata closely. Stomata functioning as hydathodes have been reported particularly of the floral region by Daumann. He noted stomata with open pores which could be closed only partly but not completely by plasmolytica, for example, on the upper side of the perigone leaves of *Magnolia*, especially on the leaf base (Daumann 1930). According to Daumann they secrete water. Likewise, on the spadix of *Anthurium digitatum*, hydathodes in the form of stomata with an open porus are found. Applying plasmolytica, the porus can only be closed to a certain extent. When the secreted liquid is removed with filter paper, a new secretion may be observed already after 25—30 minutes. The liquid secreted by the free perianth leaves through the hydathodes contains monoses and bioses (Daumann 1931). In *Arisaema consanguineum*, small droplets may be observed above the stomata of the recently opened spathe. Although the droplets apparently are secreted by the stomata, these have the normal appearance with occlusive cells containing chlorophyll and starch and with a porus which may be regulated by opening and closing. According to Steinberger-Hart (1922), the stomatal water pores of *Achillea, Impatiens*, and *Tropaeolum* are hydroactive, and in *Achillea* and *Impatiens* opening of the water pores in light could be observed (for further information see also Fahn 1979 and Napp-Zinn 1973).

In the ovary of *Jasminum nudiflorum*, Daumann (1932) found hydathodes in the form of modified stomata which secrete nectar. From these observations it results that stomata without the aid of an epithema may be enabled to secrete water. This is in accordance with the observation of Esau (1953) that in some plants the hydathodes are without epithem, and that the water moves towards the pore through ordinary mesophyll. Furthermore, Daumann reported that water (or nectar) secreting structures (such as nectaries etc.) not only secrete liquids, but under certain circumstances may also absorb water and other liquids. The often observed large size of "stomatal hydathodes" is an interesting feature in coherence with my own studies on vegetative leaves collected in Venezuelan Guiana. As is well known, in aroids, in species of *Papaver* and *Tropaeolum*, the stomata of hydathodes may adopt an enormous size. This is an obvious advantage for water secretion: the greater the porus, the lesser the resistance to infiltration.

Of most interest in this connection is the

possible function of the giant stomata on the leaves of the species studied in Venezuelan Guiana. The ecological conditions of a tropical humid forest, such as a moist, warm soil and an almost vapour-saturated atmosphere, favour the absorption of water, but do not favour transpiration. Although during the day transpiration may take place in a more or less normal way, held in equilibrium by water loss through insolation and air movement, during the night, on the contrary, transpiration may almost completely become blocked. On the other hand, continuation of the water transport is necessary, as growth is most vigorous during the night. Loss of water in a liquid form is thus the last resort for the plant. Secretion of water through the leaves — and eventually also through the stem — is a well-known phenomenon in tropical plants. Guttation may be so strong at night that a continuous rain of water drops falls from the trees or shrubs, e.g. in *Bambusa*.

It is generally agreed that guttation is caused by root pressure and root pressure occurs, because the salt concentration in the root xylem causes osmotic uptake of water and consequent development of a hydrostatic pressure in the xylem sap. Under the ecological conditions of a humid tropical forest, water absorption may greatly exceed transpiration so that the water is pushed up the xylem ducts and out through the ways of least resistance, i.e. the hydathodes. In other words, when water uptake exceeds water loss, a hydrostatic pressure is built up in the xylem and water must escape by whatever path is available. Water that exudes from hydathodes does so as a result of hydrostatic pressure developed in the xylem ducts and guttation is thus a more or less passive process, although active water secretion by hydathodes has been described also (e.g. in glands). However, guttation liquid is usually not pure water, but a solution containing a great number of dissolved substances, such as salts, sugars, etc. Fructose, glucose, arabinose, sucrose, galactose, glutamine, and P, K, Na, Ca, Mg, NO_3, NH_4, phosphate, sulphate have been reported between others. The pH measured oscillates around 6—6.75. However, the salt concentration of the liquid produced by guttation is very low. Salts are apparently taken up by the plant before the liquid is exuded. A certain inconsistency, however, exists between the periodicity of the root pressure and that of guttation. While guttation reaches its highest values during the night, root pressure shows often a periodicity with its maximum during the day time so that in certain examples a negative relation between guttation and root pressure has been observed.

As early as 1898, Molisch and Figdor independently observed that an overpressure prevails in the hydrosystem of tropical trees during the night and early morning hours which finally leads to guttation through the leaves. Braun (1983) has recently reopened the question of hydrostatic pressure and guttation. According to this author, the overpressure is due to osmotically active organic and inorganic substances secreted by the parenchymatous accessory tissue of the hydrosystem. In order to supply the crown region with the necessary mineral salts from the soil, an osmotic water shifting ("Wasserverschiebung" — translocation) takes place in the hydrosystem. Osmotically active substances are released into the vessels by the physiological activities of the accessory tissue. This leads to the uptake of water, even in the absence of transpiration, and an internal pressure is induced in the hydrosystem, which may cause the exudation of guttation liquid. Braun thus suggests that osmotically induced water transport in this way may be a mechanism of water ascent in tropical trees. This theory would evade the contradictory periodicity of root pressure and guttation found in some examples. By means of water uptake through the root and osmotic water shifting in the hydrosystem guttation sap is secreted through the leaves and possibly also through the stem. However, Braun gives no suggestion through which openings the guttation liquid leaves the leaf. Actually, we know only a few tropical plants which have ordinary hydathodes secreting guttation sap. In the bamboo, for example, the leaf tips actively secrete water during the night in such an intense way that it sounds like steady rain dripping from the leaves. In other plants, the leaf margins are provided with hydathodes, as we know from the literature.

However, the existence of some examples of hydathodes does not explain such a general phenomenon as water transport of trees when the air is vapour-saturated, and transpiration is blocked. Hydathodes or water pores have to be postulated for most of the tropical trees growing in a warm and moist climate.

The hypothesis that the giant stomata found in the material of Venezuelan Guiana function as hydathodes thus gains probability. They are the ways of least resistance and they are usually found above the veins or in their vicinity. Besides the stomatal hydathodes, glandular hydathodes may play their part and even replace the giant stomata in their function, equally as lenticels with their very large openings and fissures in the epidermis may secrete guttation sap. Finally, there may be gaps between epidermal cells which are not in complete contact with one another. Daumann (1932) observed such a "Lückenepidermis" in the ovary of *Jasminum nudiflorum.* A similar situation occurs in the flower of *Helleborus foetidus* (Bonnier 1929), in *Grevillea preissii* (Müller 1929), and in *Iris* (Daumann 1932). Gaps in the epidermis of the floral region have been described in the 19th century by many authors (see also Haberlandt 1924). Gaps in the epidermis were also observed by Roth in fruits (1977b) and in leaves (1984). In stems, water may be secreted through the bark by means of lenticels or cracks. Further ways of fast water or vapour release may be pits in the outer epidermis walls of the upper leaf epidermis in *Rheedia* sp. ("Cozoiba rebalsera"), Guttiferae, where their occurrence is very frequent. The outer cell walls of the lower leaf surface of *Drypetes variabilis*, Euphorbiaceae, seem to be irregularly perforated. Although it is not clear yet, whether real pores are present or whether the light spots in the walls only correspond to wall thinnings, they certainly increase cuticular evaporation. As generally observed in nature, there are many ways to reach the same goal. Passive stomatal hydathodes and lenticels or active glands may be very good instruments, cracks, fissures and gaps in the epidermis may be other possibilities of water secretion. However, their presence is far more frequent in plants of the humid tropical forest than had been accepted up to the present. And the ways most frequently used are probably the stomata. Not only the giant stomata situated above the veins or in their vicinity may be used as water pores. In species with stomata of different sizes particularly the larger stomata may have the same function, and even ordinary stomata may become water ways when the hydrostatic pressure in the veins is very strong.

5. OCELLI

In a dense tropical forest, light capture is a problem particularly for those plants which grow in the undergrowth. Haberlandt (1924) has drawn attention to cases where special organelles for light perception are developed. He called them "ocelli", little eyes. Actually, a variety of structures serving for light perception may be recognized. (1) There are special epidermal cells distinguished from the other epidermis cells by their size and shape. They are usually larger than the other epidermis cells and lens-shaped (plano-convex or biconvex) or more or less globular and have side walls inclined towards the inside. They rise above the epidermis level functioning as light collectors with their focus in the center of the inner wall. This type is actually very common among plants of tropical humid forests, and particularly of the cloud forest. (2) Somewhat more complicated are enlarged epidermis cells which develop cone-shaped papillas as an outgrowth at their summit corresponding to a protuberance. (3) In other cases, two superposed cells are combined to collect the light. The basal cell is larger, lens-shaped or globular and surpasses the epidermis level, while the superposed cell which lies at the summit of the basal cell is smaller having the function of a light collector. The side walls of the enlarged basal cell are inclined towards the inside, i.e. centripetally. The small cell on top of the larger one is often lens- or cone-shaped. Haberlandt considers ocelli of type 2 and 3 as modified hairs. And indeed, in some species the hairs themselves have a bulbous base which may serve as a lens. Short hairs in the form of cystoliths with strongly thickened outer walls and often cone-shaped likewise may have a refractive property and function as a lens. They usually appear as white dots on the upper (and lower) side of the green leaf blade. In all three types described, the specialized cells have the optical properties of a lens and are light collectors. The lens itself is colourless and has a transparent content.

Differences between shade and sun leaves concerning the structure of the upper epidermis cells become obvious in *Cydonia japonica* where epidermis cells of the sun leaves have a plane outer wall, whereas in the shade leaves epidermis cells are plano-convex or biconvex lenses with a convex outer wall. This is an evident reaction to the environment, according to Haberlandt 1924. In leaves of tropical humid forests, this phenomenon has hardly been studied before, but shade leaves of humid tropical forests or of the montane cloud forest frequently develop lens-shaped epidermis cells.

Different are the so-called "windows" in the epidermis, which let light pass through the photosynthetic tissue into the inside of the leaf or stem, so that they may be thought of as light channels. Such light channels are well known, e.g. of the collective genus *Mesembryanthemum*. In this case, the leaf surface is much reduced and only a small part of the leaf is exposed to insolation, for example, in the subgenera *Fenestraria* and *Lithops*, where only the leaf top is free, while the main portion of the leaf is buried in the soil. The leaf apex is transformed into a "window" composed of colourless water-storing tissue from which a centrally situated channel of colourless tissue leads into the leaf interior. Solar rays entering this channel reach the surrounding photosynthetic tissue by a reduction of their original intensity to 1/4 up to 1/10. Similarly function sclereids, for example, in the genera *Hakea* and *Capparis* where elongated sclereids penetrating the photosynthetic tissue serve as light channels. The plants mentioned above are xerophytes reducing their photosynthetically active surface to a minimum so that additional indirect irradiation of diffuse light reaching more profound layers of the mesophyll is of advantage to them. In the bifacial leaf types, the photosynthetic tissue is composed of elongated palisade cells parallel to which the light channels penetrate the leaf so that the photosynthetic cells receive additional diffuse light through their lateral walls. While light channels may compensate for a reduced leaf or stem surface, particularly in xerophytic plants, ocelli are more common in the shade leaf type. I have repeatedly drawn attention to this subject (Roth 1966, pp. 122 and 265, and 1984, pp. 17 and 446). However, light channels also occur in hygromorphic shade leaves, becoming apparent in the form of white dots on the leaf surface (see below).

Some special examples will be described in the following. Examples of ocelli are numerous, as mentioned already, and often characterize not only the upper, but also the lower surface of Angiosperm leaves in the tropical rain and cloud forest (Roth 1966, 1980; Lindorf 1979). In view of the optical conditions, especially in the cloud forest where diffuse light in the lower forest strata prevails, it is not surprising that plants growing in these stories (i.e. in the undergrowth) often possess ocelli. The patterns which ocelli create on the leaf surface may be in the form of scattered isolated cells, in the form of groups, or the entire surface may be covered with ocelli.

The upper epidermis cells are very conspicuously of a lenticular shape with convex outer walls in *Geonoma pinnatifrons*, Arecaceae, which plant grows in the undergrowth of the Venezuelan cloud forest of the Cordillera de la Costa near Rancho Grande (Lindorf 1979, Roth 1980, p. 500, Figure 3). *Besleria disgrega*, Gesneriaceae, likewise from the Venezuelan cloud forest, develops very large lenticular epidermis cells. The upper epidermis is composed of very large biconvex lenticular cells in *Psychotria aubletiana*, Rubiaceae, from the Venezuelan cloud forest. In *Heliconia hirsuta*, Musaceae, coming from the same habitat, the upper epidermis cells are transformed into very large lenticular cells with a strongly convex outer cell wall. In *Commelinopsis persicariaefolia* (Commelinaceae) from the Venezuelan cloud forest, the upper as well as the lower epidermis cells are very large having somewhat convex outer walls. Scattered very large lenticular cells with a strongly convex outer wall are found in the lower epidermis of *Phyllanthus acuminatus* (Euphorbiaceae) of the Venezuelan cloud forest. *Fittonia verschaffeltii* from the rain forest develops scattered large cells with a small cell at their apex in the upper epidermis. The small cells contain a transparent strongly refractive content. The incident light rays are concentrated on this cell which functions as a lens and are thrown on the underlying palisade parenchyma.

In *Helicostylis tomentosa*, Moraceae, from the humid forest of Venezuelan Guiana, the entire lower epidermis is composed of cells with a short papilla at the cell apex. *Rollinia multiflora* from the same habitat, has likewise a papillose lower epidermis (Roth 1984, Figures 4, 52, 55, 56, 64, 65). Examples of lower epidermis cells with papillas are plentifully found in my book "Stratification of tropical forests as seen in leaf structure" (1984). In *Catalpa*, Bignoniaceae, each lower epidermis cell is enlarged parallel to the leaf surface bearing a short papilla at its apex. Scattered large papillose cells are interspersed between very small normal cells in the lower epidermis of *Ischnosiphon* sp., Marantaceae (Roth 1984, Figure 14, lower part). Likewise in the lower epidermis of *Brunellia comocladifolia* ssp. *funckiana*, papillose cells are interspersed within normal epidermis cells. *Ischnosiphon* grows in the humid forest of Venezuelan Guiana, while *Brunellia* comes from the cloud forest at Rancho Grande. Further examples with scattered papillas from Venezuelan Guiana are *Apeiba tibourbou*, Tiliaceae, and *Vismia macrophylla*, Guttiferae (Roth 1977, Figures 22 and 26). In many cases, an irregular distribution of ocelli may be observed so that scattered isolated ocelli or small groups of them alternate with regular epidermis cells. In *Helicostylis tomentosa* and *Rollinia multiflora*, papillas form a wreath around the stomata. A pattern of white dots on the leaf surface of certain Acanthaceae is due to the sporadic formation of ocelli which are able to collect the remaining diffuse light in the forest undergrowth. As mentioned already, the light intensity which reaches the undergrowth may be less than 1% of the intensity which reaches the crown region. And it is mainly the diffuse light which plays the most important part in this stratum. Particularly in the cloud forest, the mist contributes to the increase in the diffuse light effect (Huber 1976).

Lenticular cells as light collectors thus occur on the upper as well as on the lower leaf surface, while papillas are more commonly found on the lower leaf side. To what degree they really assist in light capture has, however, never been proved as yet. Cystoliths, finally, are frequently found in Moraceae and Boraginaceae of the rain and cloud forest (see also Roth 1984). In *Cordia alliodora*, cystoliths of the upper epidermis may enter the photosynthetic tissue as far as the spongy parenchyma. In *Cordia exaltata* and *C. viridis*, cystoliths occur at the base of cone-shaped hairs with thick outer walls. These cystoliths produce a pattern of white dots on the leaf surface, acting as light channels in a similar way as sclereids, for example (see above). The optical effects of very large cystoliths in the genus *Ficus* which reach the palisade parenchyma, or of water-storing giant cells interspersed within the palisade parenchyma, as found in *Tetragastris panamensis*, or, finally, of a large water-storing hypodermis, has still to be studied. As long as the optical properties of very ordinary bifacial leaves lacking the abovementioned structures are not completely understood (see Fukshanski 1981), we can only suspect the function ocelli and similar organelles may really have.

6. PAPILLAS (Figures 1 lower, 7, 18)

As pointed out before (Roth 1980), papillas and ocelli are more frequently found on the lower than on the upper leaf surface, being originally characteristic of shade plants from the wet tropical forest. They also develop in some cultivated races of *Carica papaya* and *Manihot esculenta*, which plants are native of the humid tropical forests. Exceptions of the rule that papillas generally only occur on the lower leaf epidermis are, for example, *Virola surinamensis*, Myristicaceae, where papilla-like cells only develop in the upper epidermis, *Trichilia schomburgkii*, Meliaceae, with papilla-like protuberances on the upper and lower leaf side, and the Burseracea "Maro" where thick-walled sclerozed papillas are likewise found on the upper as well as on the lower leaf surface. Papillas may be considered not further developed small hairs, as some examples exist where papillas occur on the leaves of young trees, but unicellular hairs appear on the leaves of the adult trees or — in a few cases — vice versa. Here, the short papillas probably transform into elongated unicellular hairs.

The function of the papillas may be threefold. In certain instances, papillas surround the stomata in the form of a wreath so that a calm air chamber arises above the stomata reducing stomatal transpiration in this way. In other cases, however, they may be considered "collecting lenses" (Haberlandt 1905, 1908; Stahl 1896). Lenses of this type are particularly useful in diffuse light even when they occur on the lower leaf side. Papillas and ocelli are found in leaves of shade plants like dicotyledonous herbs and shrubs, as well as of monocotyledonous hardy annuals and bushes, but also occur in trees (Roth 1984). Papillas, furthermore, increase the leaf surface considerably (to the double or more, as an estimated value) and in this way may also increase cuticular evaporation, as the papilla-bearing cells are living. Typical papillas of shade plants are thin-walled and have a thin cuticle. Increase of evaporation through papillas could be an adaptation of tropical shade plants to the humid environment.

Papillas are found in the following species studied of the humid tropical forest of Venezuelan Guiana:

1. *Rollinia multiflora*, Annonaceae, of height category a, has papillas on the lower leaf surface (Figure 1 below). A wreath of papillas surrounds the stomata creating a "calm chamber" around them. The stomata are somewhat elevated above the surface, a peculiarity characterizing hygromorphic leaves. This feature is compensated by the protecting papillas. The thick-walled papillas with a thin cuticle are very abundant on the lower surface. The leaf is of the sun type and has some xeromorphic characters (see Roth 1984, Figures 55, 56).

2. *Vismia macrophylla*, Guttiferae, of the height 1.30 m and 2 m. The hygromorphic leaf is of the shade to medium type. Thin-walled papillas occur on the lower leaf side. Papillas are longer in the 2 m high plant than in the one of 1.30 m in height.

3. *Vismia* sp. ("Lacre amarillo"), Guttiferae, of height category a. Papillas with thick walls and a thin cuticle were only found on the lower leaf surface of a 4 m high plant, but are absent on the adult plant. Stomata are elevated above the epidermis level. An indumentum of stellate hairs occurs on the lower side which is denser in the leaf of the adult tree. The leaf is of the sun type and has some xeromorphic features.

4. *Trichilia smithii*, Meliaceae, of height category a has a leaf of the medium type with some xeromorphic characteristics. The cuticle of the lower surface is ribbed. A small papilla is found at the apex of each lower epidermis cell. In a young tree of 10 m height, the papillas are more conspicuous than in the adult tree. The leaf of the 10 m high tree is of the shade type and shows meso- to hygromorphic characters (Figure 6).

5. *Trichilia schomburgkii*, Meliaceae, of height category a. The leaf of the adult tree has no papillas, while that of a 7 m high plant shows very few papilla-like protuberances on the upper as well as on the lower leaf surface. The leaf is of the shade type and unites meso with hygro and xeromorphic characters. On the lower leaf side of the adult plant, on the contrary, extremely few unicellular hairs with thick walls develop which are partly very short. Possibly, the papillas of the young plant transform into short hairs on the adult tree.

6. "Rabo pelado", Olacaceae, of height category A. The leaf is of the sun to medium type and shows meso and xeromorphic characters. The stomata are somewhat elevated above the surface which is a hygromorphic characteristic. The papillas, up to 0.01 mm long, have thick outer walls.

7. *Pithecellobium jupunba*, Mimosaceae, of height category A, has a leaf of the xeromorphic sun type. Each lower epidermis cell is transformed

into a papillose cell. Papillas are long and thick-walled with a density of about 2000 per mm^2. They somewhat bend over the stomata, reducing transpiration in this way.

8. *Pithecellobium* c.f. *claviflorum*, Mimosaceae, of height category A, has a mesomorphic sun-type leaf. Somewhat thick-walled papillas occur on the lower surface.

9. *Piptadenia psilostachya*, Mimosaceae, of height category A, has a xeromorphic sun-type leaf. The lower epidermis cells are furnished with thick-walled papillas.

10. Unknown, Mimosaceae, of height category A, has a xeromorphic sun leaf. Very thick-walled papillas occur on the lower leaf surface.

11. *Crudia oblonga*, Caesalpiniaceae, of height category A, has a mesomorphic sun leaf with some xeromorphic characteristics. Large papillas occur on the lower surface.

12. *Andira* sp. ("Canelito negro") Papilionaceae, of height category A, has a xeromorphic sun leaf. Thick-walled papillas are found on the lower leaf surface. Additionally, sword-like hairs occur on the lower leaf side. The stomata are protected beneath the papillas.

13. *Andira retusa*, Papilionaceae, of height category A, has a xeromorphic sun leaf. Papillas with thick outer walls are present on the lower leaf surface.

14. *Hymenolobium* sp. ("Alcornoque"), Papilionaceae, of height category A. Peltate hairs form a dense indumentum on the lower leaf side, while club-shaped papillas with very thick-walled tips protect the underlying stomata. Papillas and hairs retain air tenaciously (Figures 18 and 19).

15. "Maro", Burseraceae, of height category a, has a very sclero and xeromorphic sun leaf. Thick-walled sclerosed papillas occur on both leaf sides of the adult tree. In young plants (of 4 m height), very thick-walled unicellular hairs occur on both leaf sides, but papillas are absent. Additionally, few glandular hairs are present on the lower surface of the leaves of the young plants.

16. *Ecclinusa guianensis*, Sapotaceae, of height category A, has a sclero and xeromorphic sun leaf. Small papillas are found on the lower leaf surface.

17. *Beilschmiedia curviramea*, Lauraceae, of height category a, has a shade type leaf with xeromorphic characteristics. Small papillas develop on the lower leaf surface of the adult tree.

18. *Brosimum* sp. ("Charo negro"), Moraceae, has a somewhat xeromorphic sun leaf. Papillas occur on the lower leaf surface. Additionally, unicellular hairs of medium occurrence and stalked glands are present on the lower surface.

19. *Helicostylis tomentosa*, Moraceae. A 4 m high plant has hygromorphic shade leaves. The stomata are somewhat elevated above the surface, but are surrounded by a wreath of up to 0.05 mm long papillas which create a wind screen around them (see Roth 1984, Figures 3, 4, 52, 64, 65, 95).

20. *Virola surinamensis*, Myristicaceae, of height category A, has a xeromorphic sun leaf. Papilla-like cells occur on the upper leaf surface, while the lower surface is covered by large, thick-walled stellate hairs.

Papillas which surround and protect the stomata, furnishing a windless chamber above them and reducing transpiration in this way, occur in the following species: *Rollinia multiflora, Helicostylis tomentosa, Pithecellobium jupunba, Crudia oblonga, Andira* sp., and *Hymenolobium* sp. Papillas together with hairs occur on the lower leaf side of *Andira* sp. ("Canelito negro"). Papillas together with an indumentum of short hairs are observed in the lower epidermis of *Vismia* sp. ("Lacre amarillo"). Papillas together with a dense indumentum are conspicuous in *Hymenolobium* sp. ("Alcornoque montañero"). Papillas, unicellular hairs and stalked glands are characteristic of *Brosimum* sp. ("Charo negro") and, finally, unicellular hairs, stalked glands and long papillas (up to 0.05 mm long) occupy the lower leaf surface of *Helicostylis tomentosa*. In the abovementioned examples, papillas may — at least partly — function as a kind of protection reducing stomatal transpiration.

Very thick-walled papillas are found in *Rollinia multiflora*, in the leaves of the young trees of *Vismia* sp. ("Lacre amarillo"), in the "Rabo pelado", in *Pithecellobium jupunba*, in *Pithecellobium claviflorum*, in *Piptadenia psilostachya*, in an unknown Mimosacea, in *Andira* sp. ("Canelito negro"), in *Andira retusa*, in *Hymenolobium* sp. ("Alcornoque montañero"), and in the "Maro". Most of these species are represented by high trees, eight of height category A. The thick walls of the papillas may somewhat reduce evaporation, but in certain cases they may even have an optical effect, particularly when the tips of the papillas are thickened. A lenticular effect may arise in this way, e.g. in *Helicostylis tomentosa*.

Papillas surrounding the stomata may be especially effective when the stomata are elevated above the surface. This hygromorphic character is thus compensated in high trees by a xeromorphic peculiarity equally as in certain xeromorphic leaves with crypts (e.g. of *Nerium oleander*) in which the stomata are elevated above the surface, combining hygromorphic with xeromorphic characteristics that point in different directions.

Differences between leaves of young and adult trees suggest a pro- or regressive drift in the papilla formation from small hygromorphic plants towards high trees, expressing in this way the stratification of the forest and reflecting its special microclimate. This signifies that papillas are apparently a useful peculiarity in hygromorphic shade leaves, but either disappear in the meso- or xeromorphic leaves of the adult trees or transform into long hairs reducing transpiration in this way. In *Trichilia smithii*, papillas are more conspicuous in the young than in the adult tree. No papillas are found in the adult trees of *Trichilia schomburgkii*, while few papilla-like protuberances occur on the upper as well as on the lower leaf surface of the young plants. In *Vismia* sp. ("Lacre amarillo") papillas are present only on the young plant, but absent in the adult one. Papillas thus appear to be more characteristic of young plants which tend towards the hygromorphic shade leaf type. Only seldom the reverse is realized: Small papillas only occur on the leaves of the adult trees of *Beilschmiedia curviramea*, but not on those of the young plants.

On the other hand, papillas very much increase the leaf surface. They may augment the leaf surface by the double or more. Papillas are living cells. As they are frequently covered by a thin cuticle, they increase cuticular evaporation. This may be a desired effect in a very humid tropical environment. Furthermore, in the hygromorphic leaf type and in leaves of young trees the papillas are generally thin-walled. In the xeromorphic leaf type, on the other hand, papillas usually result thick-walled, and may even elongate into long hairs. For example, in *Vismia macrophylla*, papillas gain in length towards the canopy. In this case, they have probably the function of transpiration reduction.

We thus may perceive a modification of function of the papillas accompanying forest stratification: With the increasing height of the trees, the short living papillas may transform into long dead hairs which protect the surface against excessive transpiration, especially when they surround the stomata: a windless chamber is thus produced around the stomata and air is tenaciously retained above them. Modification of function thus may occur due to the changing microclimate in the different strata of the forest.

Papillas may thus have different functions. They may have an optical effect replacing ocelli, particularly when their tips are thick-walled and lens-shaped. Another function of papillas is the protection of stomata, when they surround the stomata in the form of a wreath, serving as a wind screen. Papillas with a thin cuticle considerably increase the leaf surface and consequently the cuticular evaporation. And finally, air is tenaciously retained between the papillas (e.g. in *Hymenolobium* sp.), a great advantage for plants living in a wet, often almost vapour-saturated atmosphere. Papillas may thus function in the same way as surface sculpturations (see next section).

7. SURFACE SCULPTURING
(Figures 8—17, 19, 20, 22)

A smooth surface of the upper and lower leaf epidermis is the usual picture presented by plants growing in a temperate climate. Cuticular ribbing or wall protuberances in the epidermis are better known of petals than of vegetative leaves. Where leaves have to be protected from excessive tran-

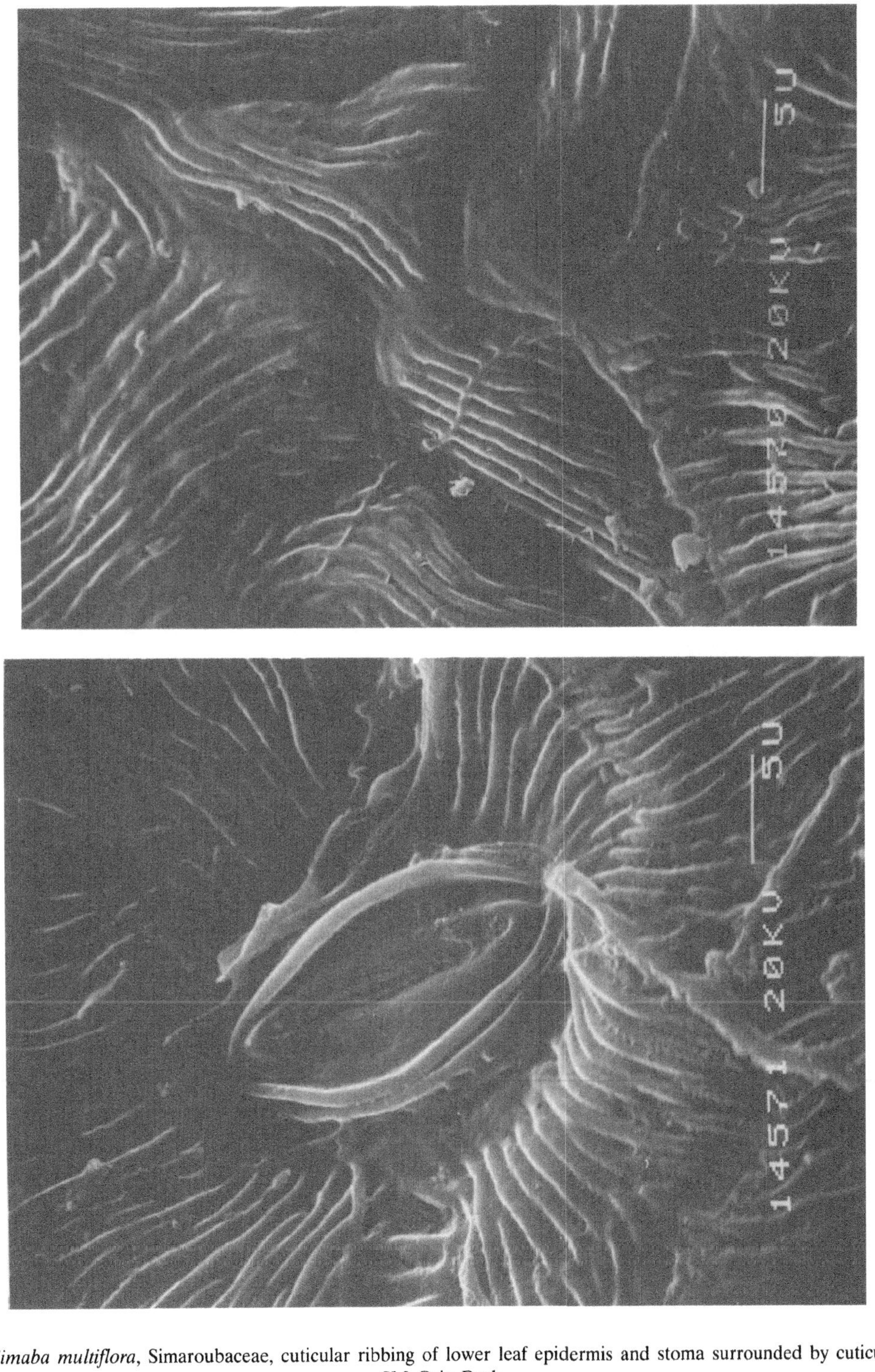

Fig. 8. Simaba multiflora, Simaroubaceae, cuticular ribbing of lower leaf epidermis and stoma surrounded by cuticular folds. SM. Orig. Roth.

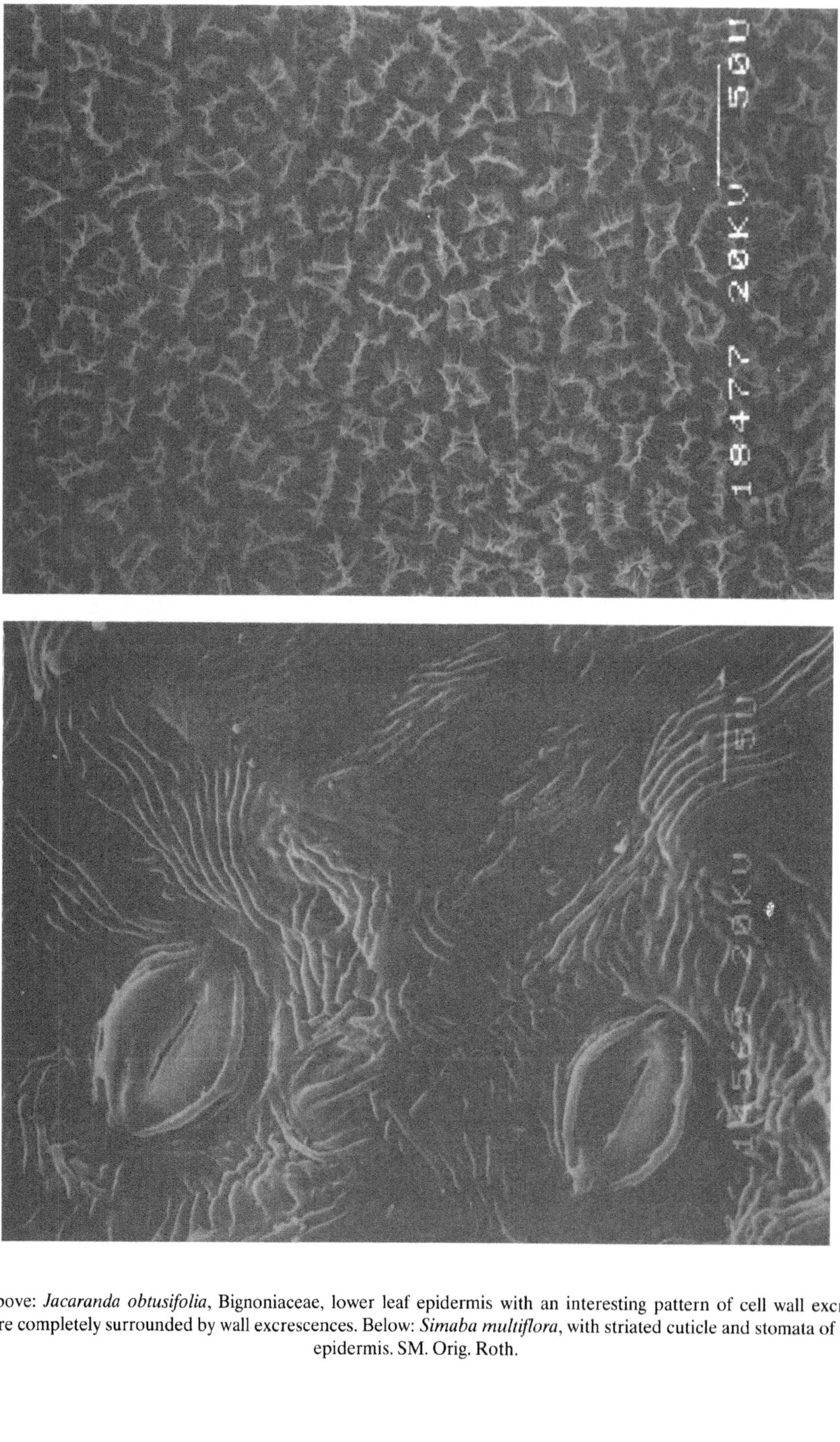

Fig. 9. Above: *Jacaranda obtusifolia*, Bignoniaceae, lower leaf epidermis with an interesting pattern of cell wall excrescences. Stomata are completely surrounded by wall excrescences. Below: *Simaba multiflora*, with striated cuticle and stomata of lower leaf epidermis. SM. Orig. Roth.

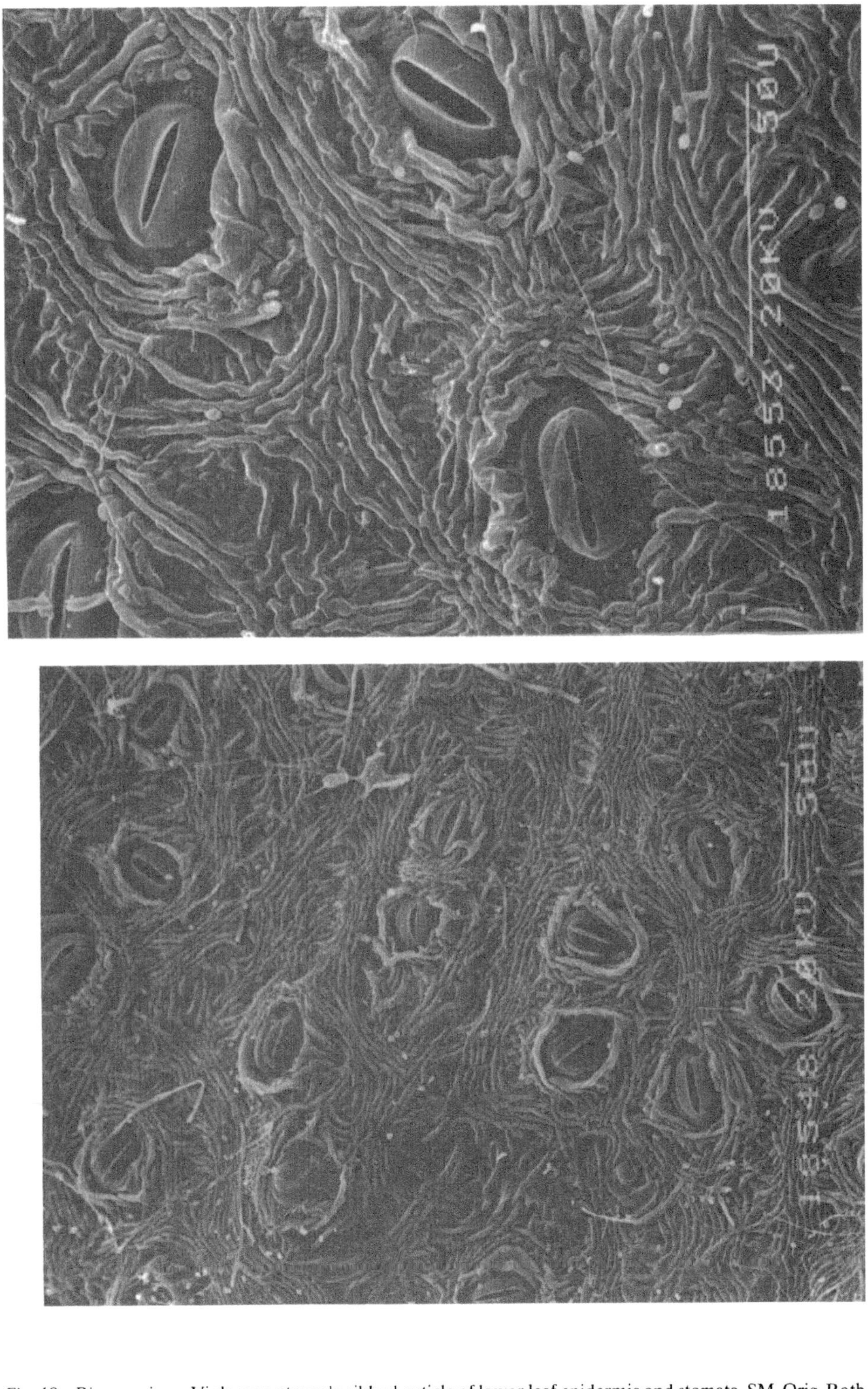

Fig. 10. *Rinorea riana*, Violaceae, strongly ribbed cuticle of lower leaf epidermis and stomata. SM. Orig. Roth.

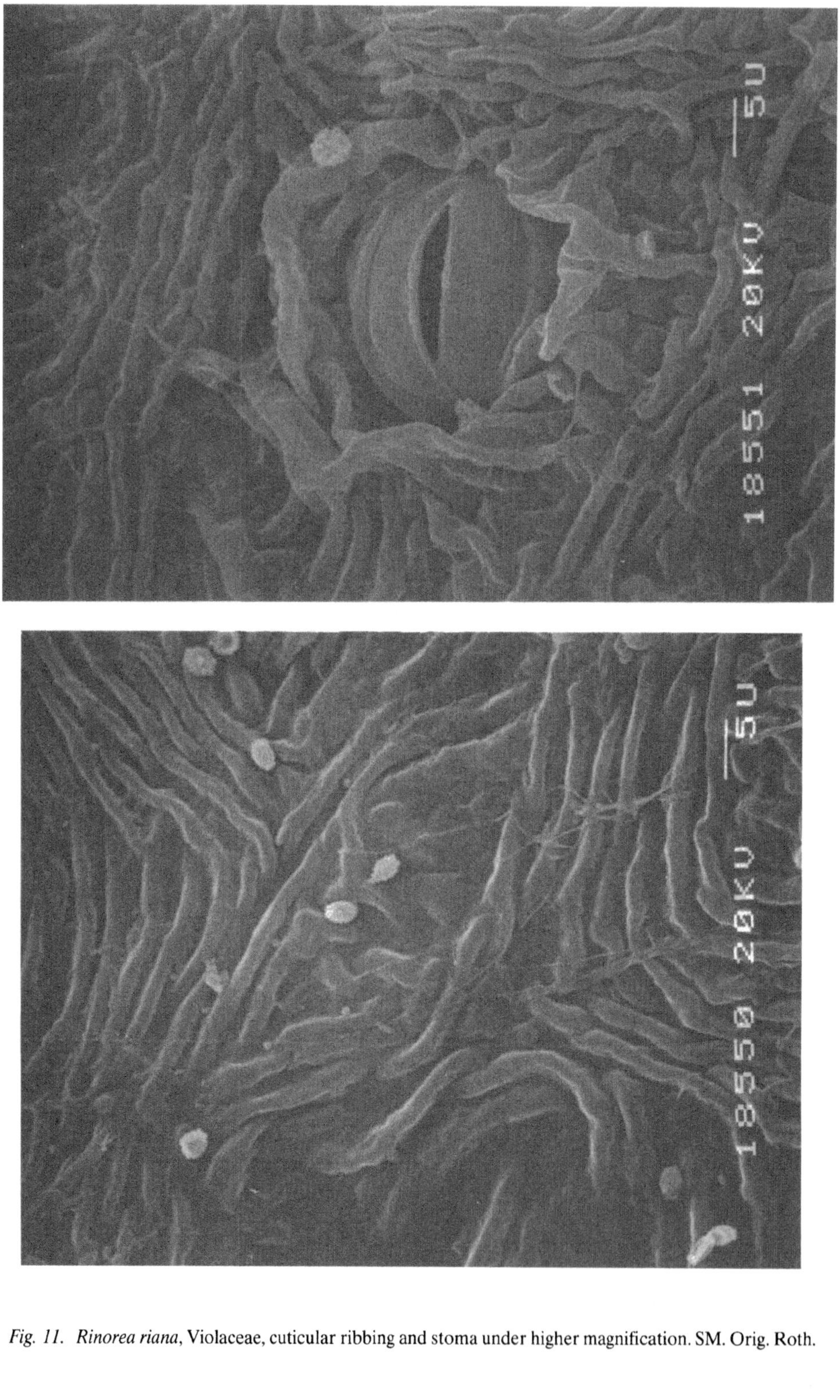

Fig. 11. Rinorea riana, Violaceae, cuticular ribbing and stoma under higher magnification. SM. Orig. Roth.

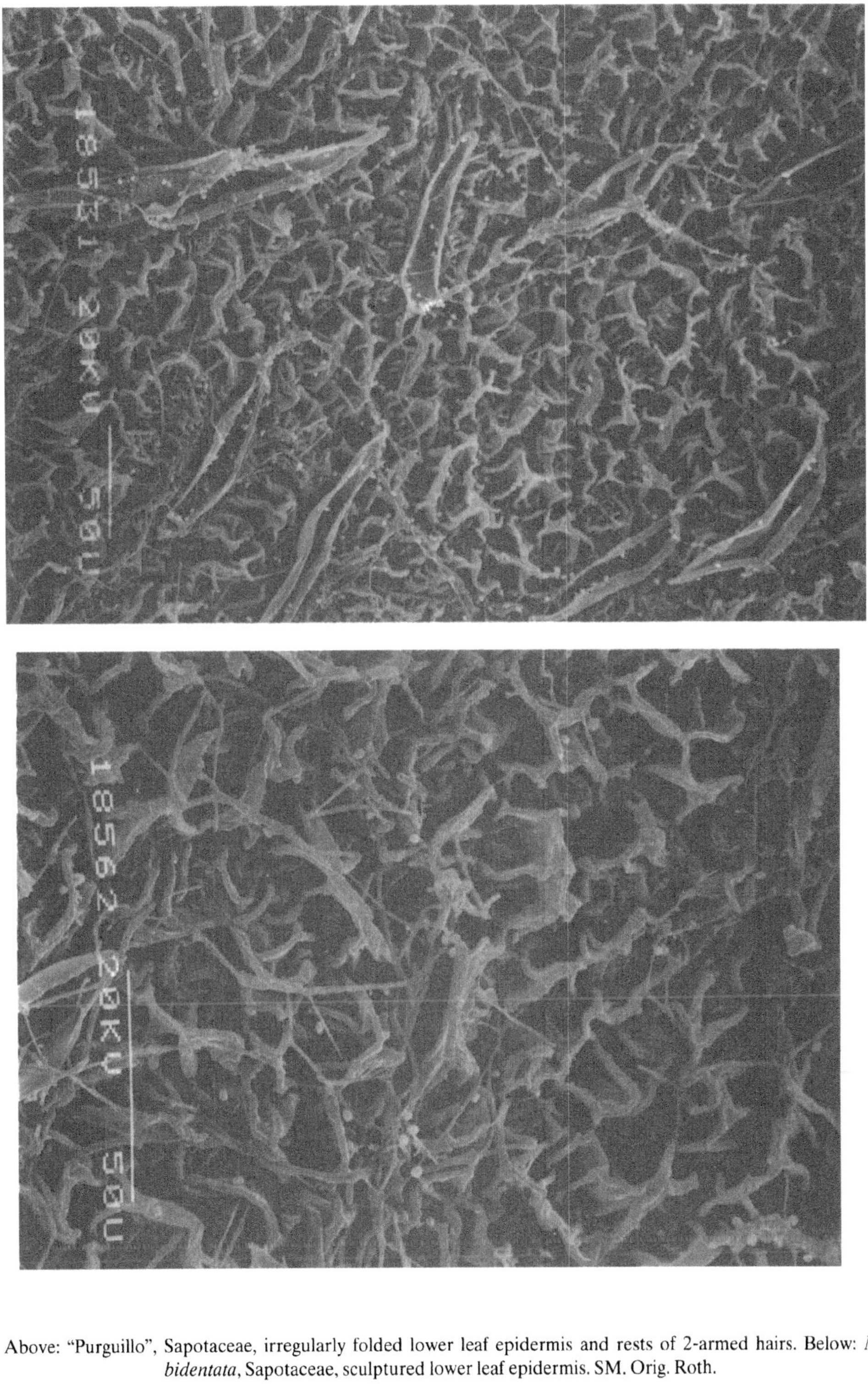

Fig. 12. Above: "Purguillo", Sapotaceae, irregularly folded lower leaf epidermis and rests of 2-armed hairs. Below: *Manilkara bidentata*, Sapotaceae, sculptured lower leaf epidermis. SM. Orig. Roth.

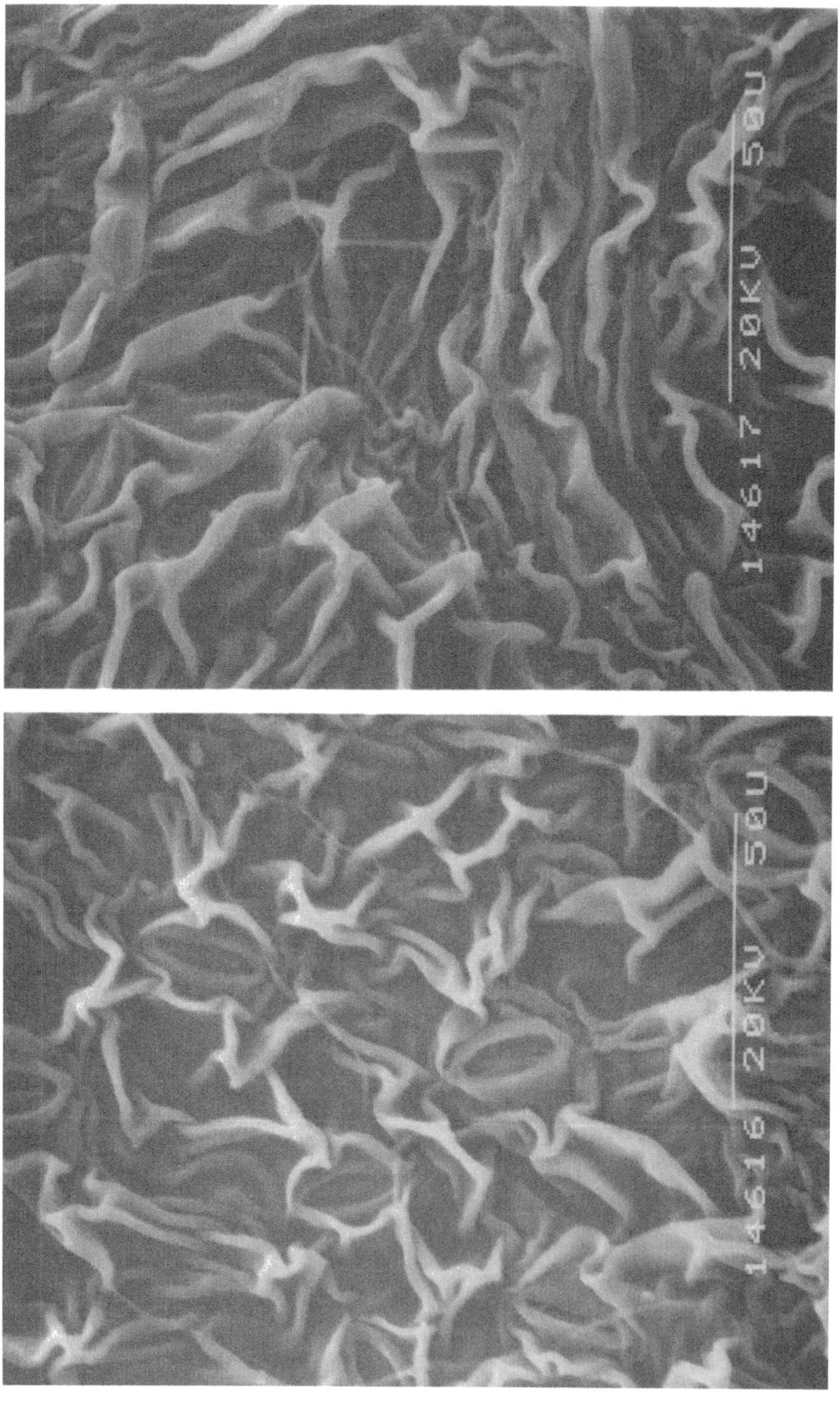

Fig. 13. Ficus sp., Moraceae, liana, lower leaf epidermis with irregular cuticular folds. SM. Orig. Roth.

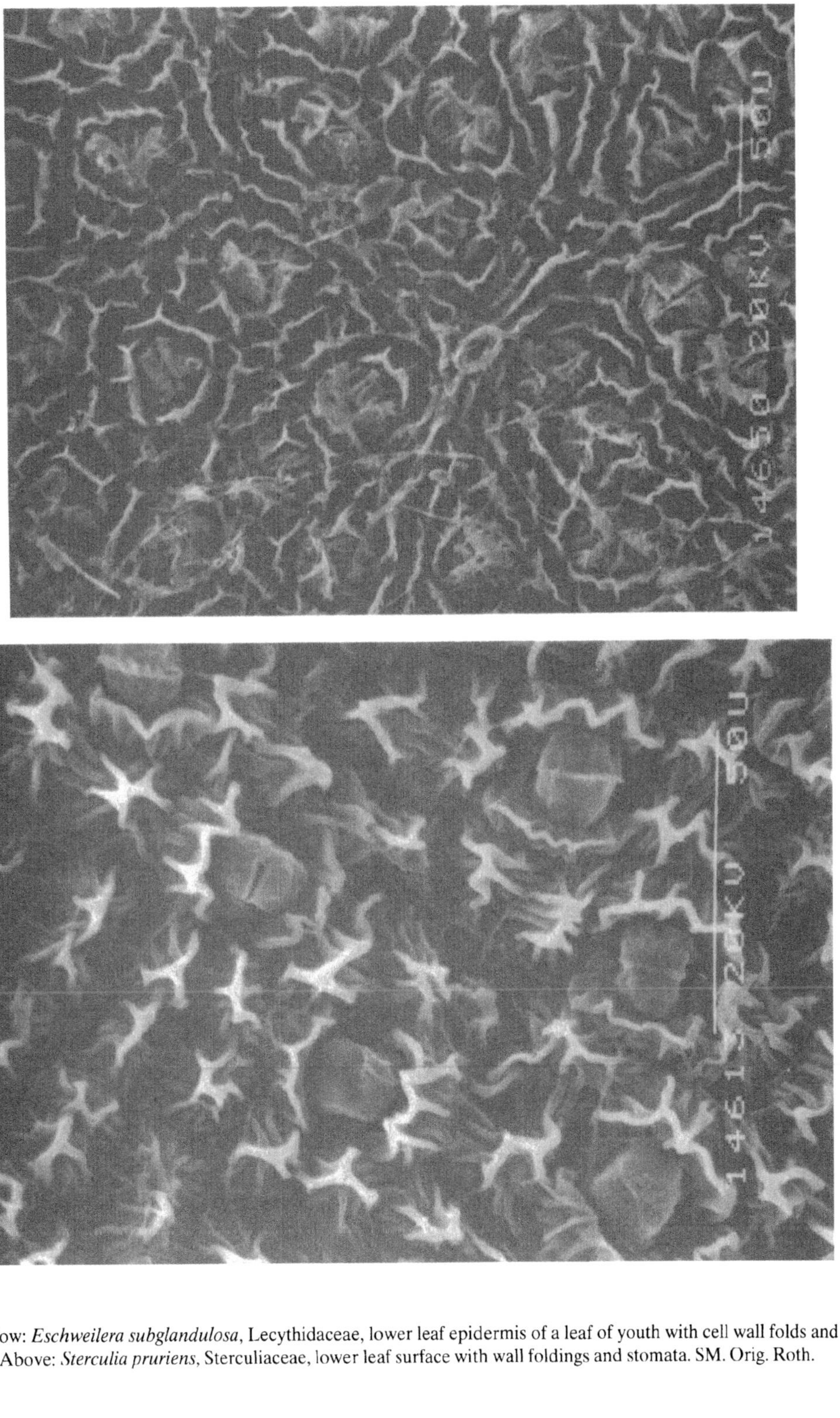

Fig. 14. Below: *Eschweilera subglandulosa*, Lecythidaceae, lower leaf epidermis of a leaf of youth with cell wall folds and stomata. Above: *Sterculia pruriens*, Sterculiaceae, lower leaf surface with wall foldings and stomata. SM. Orig. Roth.

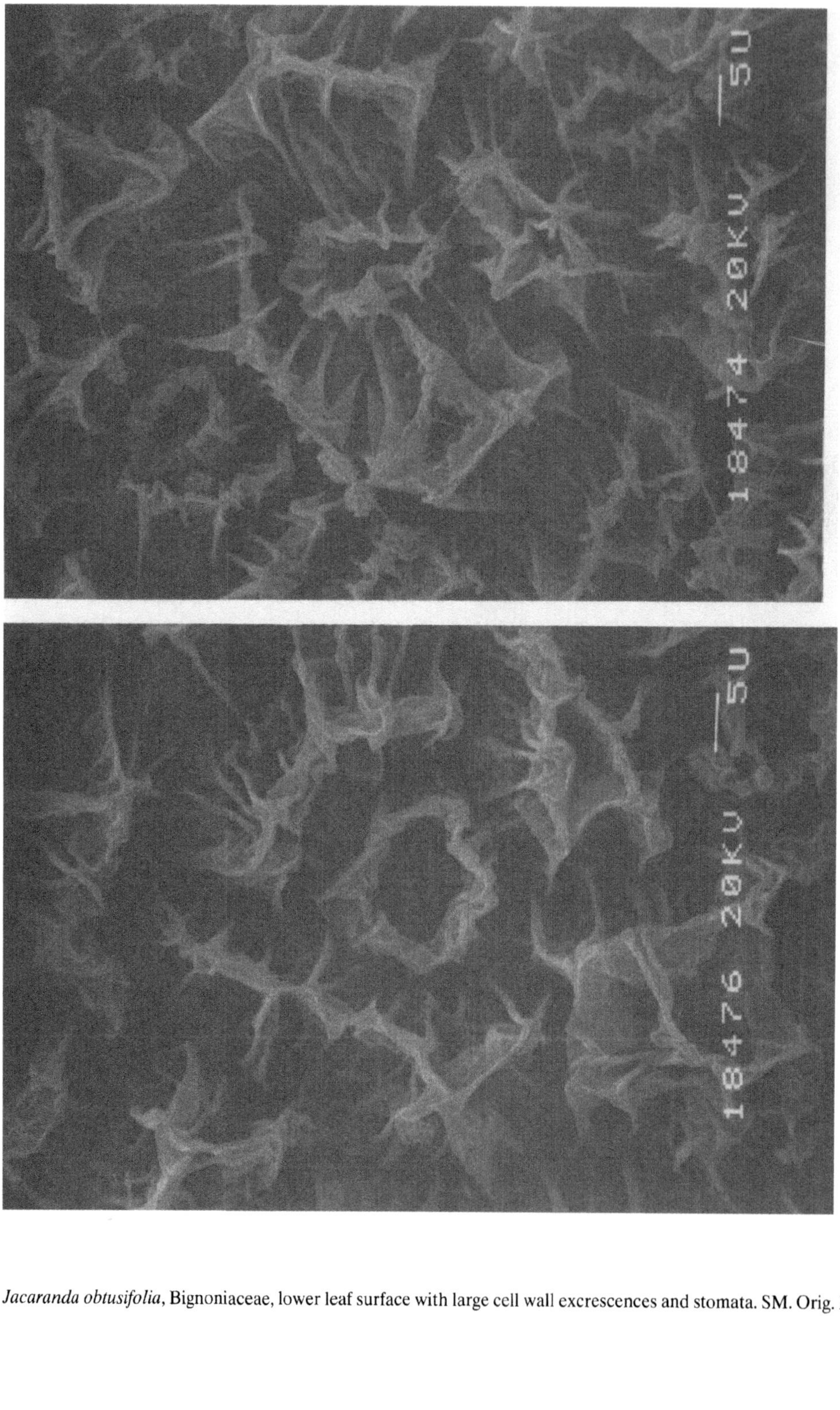

Fig. 15. Jacaranda obtusifolia, Bignoniaceae, lower leaf surface with large cell wall excrescences and stomata. SM. Orig. Roth.

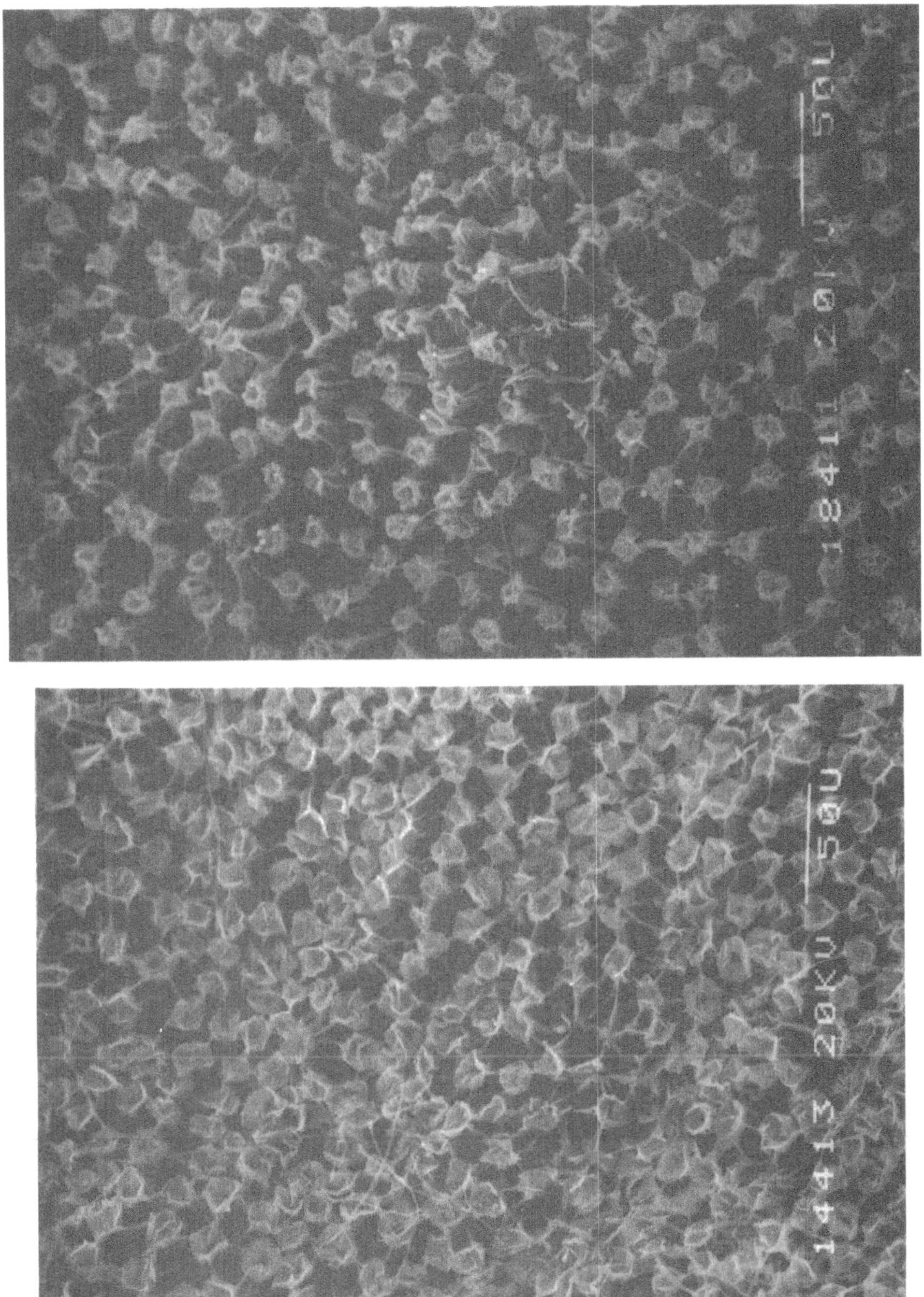

Fig. 16. A pattern of cell wall excrescences on the lower leaf epidermis very much alike in two different species belonging to two very different families. Below: *Aspidosperma excelsum*, Apocynaceae — Above: *Diospyros melinonii*, Ebenaceae. SM. Orig. Roth.

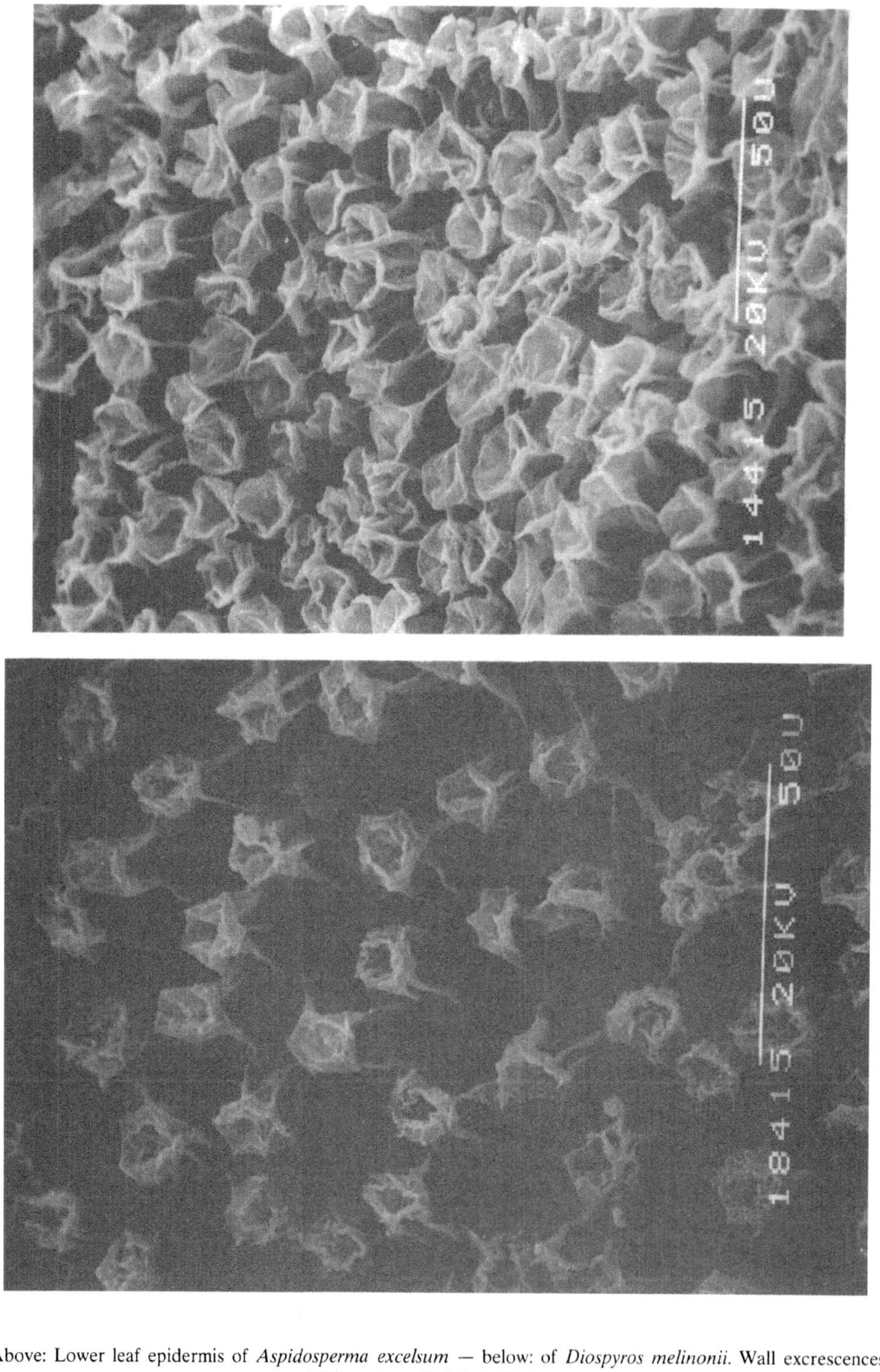

Fig. 17. Above: Lower leaf epidermis of *Aspidosperma excelsum* — below: of *Diospyros melinonii.* Wall excrescences about 0.02 mm long. SM. Orig. Roth.

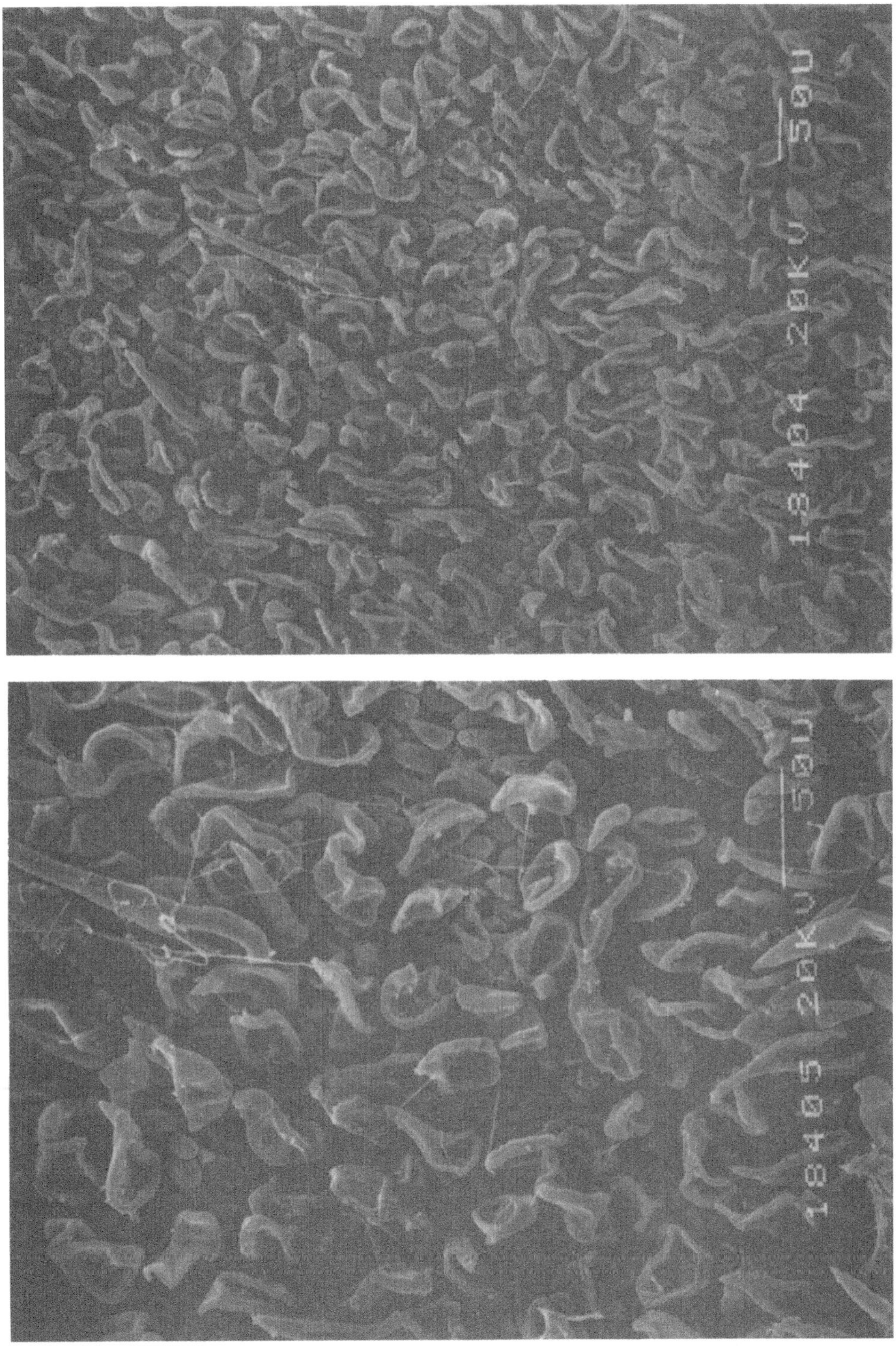

Fig. 18. Hymenolobium sp., Papilionaceae, lower leaf epidermis with papillas (shorter) and peltate hairs forming a dense indumentum. SM. Orig. Roth.

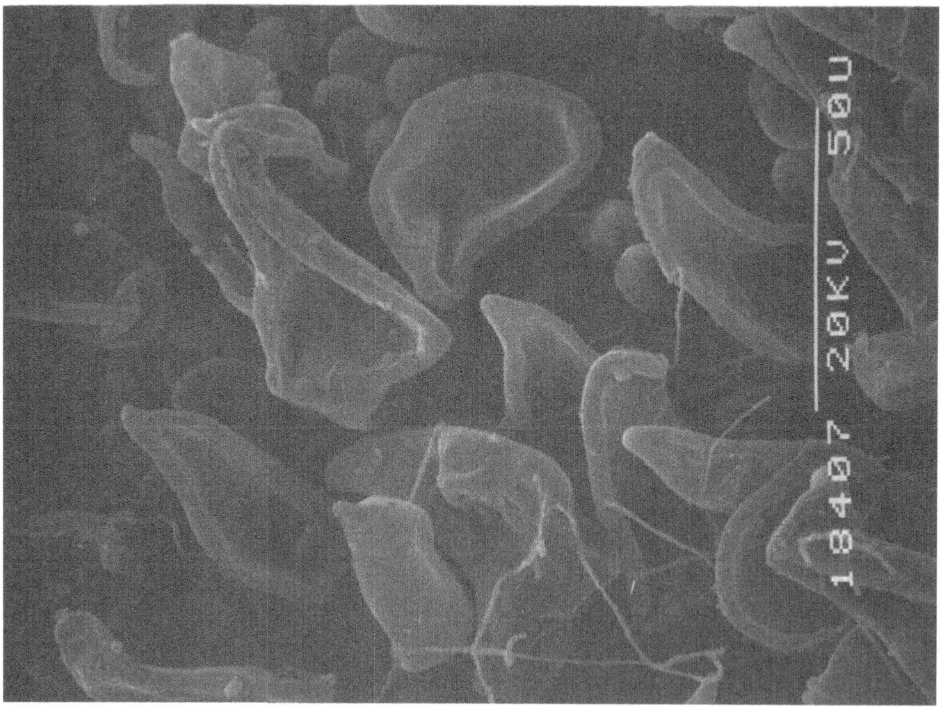

Fig. 19. Above: *Aspidosperma excelsum*, wall excrescences of lower leaf epidermis under high magnification. Below: *Hymenolobium*, peltate hairs and papillas covering completely lower leaf surface. SM. Orig. Roth.

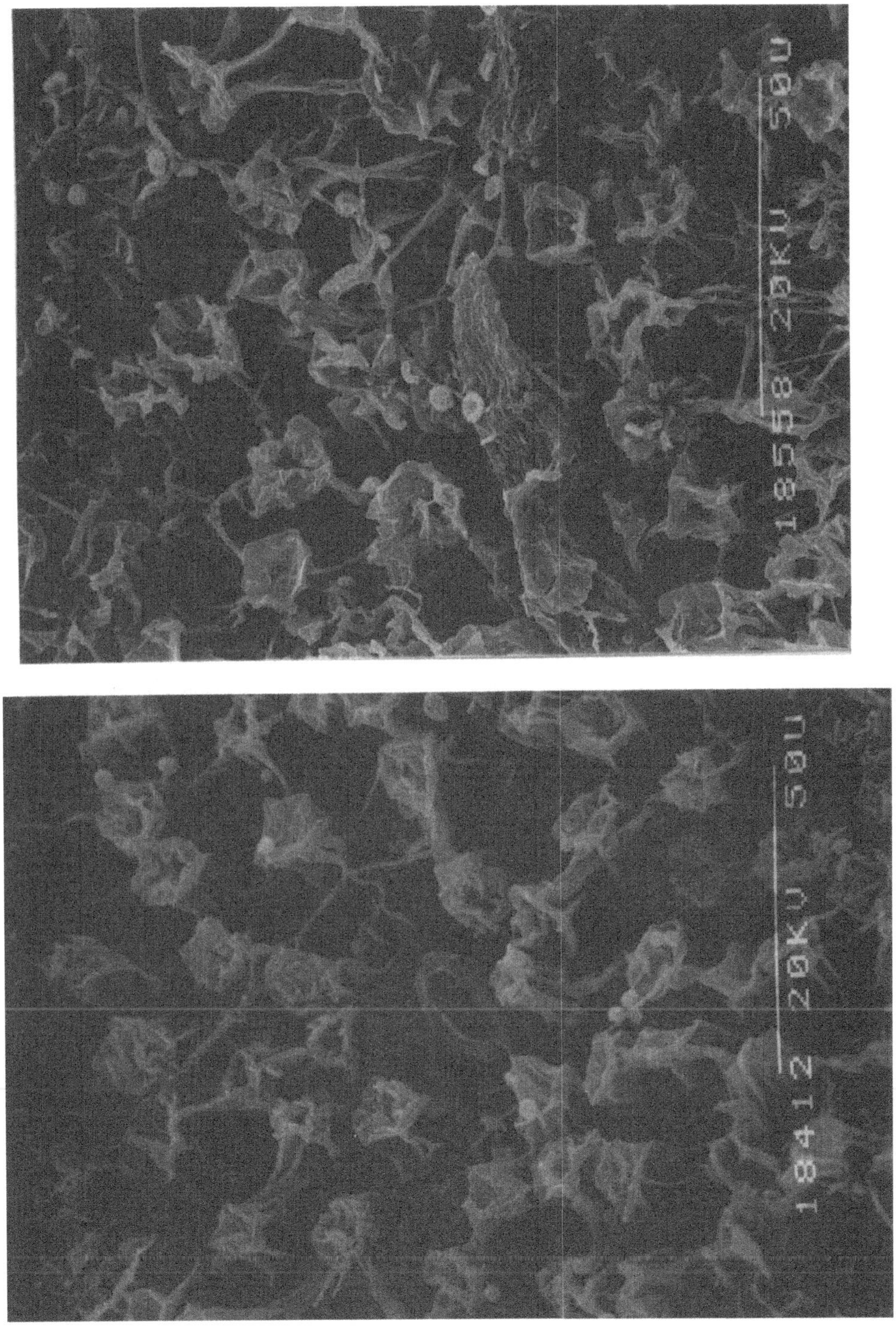

Fig. 20. Above: *Aspidosperma marcgravianum*, Apocynaceae, lower leaf epidermis with hairs and wall excrescences, a pattern somewhat different and less regular than that of Aspidosperma excelsum. Below: *Diospyros melinonii*, lower leaf epidermis with stoma. SM. Orig. Roth.

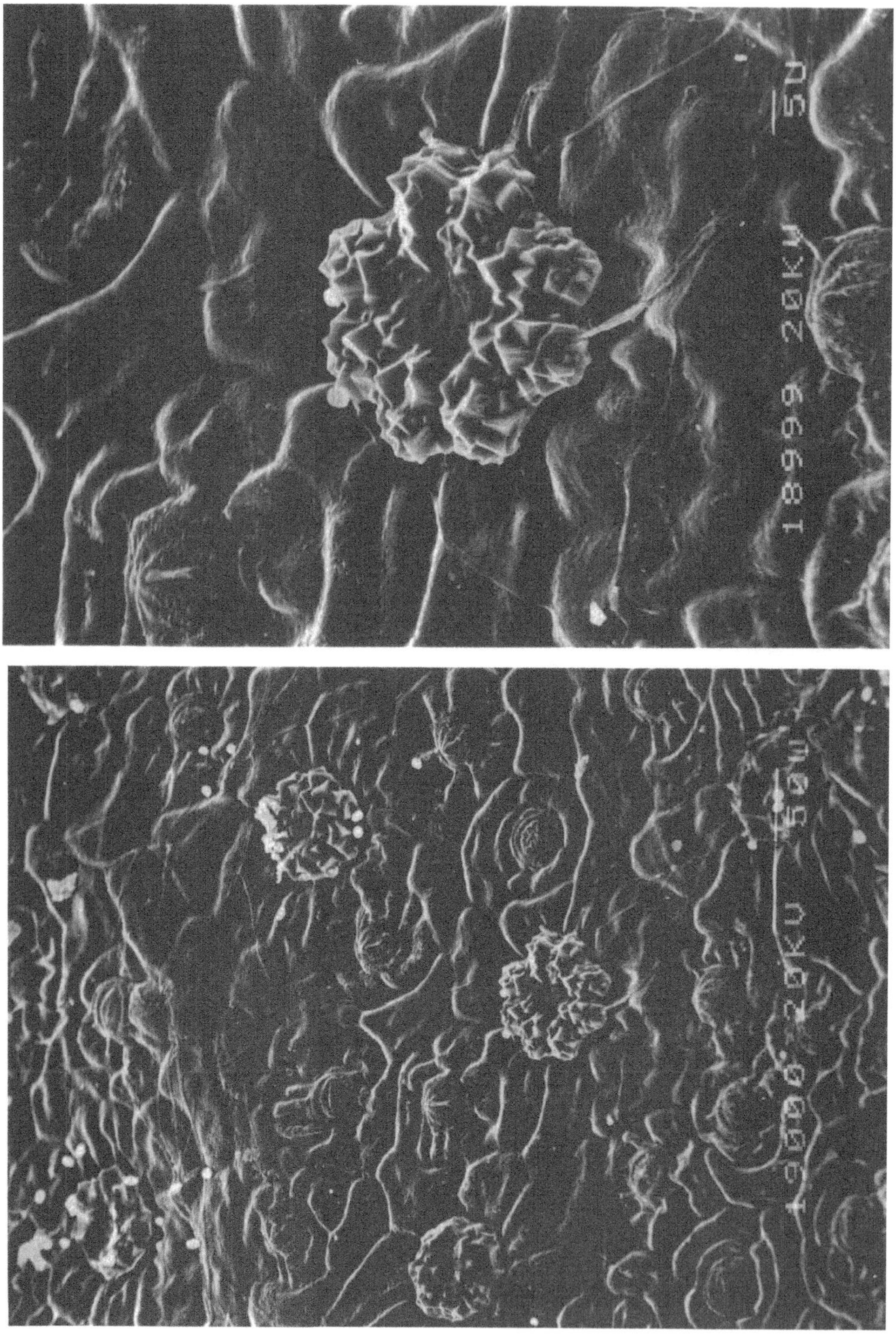

Fig. 21. Tabebuia stenocalyx, Bignoniaceae, lower leaf epidermis with stomata and smaller headed glands, each head cell containing a rhomboid crystal of Ca oxalate. SM. Orig. Roth.

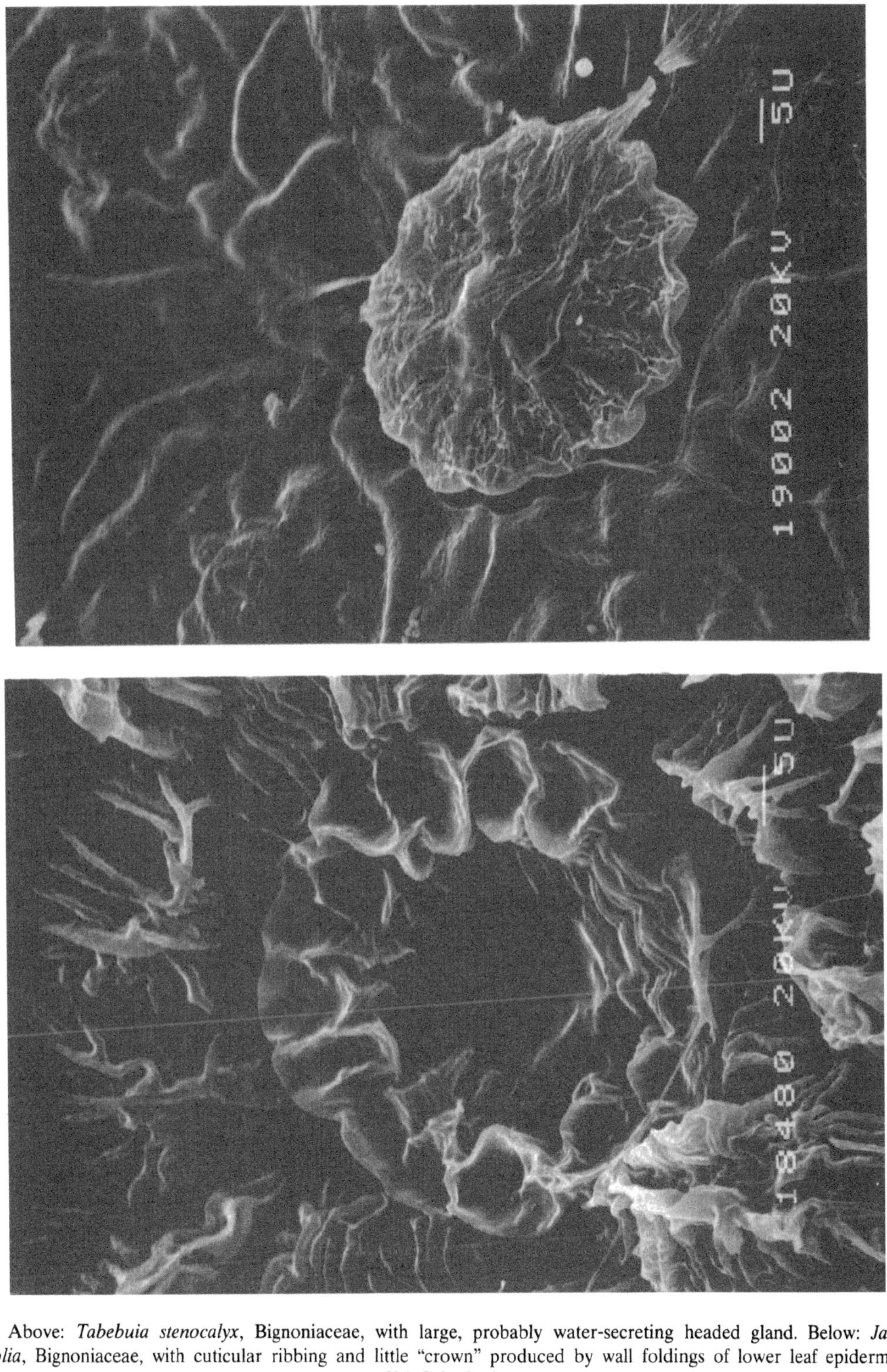

Fig. 22. Above: ***Tabebuia stenocalyx***, Bignoniaceae, with large, probably water-secreting headed gland. Below: *Jacaranda obtusifolia*, Bignoniaceae, with cuticular ribbing and little "crown" produced by wall foldings of lower leaf epidermis cells. SM. Orig. Roth.

spiration, water loss and strong insolation, they develop hairs of different kinds. A phenomenon completely different from hair formation is the "sculpturing" of leaf surfaces, as I called it, which becomes apparent in very distinct ways. A very weak sculpturing is perceptible in a cuticular granulation or a slight cuticular ribbing. Sculpturing is, however, more pronounced when the surface appears "wrinkled" or where irregular folds and ledges develop in the walls or, finally, scales and protuberances emerge. Elevations of this type may considerably rise above the surface, and scarcely have been described of leaves of temperate regions. This surface sculpturing is usually more conspicuous on the lower than on the upper leaf surface. A certain structuration of the surface is, however, conspicuous in many leaves of the material studied in Venezuelan Guiana. In the following peculiar examples are described.

1. *Eschweilera* c.f. *trinitensis*, Lecythidaceae, shows a cuticular ribbing on the lower leaf surface, while an occasional ribbing is only observed above the nerves of the upper side (Figure 6). The leaf is of the xeromorphic sun type and the tree belongs to height category A.

2. *Lecythis davisii*, Lecythidaceae, has a mesomorphic sun leaf. The tree is of height category A, and a ribbing of the leaf surface is only apparent above the guard cells and the adjoining cells of the stomata.

3. *Eschweilera subglandulosa*, Lecythidaceae, (Figure 1.7) of height category A, has a very xeromorphic sun leaf. The lower leaf surface is ribbed with somewhat curved ribs above the normal epidermis cells, but with straight parallel ribs along the veins. Excrescences occur on leaves of youth (Figure 14).

4. *Couratari multiflora*, Lecythidaceae, of height category A, has a mesomorphic sun leaf. The lower leaf side shows a weak ribbing above the nerves. But ribbing is more pronounced around the giant stomata, the ribs radiating from them in different directions.

5. *Eschweilera* sp. ("Majaguillo erizado"), Lecythidaceae, is of height category A and has a xeromorphic sun leaf. On the lower leaf side a weak ribbing becomes apparent above the nerves.

6. *Trichilia propingua*, Meliaceae, has an upper leaf surface which is somewhat granulated. The lower leaf surface, on the other hand, is partially ribbed and sculptured, especially around the stomata. The leaf is of the xeromorphic sun type, and the tree belongs to height category a.

7. *Trichilia smithii*, Meliaceae, of height category a, has a xeromorphic sun leaf. The upper leaf side appears to be covered with small sticks (of wax?), while the lower side has cuticular ribs (Figure 6).

8. *Trichilia schomburgkii*, Meliaceae, of height category a and with a xeromorphic shade leaf, has a ribbed cuticle on the lower leaf surface.

9. *Guarea schomburgkii*, Meliaceae, of height category a and with a mesomorphic shade leaf, has a strongly ribbed cuticle on the lower side.

10. *Simarouba amara*, Simaroubaceae, is of height category A and has a leaf of the xero and scleromorphic sun type. The surface of the lower leaf side is ribbed.

11. *Simaba multiflora*, Simaroubaceae, of height category A, has a mesomorphic sun leaf. On the lower leaf surface ribbing of the cuticle may be observed.

12. *Tapirira guianensis*, Anacardiaceae, with a xeromorphic sun leaf, is of height category A. The lower leaf side shows straight ribs, while the upper strongly ribbed leaf side has curved ribs.

13. *Aspidosperma excelsum*, Apocynaceae (Figures 1.1 and 16, 17, 19), with a xero and scleromorphic sun leaf is of height category A. Protuberances of a special kind with a "head" at the apex of each cell occur on the lower leaf side. The protuberances give rise to a calm front cavity above the guard cells of the lower leaf side, representing an effective protection against excessive transpiration (see Roth 1984, Figures 12, 47, 88).

14. *Aspidosperma marcgravianum*, Apocynaceae, of height category A, has a scleromorphic sun leaf. 0.02 mm long protuberances with a cutinized "head" protecting the stomata are found on the lower leaf side (Figure 20).

15. *Rheedia* sp., "Cozoiba rebalsera", Guttiferae, belongs to height category A and has a xeromorphic leaf of the medium type. While the upper leaf surface is only somewhat sculptured,

the lower leaf side is strongly sculptured or ribbed, much more than the upper side.

16. *Symphonia globulifera*, Guttiferae, of height category A and with a xeromorphic sun leaf, has a ribbed upper cuticle.

17. *Mabeaa taquiri*, Euphorbiaceae, (Figure 1.5) has a xeromorphic sun leaf. The lower leaf side shows a very pronounced sculpturation, partly with cutinized undulated wall foldings (see Roth 1984, Figure 57).

18. *Sandwithia guayanensis*, Euphorbiaceae, of height category a, has a mesomorphic shade leaf. The cuticle of the lower leaf side is weakly ribbed, especially around the stomata.

19. *Drypetes variabilis*, Euphorbiaceae, of height category A, has a xeromorphic sun leaf. The upper leaf surface is granulated, while the cells of the lower leaf surface appear as if they were perforated. They probably develop very thin spots in the walls.

20. *Mabea piriri*, Euphorbiaceae, of height category a, has a xeromorphic sun leaf. On the lower leaf surface, wall protuberances up to 13 μm in height occur which grow densely together, completely covering the stomata (see Roth 1984, Figures 7, 59).

21. *Hieronyma laxifolia*, Euphorbiaceae, of height category A, has a xeromorphic sun leaf. Large "peltate" star-shaped scales of a low density occur on the upper as well as on the lower leaf surface.

22. *Conceveiba guayanensis*, Euphorbiaceae, of height category a, has a mesomorphic shade leaf. The lower leaf surface is striated.

23. *Pera schomburgkiana*, Euphorbiaceae, of height category A, has a mesomorphic sun leaf. "Peltate" star-shaped scales occur at a very low density on the lower leaf surface.

24. *Ocotea nicaraguensis*, Lauraceae, of height category A, has a xeromorphic sun leaf. The lower leaf surface is slightly striated, especially above the nerves.

25. *Beilschmiedia curviramea*, Lauraceae, of height category a, has a xeromorphic shade leaf. The lower leaf surface is strongly sculptured with small protuberances, as seen is transverse section, while the upper surface is smooth (Figure 1.8).

26. *Aniba excelsa*, Lauraceae, of height category A, has a xeromorphic sun leaf. Only a weak ribbing above the guard cells may be observed on the lower leaf side.

27. *Ocotea martiana*, Lauraceae, of height category A, has a mesomorphic shade leaf. The lower leaf surface is ribbed.

28. *Jacaranda obtusifolia*, Bignoniaceae, (Figure 1.2) of height category A, has a xeromorphic sun leaf. While the upper leaf surface is only weakly granulated or ribbed, scale-like protuberances completely cover the lower surface (Figures 9, 15 and 22) (see also Roth 1984, Figures 48, 49).

29. *Mouriria sideroxylon*, Melastomaceae, of height category A, has a xeromorphic leaf of the medium type. A partial ribbing of the cuticle may be observed on the lower leaf side.

30. *Mouriria huberi*, Melastomaceae, of height category A and with a xeromorphic leaf of the medium type, shows granulation on the upper as well as on the lower leaf surface.

31. "Saquiyak", Melastomaceae, of height category aa and with a xeromorphic shade leaf, has a ribbed upper leaf surface.

32. *Mouriria sideroxylon*, Melastomaceae, of height category A and with a xeromorphic leaf of the medium type, has an upper leaf surface which is granulated and partly ribbed. The lower surface is only somewhat granulated.

33. *Lacistema aggregatum*, Lacistemaceae, of height category a, has a mesomorphic sun leaf. Upper and lower leaf surface are equally well granulated.

34. *Zinowiewia australis*, Celastraceae, of height category A, has a mesomorphic leaf of the medium type. The upper leaf side is weakly granulated, while the lower leaf side is ribbed, and shows circular folds around the stomata.

35. *Cordia alliodora*, Boraginaceae, of height category A, has a mesomorphic sun leaf. The lower leaf side is ribbed.

36. *Cordia exaltata*, Boraginaceae, of height category A with a hygromorphic sun leaf has a strongly ribbed lower leaf surface.

37. *Erisma uncinatum*, Vochysiaceae, of height category A, has a xeromorphic sun leaf. The

upper leaf surface shows a very slight granulation, while on the lower leaf surface strong granulation is combined with sculpturing and ribbing.

38. *Vochysia lehmannii*, Vochysiaceae, of height category A, has a mesomorphic sun leaf. The lower leaf surface is somewhat granulated.

39. *Qualea dinizii*, Vochysiaceae, of height category A, has a xeromorphic sun leaf. The upper and lower leaf side show a weak ribbing which becomes stronger above the nerves. However, the cells surrounding the giant stomata are strongly ribbed in a way that the ribs radiate from them (Figures 1, 4).

40. *Duroia* sp., "Conserva", Rubiaceae, of height category aa, has a xeromorphic sun leaf. The lower leaf surface is slightly ribbed.

41. *Amaioua guianensis*, Rubiaceae, of height category a with a mesomorphic sun leaf, has a ribbed lower leaf surface.

42. *Faramea torquata*, Rubiaceae, of height category a, has a mesomorphic shade leaf. The upper surface shows a ribbed cuticle.

43. *Banara nitida*, Flacourtiaceae, of height category A, has a xeromorphic sun leaf. A very weak sculpturing is visible on the upper leaf surface.

44. *Capparis guaguaensis*, Capparidaceae (Figure 1, 3) of height category a, has a xeromorphic sun leaf. Scales or crests arise from the top of each cell extending towards the cell border in the form of a striation on the lower leaf surface. These scales completely cover the leaf surface so that also the stomata are partly covered, being well protected (see Roth 1984, Figures 5, 6, 11, 13, 62, 63).

45. *Capparis amplissima*, Capparidaceae, of height category A, has a xeromorphic sun leaf. The lower leaf surface shows a slight cuticular sculpturing.

46. *Crataeva tapia*, Capparidaceae, of height category aa, has a hygromorphic shade leaf. The upper leaf surface is striated.

47. *Paypayrola longifolia*, Violaceae, of height category a, has a hygromorphic shade leaf. The cuticle of the upper leaf side is granulated, while that of the lower leaf side is ribbed.

48. *Rinorea riana*, Violaceae, of height category aa with a hygromorphic shade leaf has a granulated upper leaf surface, while the lower one is ribbed (Figures 10, 11).

49. "Gaspadillo blanco", Violaceae, of height category aa and with a hygromorphic shade leaf, has a slightly granulated upper leaf surface, while the lower leaf side is ribbed.

50. *Sterculia pruriens*, Sterculiaceae, of height category A and with a xeromorphic sun leaf, shows wall foldings and excrescences on the lower leaf surface (Figure 14) (see also Roth 1984, Figure 60).

51. *Apeiba echinata*, Tiliaceae, of height category A and with a mesomorphic shade leaf, has a ribbed cuticle on the upper leaf side. On the lower leaf side, stellate hairs and glands occur.

52. *Byrsonima aerugo*, Malpighiaceae, of height category A and with a xeromorphic sun leaf, has a sculptured upper and lower leaf surface.

53. *Tetragastris* or *Dacryodes*, "Aracho blanco", Burseraceae, of height category a and with a xero-scleromorphic sun leaf, has a granulated lower cuticle.

54. "Maro", Burseraceae, of height category a, with a mesomorphic sun leaf, has a strongly ribbed upper leaf surface, while hairs and glands are found on the lower leaf side.

55. "Rabo pelado", Olacaceae, of height category A and with a mesomorphic sun leaf, has a granulated cuticle on the upper as well as on the lower leaf side.

56. *Meliosma herbertii*, Sabiaceae, of height category a and with a xeromorphic leaf of the medium type has a granulated upper leaf surface, while the lower leaf side is ribbed or granulated.

57. *Pouteria* c.f. *trilocularis*, Sapotaceae, of height category a, has a xeromorphic sun leaf. The cuticle of the lower leaf side is ribbed.

58. *Pouteria venosa*, Sapotaceae, of height category A, has a xeromorphic sun leaf. The surface is ribbed on the lower leaf side.

59. (a). *Manilkara bidentata*, Sapotaceae, of height category A, has a xeromorphic sun leaf. The upper and lower leaf surface are strongly sculptured, on the adult tree. (b). *Manilkara bidentata*.

The leaves of a young tree of only 6 m in height are of the xeromorphic medium type. The upper leaf surface is only granulated, while the lower one is strongly sculptured (Figure 12).

60. (a). *Chrysophyllum auratum*, Sapotaceae, of height category A, has a xeromorphic sun leaf. The lower leaf surface is sculptured with "knobs", on the adult tree. (b). *Chrysophyllum auratum*, Sapotaceae. The leaves of a young 7 m high tree are of the less xeromorphic medium type. Their lower leaf surface is strongly ribbed.

61. "Purguillo", Sapotaceae, of height category A, with a xeromorphic sun leaf, has an upper leaf surface which is granulated and sculptured, while the lower surface is strongly folded.

62. (a). *Diospyros melinonii*, Ebenaceae. The leaves of a tall tree are of the sclero-xeromorphic sun type. Their lower surface is strongly ribbed. (b). The leaves of a young 9 m high tree are of the xeromorphic sun type. On the lower surface folded wall excrescences arise from the top of each epidermis cell, reaching up to 0.02 mm in height (Figures 16, 17, 20).

63. *Virola sebifera*, Myristicaceae, of height category A, has a xeromorphic sun leaf. The lower leaf surface is sculptured with little "teeth".

64. *Virola surinamensis*, Myristicaceae, of height category A, and with a xeromorphic sun leaf, develops a very peculiar lower epidermis with strongly thickened cell walls. The epidermis cells form protrusions under which the stomata are hidden (Figure 4.7). In surface view, roundish cells may be observed between which gaps of different shape and size appear. Within the gaps air is held tenaciously.

65. *Iryanthera lancifolia*, Myristicaceae, with a xeromorphic sun leaf, belongs to height category A. The upper leaf surface is only granulated, while the lower surface is strongly ribbed.

66. *Ficus* sp., "Matapalo", Moraceae, a liana, has hygromorphic leaves of the medium type. Cutinized wall folds occur on the lower leaf side (Figure 13).

Of the 232 species studied of Venezuelan Guiana, 66 or 28.4% show certain surface sculpturation. Of these species, 65% belong to height category A, 26.9% to height category a, and 7.9% to height category aa.

Of the possible combinations of leaf types all nine are found:

53.8% of the species have xeromorphic sun leaves
7.6% " " " " xeromorphic leaves of the medium type
3.0% " " " " xeromorphic shade leaves

15.3% " " " " mesomorphic sun leaves
1.5% " " " " mesomorphic leaves of the medium type
9.2% " " " " mesomorphic shade leaves

1.5% " " " " hygromorphic sun leaves
1.5% " " " " hygromorphic leaves of the medium type
6.1% " " " " hygromorphic shade leaves

Surface sculpturing or ribbing generally occurs mainly on the lower surface or is more pronounced on the lower than on the upper surface. Rarely do both sides show the same degree of surface sculpturing or ribbing (e.g. in *Qualea dinizii*) and only in very few cases the opposite relation could be observed, i.e. a stronger surface sculpturing on the upper than on the lower surface (e.g. in *Trichilia smithii* and *Mouriria sideroxylon*). A stronger ribbing or sculpturation is frequently observed above or around the guard cells of the stomata and above the nerves, or ribbing only occurs around the stomata and above the nerves. Ribbing frequently radiates from the stomata as a center, but passes parallel to the extension of the nerves.

A slight granulation may be considered the weakest type of surface sculpturation. Ribbing gives the surface more structuration, and is increased in surface folding. Crests and scales, finally, represent the most pronounced types of surface sculpturing. An explanation for the more frequent occurrence of any type of sculpturing on the lower leaf side is the fact that the upper leaf surface has to remain smooth for three reasons: water drainage is more rapid on a smooth surface, any surface sculpturation would interfere with light absorption, and epiphytes attach more easily to a rough surface. Consequently, the lower leaf surface *may* develop sculpturation, whether it is advantageous for the leaf or not.

Interesting are cases where the leaf of a young tree shows a sculpturation different from that of an adult tree. In *Manilkara bidentata*, the upper as well as the lower surface of an adult tree is strongly sculptured, while in the leaves of a young tree the upper surface is only granulated. Surface sculpturing is thus reduced on the upper leaf side of young trees, probably for the abovementioned reasons. In *Chrysophyllum auratum*, the lower leaf surface of an adult tree is heavily sculptured with little "knobs" on top of the epidermis cells, while it is only ribbed in leaves of young trees. Also here, a reduction of the sculpturing in leaves of young trees. In *Diospyros melinonii*, however, the lower leaf surface of a tall tree is strongly ribbed, while that of a young tree develops wall excrescences arising from the top of each epidermis cell and reaching up to 0.02 mm in height. On the lower leaf side of young trees of *Eschweilera subglandulosa*, protuberances are produced by wall foldings which partly surround the stomata. They are absent on the leaves of adult trees.

7.1. Possible function of the epidermis sculpturation

It is difficult to explain the possible function of this surface sculpturing. As almost one third of the species studied shows some kind of surface structuration, it is suggested here that some advantage should be the reason. Striking is the fact that the upper surface is usually smooth. On a smooth surface water runs off more easily, whereas on a sculptured surface the danger is increased that epiphytes and parasites, such as fungi, algae, mosses etc., colonize it. A quick drainage of the upper leaf surface is necessary to ensure optimal photosynthetic processes. Water drops or a water film would change light incidence to the worse. And the sculpturation itself could change light absorption by the chloroplasts. A smooth upper leaf surface is thus more advantageous.

On the lower leaf surface, scales and protrusions of the epidermal cells which surround the stomata have almost certainly the effect of transpiration reduction. A calm chamber or wind screen above the stomata results. Examples of this type are the following.

Eschweilera subglandulosa with a very xeromorphic sun leaf develops very thick cell walls on the lower epidermis surface. Stomata density is medium, but the stomata possess pronounced cuticular "cornets" or ledges so that a calm air chamber is created. The protuberances and wall foldings on the lower side of leaves of young trees may have the same function. *Capparis guaguaensis* with a sun leaf that develops some xeromorphic characteristics shows a relatively low stomata density. The stomata are sunken below the surface. Additionally, "peltate" scales on the lower leaf surface partly cover the stomata, reducing transpiration in this way. *Mabea piriri* has a sun leaf with developing xeromorphic features. The lower epidermis cells have somewhat thickened outer walls. Stomata which occur very frequently are hidden below protuberances. The excrescences which completely cover the lower surface help reduce transpiration. *Aspidosperma* (both species) has a xero- and scleromorphic sun leaf. The lower epidermis cells have very thick outer walls. Stomata density is relatively lower and stomata are sunken below the surface. Protuberances with "heads" develop on the lower leaf surface to protect the stomata. *Jacaranda obtusifolia* with a xeromorphic sun leaf shows an elevated stomata density. Scale-like protuberances which sometimes form little "crowns" on top of the cells completely cover the lower surface. The stomata which are surrounded by the scales in a collar-like manner are protected from excessive transpiration in this way. *Sterculia pruriens* with a very xeromorphic sun leaf and a high stomata density develops wall foldings on the lower leaf surface. These wall protrusions surround the stomata supplying an effective wind screen. In the abovementioned examples there is no doubt that the surface sculpturing on the lower leaf side protects the stomata in reducing transpiration. In xeromorphic sun leaves which are exposed to insolation and wind, this mechanism seems to be beneficial.

Another peculiarity of strong surface sculpturing is that it tenaciously retains air in the gaps between the protuberances and especially above the stomata when excrescences surround them. Strong surface sculpturing thus drastically reduces the wetting capacity of the surface. This may be an important function of the surface structuration in a very moist, almost vapour-saturated environment.

The strongly sculptured lower epidermis with scales and crests has furthermore an isolating effect. By subdivision of the surface into small compartments which hold the air, comparable to those of a foam material or a sponge, temperature exchange is considerably reduced. Likewise foils with a ruffled and crumpled surface, as, for example, a dry shrivelled skin, have a good isolating effect. Both types are realized in tropical leaves. In spite of the protecting effect against overheating, respiration is ensured, as the compartments on the leaf surface are in communication with the open air. The sculptured leaf surface thus represents a very perfect technical construction.

Any sculpturing or formation of protuberances and projections increases the surface. It would also increase cuticular evaporation, if cutinization is not markedly augmented. Of seeds with a rough surface it is known that they are more sensitive to water loss. An increasing cuticular evaporation could also be beneficial for plants growing in a humid environment (see also the paragraph dealing with giant stomata).

A certain ribbing around the stomata in a radiating manner may cause the water to flow off more rapidly so that the gaseous exchange is not disturbed.

Another aspect of surface sculpturing could possibly be light capture. As surface sculpturing is more frequently observed on the lower than on the upper surface and not so seldom on the shade leaf type, absorption of light may have some effect. In modern technology, solar chambers which capture the solar energy often have a surface similar to that found on the lower leaf surface of the above-mentioned plant species. It is a rough surface with small irregular protrusions. However, this remark is only a vague supposition.

8. SUMMARY

Although structural botany as seen under the light microscope appears no longer to be "modern", having been overtaken by molecular biology, the tropical flora, which is structurally almost unknown, yet continues to confront us with structural and physiological problems that are unsolved to the present day. Already at the end of the last century, Haberlandt, Stahl, Areschoug and other renowned botanists had pointed out the presence of ocelli, water pores, fissures in the epidermis, etc., but the real function of these structures has never been demonstrated experimentally. Only one structural problem has been seized continually and with insistence: whether the drip-tip is really advantageous for the rapid drainage of leaves in a wet tropical forest. In the present discussion there is one large complex of problems included dealing with water conduction in a tropically humid, almost vapour-saturated environment, and another one which concerns light capture in a dense undergrowth where only 1/100 of the light reaching the crown region, or an even smaller fraction of it, is sufficient to maintain photosynthesis. The question of water translocation in a very humid atmosphere has recently been reopened by Braun (1983) in postulating a hydrostatic overpressure in the hydrosystem to maintain the water flow, when transpiration is blocked. According to Braun, this overpressure is accomplished by secretion of osmotically active substances through the accessory tissue of the hydrosystem into the vessels and consequent water uptake by the roots which causes exudation of guttation liquid. The ways through which water may be secreted passively are necrotic holes, fissures in the epidermis, the large holes of lenticels, the giant stomata above the veins or in their close vicinity, as has been shown in the present paper — being the most rapid path ways for exudation on emergency. The slower way leads through glandular hairs or hydathodes with an epithema which secrete water actively. Braun compares the situation of tropical trees in a vapour-saturated atmosphere when transpiration is blocked, with that of still leafless plants growing in temperate regions during spring time when guttation is called forth by an injury ("Blutungssaft"). The accessory tissues of trees from tropical regions with macroporous wood are especially well developed and are probably functioning almost throughout the year. Braun, therefore, concludes that the hydrosystem of tropical trees is under overpressure when transpiration is reduced to a minimum and that an osmotic water translocation ("Wasserverschiebung") represents an additional way of water conduction in tropical trees besides transpiration.

Fissure formation has been observed on the lower as well as on the upper leaf epidermis. Fissures may either originate in a schizogeneous way, possibly preceded by surface tensions, or they may arise through decomposition of wall material in a lysigeneous way. Both ways may be connected by transitional stages. Later on, the surrounding cells may follow with periclinal divisions so that a lenticel is originated.

Lenticels, in their turn, occur on the upper as well as on the lower leaf surface, but are more frequent on the lower side, usually developing around a stoma. Lenticel formation starts with meristematic activity around a single stoma or a group of them, often preceded by radial extension of the cells surrounding the stoma. Periclinal divisions, however, not only occur in the epidermis itself, but also the subepidermal cells surrounding the respiratory cavity are induced by the meristematic activity so that finally cell rows radiate from the stoma as a center in all directions. The stoma finally obliterates to give way to a large hole. An unusual type of lenticel formation differing from the ordinary one above described, has been observed on the upper leaf side in the absence of stomata. In this case, lenticels either develop around a group of epidermis cells which may be distinguished by their staining affinity different from ordinary epidermis cells or meristematic activity may start around a schizogeneously developed intercellular space or around a necrotic spot where dissolution of cell walls took place. Even around the base of an abscised hair, around an injury or an insect sting, meristematic activity may begin with periclinal divisions. It is possible that surface tensions are responsible for the starting cell elongation and the following meristematic activity. Similar process have been observed around stone cells, oil cells, glands, or idioblasts of different kinds, which may function as obstacles during growth in surface of the epidermis (see also Roth 1981).

The size of lenticels oscillates between 0.1 and 0.65 mm in diameter, the central hole reaching 0.02 up to 0.1 mm in diameter, and the height of the lenticel above the leaf surface may attain 0.05 mm. Lenticel density is usually low and a density of 0.8 lenticels per mm^2 is already elevated, but densities of 13—27 lenticels per mm^2, which have been observed on leaves of certain Lecythidaceae, are already comparable to stomata densities. In the present material, lenticels mainly occur on leaves of very high trees which are of the sun type and have xeromorphic characteristics. Mesomorphic shade leaves rarely bear lenticels. The usual function of the lenticels is the gaseous exchange. However, cork warts which serve as hydathodes have been observed by Areschoug (1902), particularly in mangrove plants. They are supposed to secrete salts dissolved in the water. In the present material, not infrequently lenticels have been found above the nerves and their water-secreting function is very probable, consequently. This assumption is supported by the fact that several lenticel-bearing species additionally possess large stomata which may have the same water-secreting function (see below).

Up to the present, two types of hydathodes are well known: Glandular trichomes, on the one hand, and epithema hydathodes, on the other. The epithema, often a chlorophyll-free parenchyma penetrated by intercellular spaces, is covered by an epidermis, and the guttation water is secreted actively through modified stomata, fissures or holes. A third type, suspected to serve as a passive water way when transpiration is blocked, is represented by "giant-stomata", found in the material studied of Venezuelan Guiana. A large proportion of species (about 33%) show a great variation in stomatal size of the leaves. The greatest difference measured between small and large stomata of one and the same leaf was 54 μm, an enormous deviation considering the mean size of stomata of about 20—25 μm. The so-called giant-stomata are distinguished from the ordinary stomata by the following characteristics: By their larger size, their scarcer occurrence, their possible affinity to stains, their preferred position above veins or in close vicinity of them, and by their possible wide open porus. The size of giant-stomata has, however, to be seen in relation with the ordinary stomata of the same leaf. There is no absolute size which would define the size of giant-stomata, but only their relative size in comparison with the normal stomata which usually prevail in number. About 15% of the species studied show giant stomata in the above-mentioned sense belonging to 17 different plant families some of which are considered more primtive (such as the Burseraceae, Guttiferae, Lecythidaceae), while more advanced families are in the minority (e.g. the Euphorbiaceae). Giant-stomata occur with more frequency in high trees with a xeromorphic sun leaf type. In certain examples it is quite obvious that the stomata are laterally extended by surface tensions so that "ring-pores" originate. A total width of the stomatal apparatus, including the companion cells, of up to 130 μm has been observed. The largest stomata usually occur above the nerves or in their close vicinity. Ring-pores with a wide open porus occur, for example, in *Eschweilera subglandulosa* at a density of 9 per mm^2. In *Cordia alliodora*, the giant-stomata reach a density of 22—54 per mm^2, while the complete stomata density amounts to 400 per mm^2. Surprisingly, radial cell extension of neighbouring cells which surround the stomata, may precede formation of giant-stomata and even periclinal divisions in the radially extended cells may occur in a similar way as observed during lenticel formation. It is, therefore, not improbable that certain types of giant-stomata later transform into lenticels.

Glandular hairs with a foot, a stalk and a uni- or pluricellular head occur in about 8% of the species studied, being more frequently found on the lower leaf side. It is most probable that they function as hydathodes, secreting water actively. Giant glands with a head diameter of up to 180 μm were observed in *Centrolobium paraense*.

Stomata which are distinguished by their larger size, their wide open aperture and their special shape ("ring-pores") have been described before in fruits. Likewise epithema hydathodes often possess stomata differing from the ordinary stomata by their larger size, their wide open porus and their inability to open and close the porus. In Araceae, species of *Papaver* and *Tropaeolum*, the stomata of hydathodes may adopt an enormous size. This is an obvious advantage for water secretion: the greater the porus, the lesser the resistance to infiltration. In the floral region, however, Daumann found stomata which function as hydathodes in the absence of an epithema. It is, therefore, most probable that the giant-stomata observed in the leaf material of Venezuelan Guiana serve as water ways or hydathodes, secreting water directly without the aid of an epithema. This suggestion is supported by the position of the stomata above the veins or in their close vicinity. A special epithema may possibly be replaced by a loose mesophyll rich in intercellular spaces. Water collecting within the intercellular spaces below the substomatal chamber may later be secreted through the giant-stomata, an assumption which becomes more probable when we consider the problems arising for the plant in a vapour-saturated atmosphere and the consequent lack of transpiration during the night.

Guttation may thus be the last resort of the plant to maintain the water flow. In some nights guttation is so strong that a continuous rain of water drops comes from the trees and shrubs. However, ordinary hydathodes with an epithema are only known of a few tropical plants, and in the leaf material studied of Venezuelan Guiana no hydathodes of this type have been discovered up to the present. But the hydrostatic pressure within the hydrosystem of the plant may become so high that injection with water of the intercellular spaces becomes a consequence and water is secreted through whatever way is available and offers least resistance. Large stomata above the veins, lenticels and fissures in the epidermis may be mentioned in the first place. The energy necessary for water secretion is supplied by a hydrostatic overpressure in the hydrosystem which is generated by osmotically active organic and inorganic substances secreted by the parenchymatous accessory tissue of the hydrosystem. These substances released into the vessels lead to water uptake by the roots, even in the absence of transpiration so that an internal overpressure is induced, according to Braun (1983). Water is consequently secreted through the stomata by guttation. Braun calls this an osmotic water shifting. Stomatal hydathodes (giant-stomata) may, therefore, be considered very important passive regulators of the water content of the leaves and have to be postulated for most of the tropical trees growing in a warm and moist climate. Last but not least, stomata of a very large size (about 1/3 larger than normal) have been observed already in the eocene in Rhodomyrtus sinuata Band. Hofmann (1934) considers them as hydathodes which help to regulate the translocation of nutritive substances by guttation.

In a dense tropical forest, light capture is a problem particularly for those plants which grow in the undergrowth. Haberlandt has drawn attention to cases where special organelles for light perception are developed. Certain epidermis cells may be distinguished from ordinary epidermis cells by their size and shape, being globular or plano-convex or even biconvex, and rising above the epidermis level by their larger size. As their cell content is transparent, they may function as light collectors and, therefore, were called ocelli, little eyes, by Haberlandt. They may be interspersed singles between other epidermis cells, appear in small groups or the entire epidermis may consist of ocelli. Besides, ocelli may be modified hairs, either in the form of papillae when the cells form a little outgrowth on the apex, or two superposed cells may be combined in a light collector. Even short hairs in the form of cystoliths with strongly thickened outer walls may have refractive properties. Lenticular cells as light collectors occur on the upper as well as on the lower leaf surface, while papillas are more commonly found on the lower leaf side. An upper or lower epidermis consisting in lenticular cells or — at least — with interspersed lenticular cells has often been observed in the material studied of Venezuelan Guiana as well as in that of the cloud forest of Rancho Grande. To what degree they really assist in light capture has, however, never been proved experimentally.

Cystoliths on the upper epidermis may penetrate the photosynthetic tissue as far as the spongy parenchyma, producing a pattern of white dots on the leaf surface and acting as light channels in a similar way as sclereids in xerophytes. The optical effects of very large cystoliths as in the genus *Ficus*, or of water-storing giant cells interspersed within the palisade parenchyma as found in *Tetragastris panamensis*, or of a large water-storing upper hypodermis, has still to be studied. Of course, there are many more rare structures in tropical leaves the optical effect of which is not yet understood, such as layers of colourless cells alternating with layers of photosynthetic tissue, running parallel to the leaf surface (e.g. in *Sloanea guianensis*).

Papillas, originally characteristic of shade plants from the wet tropical forest, may have three different functions: When they surround the stomata in the form of a wreath, as for example, in *Helicostylis tomentosa*, they may reduce transpiration. In certain cases, they may be considered collecting lenses, particularly when their outer walls are strongly thickened, and finally, they increase the leaf surface significantly and in this way may augment the cuticular evaporation, when the cuticle remains thin. Differences in the papilla formation between leaves of young and adult trees suggest a pro- or regressive drift leading from small hygromorphic plants towards high xeromorphic trees and expressing in this way the stratification of the forest in relation to the microclimate. Papillas are apparently a useful peculiarity in hygromorphic shade leaves, but either disappear in the meso- and xeromorphic leaves of the adult trees or transform into long hairs reducing transpiration in this way. Increase of cuticular evaporation may be a desired effect in a very humid tropical environment. In the hygromorphic leaf type and in leaves of young trees, the papillas are usually thin-walled, while in the xeromorphic leaf type, they usually result thick-walled and may even elongate into hairs. Papillas may gain in length towards the upper forest strata. With the increasing height of the trees, the short living papillas may transform into long dead hairs which protect the leaf surface against excessive transpiration. Modification of function thus occurs due to the different microclimate in the various forest strata. Finally, air is tenaciously retained between short papillas, an advantage for plants living in a wet, almost vapour-saturated atmosphere.

Surface sculpturing such as cuticular granulation or ribbing, in the form of wrinkles, of irregular folds and ledges or in a very pronounced form as scales and protuberances, is characteristic of about one third of the plants studied in Venezuelan Guiana. It is usually more conspicuous on the lower than on the upper surface. Scales and protrusions around stomata generate a wind screen above the stomata. On the other side, air is tenaciously retained above the stomata so that the wetability is much reduced. It is certainly an important peculiarity of surface sculpturation to retain air in the gaps between the protuberances. Strong surface sculpturing drastically reduces the wetting capacity of the surface. This seems to be an important function in a very moist, almost vapour-saturated environment. On the other hand, crests and protuberances increase the surface and herewith also augment cuticular evaporation, if cutinization is not markedly strngthened. Seeds with a rough surface, for example, are more sensitive to water loss. An increased cuticular evaporation could also be beneficial for plants growing in a humid environment. A strongly sculptured epidermis with scales and crests has furthermore an isolating effect. By subdivision of the surface into small compartments which hold the air, temperature exchange is considerably reduced. A certain ribbing around the stomata in a radiating manner may accelerate water drainage. A final aspect of surface sculpturing could be light capture which may be increased by a surface with irregular protrusions particularly in diffuse light.

Peculiar structures of tropical plants, especially of the humid tropical forest and the montane cloud forest thus confront us with a majority of problems concerning water transport, transpiration and guttation, on the one hand, and light capture or isolating effects, on the other, none of which has ever been attacked experimentally. Although certain trends have been observed in the humid tropical forest leading from the hygromorphic shade leaf type in the undergrowth towards the more xeromorphic sun leaf type in the crown region, we have to bear in mind that the leaf of a given species not necessarily has to unite all possible xeromorphic or hygromorphic characteristics at the same time, but that xeromorphic peculiarities may be combined with hygromorphic qualities and that even antagonistic properties present at the same time may balance the economy of plants. The peculiarity of stomata elevated above the surface to increase transpiration, a hygromorphic characteristic, may, for example, be compensated by the position of the stomata within a crypt to reduce transpiration, a xeromorphic character. Or another example: The large size of stomata in leaves of the undergrowth is compensated by their lower number, and vice versa, the high stomata density of leaves in the crown region is balanced by their small size. There is no strict, absolute or final solution detectable in the structure of plants, but only combination of characters must be well balanced.

REFERENCES

Areschoug, F. W. C. (1902): Untersuchungen über den Blattbau der Mangrove-Pflanzen. Bibl. Bot., Heft 56.

Arzt, Th. (1962): Beobachtungen an Fruchtknoten und an der reifen Frucht der Tulpe. Ber. Oberhess. Ges. Nat.-u. Heilk. Giessen, Naturwiss. Abt. 32: 38—59.

Biebl, R. and H. Germ (1950): Praktikum der Pflanzenanatomie. Springer, Wien.

Bone, R. A., D. W. Lee, & J. M. Norman (1984): Leaf epidermal cells function as lenses in tropical rain forest shade plants. Appl. Optics 24: 1408—1412.

Braun, H. J. (1983): Zur Dynamik des Wassertransportes in Bäumen. Ber. Dtsch. Bot. Ges. 96: 29—47.

Burtt, L. (1978): Notes on rain forest herbs. Gard. Bull. Singapore 29: 37—49.

Daumann, E. (1930): Das Blütennektarium von Magnolia und die Futterkörper in der Blüte von Calycanthus. Planta 11: 108—116.

Daumann, E. (1931): Nektarabscheidung in der Blütenregion einiger Araceen. Planta 12: 38—48.

Daumann, E. (1932): Über postflorale Nektarabscheidung. Beih. Bot. Cbl. 49: 720—734.

Daumann, E. (1935): Die systematische Bedeutung des Blütennektariums der Gattung Iris. Beih. Bot. Cbl. 53: 525—625.

Eames, A. J. (1961): Morphology of the Angiosperms. McGraw-Hill, New York — Toronto — London.

Eames, A. J. and L. H. MacDaniels (1947): An introduction to Plant Anatomy. McGraw-Hill, New York and London.
Esau, K. (1953): Plant Anatomy. J. Wiley & Sons, New York, London, Sydney.
Fahn, A. (1979): Secretory tissues in plants. Academic press, London, New York, San Francisco.
Figdor, W. (1898): Untersuchungen über die Erscheinung des Blutungsdruckes in den Tropen. — Sitzungsber. math.-nat. Cl. Kaiserl. Akad. Wiss. Wien CVII, Abt. 1.
Fischer, M. (1921): Beobachtungen über den anatomischen Bau der Früchte und über ein inneres Ausscheidungssystem in denselben bei den Kulturrassen und Varietäten von Capsicum. Zt. Allg. Österr. Apotheker-Ver. 59: 83—86 and 93—94.
Fischer, M. (1929a): Ergebnisse der Spaltöffnungsforschung an Früchten. Biol. Zbl. 49: 221—251.
Fischer, M. (1929b): Beiträge zur Kenntnis der Spaltöffnungsapparate an Früchten und zur Durchlüftung der Hohlfrüchte. Beih. Bot. Cbl. 45: 271—389.
Fukshanski, L. (1981): Optical properties of plants. Invited review lecture.
Gaulhofer, K. (1908): Über die anatomische Eignung der Sonnen- und Schattenblätter zur Lichtrezeption. Ber. Dtsch. Bot. Ges. 26 a. 484—494.
Givnish, T. J. (1984): Leaf and canopy adaptations in tropical forests. Eds. E. Medina, H. A. Mooney, & C. Velásquez. "Physiological ecology of plants of the wet tropics. Pp. 51—84. Junk, The Hague.
Haberlandt, G. (1905): Die Lichtsinnesorgane der Laubblätter. W. Engelmann, Leipzig.
Haberlandt, G. (1908): Über die Verbreitung der Lichtsinnesorgane der Laubblätter. SB Kaiserl. Akad. Wiss. Wien, math.-nat. Kl. 117: 621—635.
Haberlandt, G. (1924): Physiologische Pflanzenanatomie. 6. Aufl. W. Engelmann, Leipzig.
Hofmann, Elise (1934): Palaeohistologie der Pflanze. Springer, Wien.
Huber, O. (1976): Pflanzenökologische Untersuchungen im Gebirgsnebelwald von Rancho Grande (Venezolanische Küstenkordillere). Dissertation Univ. Innsbruck.
Johnson, Marion A. (1937): Hydathodes in the genus *Equisetum*. Bot. Gaz. 98: 598—608.
Klepper, B. & M. R. Kaufmann (1966): Removal of salt from xylem sap by leaves and stems of guttating plants. Plant Physiol. 41: 1743—1747.
Lee, D. W. (1986): Unusual strategies of light absorption in rain forest herbs. In: on the economy of plant form and function, Ed. T. J. Givnish. Pp. 105—131. Cambridge University Press, Cambridge.
Lindorf, Helga (1979): Estructura foliar de 15 monocotiledóneas de sombra del bosque nublado de Rancho Grande. Trabajo de ascenso, U. C. V., Caracas.
Maercker, Uta (1965): Das Phänomen der Frostinfiltration bei *Ilex aquifolium*. Zeitschr. f. Pflanzenphys. 53: 162—168.
Metcalfe, C. R. and L. Chalk (1950 & 1965): Anatomy of the Dicotyledons. Vols. I and II. Clarendon Press, Oxford.
Molisch, H. (1898): Über das Bluten tropischer Holzgewächse im Zustand völliger Belaubung. — Ann. Jard. Bot. Buitenzorg, 2. Suppl.: 23—32.
Müller, L. (1929): Über Bau und Nektarausscheidung der Blüte von *Grevillea preissii* Meisn. Biologia generalis 5.
Napp-Zinn, K.: Anatomie des Blattes II. Encyclopedia of plant anatomy. Borntraeger, Berlin — Stuttgart 1973.
Paturi, F. (1974): Geniale Ingenieure der Natur. Econ, Düsseldorf — Wien.
Pyykkö, Maire (1979): Morphology and anatomy of leaves from some woody plants in a humid tropical forest of Venezuelan Guayana. Acta Bot. Fennica 112: 1—41.
Roth, Ingrid (1966): Anatomía de las plantas superiores. 2nd. ed. 1976. EBUC, Caracas.
Roth, Ingrid (1977a): Anatomía y textura foliar de plantas de la Guayana Venezolana. Acta Bot. Venez. 12: 79—146.
Roth, Ingrid (1977b): Fruits of Angiosperms. Encyclopedia of plant anatomy, Bd. X, T. 1. Borntraeger, Berlin — Stuttgart.
Roth, Ingrid (1980): Blattstruktur von Pflanzen aus feuchten Tropenwäldern. Bot. Jahrb. Syst. 101: 489—525.
Roth, Ingrid (1981): Structural patterns of tropical barks. Encyclopedia of plant anatomy, eds. H. J. Braun, S. Carlquist, P. Ozenda & I. Roth. Borntraeger Berlin — Stuttgart.
Roth, Ingrid (1984): Stratification of tropical forests, as seen in leaf structure. Junk, The Hague — Boston — Lancaster.
Roth, Ingrid (1987): Stratification of a tropical forest as seen in dispersal types. Tasks for vegetation science 17. Ed. H. Lieth. Dr. W. Junk Publ., Dordrecht — Boston — Lancaster.
Roth, Ingrid & Ingrid Clausnitzer (1972): Desarrollo de los hidátodos en *Sedum argenteum*. Acta Bot. Venez. 7: 207—217.
Roth, Ingrid, Tatiana Mérida and Helga Lindorf (1986): Morfología y anatomía foliar de plantas de la selva nublada. La selva nublada de Rancho Grande Parque Nacional "Henri Pittier". El ambiente físico, Ecología vegetal y anatomía vegetal. Edit. O. Huber. Acta Cient. Venez., Caracas.
Sauer, H. (1933): Blüte und Frucht der Oxalidaceen, Linaceen, Geraniaceen, Tropaeolaceen und Balsaminaceen, vergleichend-entwicklungsgeschichtliche Untersuchungen. Planta 19: 417—481.
Stahl, E. (1896): Über bunte Laubblätter. Ein Beitrag zur Pflanzenbiologie. Ann. Jard. Bot. Buitenzorg 13: 137—216.
Steinberger-Hart, A. L. (1922): Über Regulation des osmotischen Wertes in den Schliesszellen von Luft- und Wasserspalten. Biol. Zbl. 42: 405—419.
Volkens, G. (1883): Über Wasserausscheidung in liquider Form an den Blättern höherer Pflanzen. Jahresber. K. Bot. Gartens, Bot. Mus., Berlin, 2: 166—209.
Zimmermann, W., and H. Bachmann-Schwegler (1962): Morphologie und Anatomie von *Pulsatilla*. Flora 152: 315.

FINAL REMARKS

The content of this book confirms that the microclimatic gradient in a tropical forest not only affects the inner leaf structure, but also leaf surface area and leaf morphology. The contribution by Rollet confirms the statements of Roth (1984) and Roth and Mérida (1971) that the length/width ratio of leaves decreases from the undergrowth towards the canopy, leaves becoming less elongated and less lanceolate but shorter and broader. Drip tips prevail in the undergrowth becoming scarcer towards the canopy. Within the saplings only 2.5/10 are without drip-tips, the proportion changes to 4/10 between 5 m and 20 m height, and to 5/10 above 20 m. Furthermore, the mean length of drip-tips decreases with the increasing height of the trees. Particularly, trees above 20 m in height show a sudden drop in drip-tip length. Another phenomenon is that leaf size changes during the lifetime of plants. Fully developed leaves of juvenile or young plants (small trees) are commonly larger than leaves of adult trees. Likewise the length/width ratio of leaves changes during the life time. In young trees, the leaves may be long and narrow, but rather oblong when the tree is adult (e.g. in *Rollinia multiflora*). In general, leaf length/width ratio is higher in juvenile than in adult trees. This phenomenon seems also to depend on the microclimatic gradient in the forest. Concerning leaf size of different species in the distinct forest strata, Rollet noted a certain tendency towards an increase of the mean leaf area or mean leaflet area from the canopy trees towards the shrubby layer and the undergrowth. Leaf thickness obviously varies with the age of the tree. It is roughly twice as thick in the adult than in the juvenile stage. Between 1 m and 10 m height the mean leaf thickness amounts to 144 microns, between 10 m and 20 m height it increases to 205 microns, and above 20 m to 235 microns. Leaf density 1 g/1 cm^2 only slightly increases from 10 m to 29 m or above, but below 10 m leaf density decreases significantly. Leaf consistence is a character fairly correlated with density.

Although the proportion of compound leaves is lower in montane rainforests than in semideciduous tropical humid forests, a relatively high proportion of compound leaves is obvious in tropical humid forests. In the humid forest of Venezuelan Guiana, Rollet observed a tendency of enrichment in compound leaves from the lower forest strata to the canopy. Compound leaves are more deciduous and occur more frequently among light demanding species. There is thus a clear gradation visible in the forest concerning the quantified parameters of leaf length/width ratio, size, thickness and density from the undergrowth towards the forest top. These gradual changes have their roots in the microclimatic gradient of the forest which changes in humidity, light intensity, air movement, and temperature from top to bottom.

Venation studied by Högermann is a part of the inner leaf structure. Högermann tried to find relations between venation density of leaves and height of trees. Comparing the three forest strata with one another she found a certain tendency of leaves of adult trees to increase the number of secondary veins in the lamina, as well as the vein length/cm^2 in the upper stratum. Her material is too small for general conclusions, as she only used three families (Sapotaceae, Lauraceae, and Euphorbiaceae). Comparing the leaves of young with those of adult trees, she noted that a larger part of the lamina in leaves of youth is supplied by a secondary vein than in those of adult trees. Distances between secondaries are thus larger in leaves of youth, as compared with leaves of adult trees (see also Roth 1984). About 66% of the leaves of adult trees show a higher venation density/cm^2 than leaves of young treelets. In about

B. Rollet et al.: Stratification of Tropical Forests as seen in Leaf Structure, Part 2, pp. 239–240.

77% of the leaves of adult trees the gross conductivity volume is higher than in leaves of young trees. Furthermore, a higher number of vein islets as related to the venation length is observed in leaves of adult trees (90%). The quotient venation length/cm^2 to number of vein islets + veinlet terminations indicates a considerably higher venation density in leaves of adult trees (71%) than in leaves of young trees. Similar results are obtained when the quotient venation length/cm^2 to number of vein islets + ramifications + veinlet terminations is considered. This shows that the vertical gradient in the forest expresses itself also in the leaf venation. Leaves of young trees (corresponding to leaves of smaller *treelets*) show a lesser venation density than leaves of grown up adult trees.

Changes in the inner leaf structure from the lower tree and shrub stratum towards the crown region have been studied before by Roth (1984). About 50 different anatomical characters were investigated and related to the tree height. It could be shown that not only the size and number of stomata as well as the photosynthetic apparatus profoundly change from the undergrowth towards the tree top, but also many other parameters. Leaf structure adapts to the decreasing light intensity in the undergrowth as well as to the increasing drought in the crown region. With the studies of leaf size and morphology and those of leaf venation as related to the tree height, the picture is rounding off more and more. In the future, more emphasis should be laid on a comparison between the herbaceous layer and the undergrowth, on the one hand, and high trees, on the other. With the studies of Rollet and Högermann, the results of Roth (1980, 1984) and Roth and Mérida (1971) could thus be confirmed and extended.

There is actually no other area in the tropical forest which is structurally so well known in its number of species, number of individuals and height categories, as well as in its leaf morphology and anatomy, bark structure, dispersal types, and wood anatomy than the humid tropical forest of Imataca in Venezuelan Guiana.

INGRID ROTH

INDEX OF COMMON PLANT NAMES

INDEX OF SCIENTIFIC PLANT NAMES

GPSR Compliance

The European Union's (EU) General Product Safety Regulation (GPSR) is a set of rules that requires consumer products to be safe and our obligations to ensure this.

If you have any concerns about our products, you can contact us on ProductSafety@springernature.com

In case Publisher is established outside the EU, the EU authorized representative is:

Springer Nature Customer Service Center GmbH
Europaplatz 3
69115 Heidelberg, Germany

Batch number: 10800581

Printed by Printforce, the Netherlands